Texts in Applied Mathematics 30

Springer
New York
Berlin
Heidelberg
Barcelona
Budapest
Hong Kong
London
Milan
Paris
Singapore
Tokyo

Claude Gasquet†
Université Joseph Fourier (Grenoble I)

Patrick Witomski
Directeur du Laboratoire LMC-IMAG
Tour IRMA, BP 53
38041 Grenoble, Cedex 09
France

Translator

Robert Ryan
12, Blvd. Edgar Quinet
75014 Paris
France

†*Deceased.*

Mathematics Subject Classification (1991): 42-01, 28-XX

Library of Congress Cataloging-in-Publication Data
Gasquet, Claude.
Fourier analysis and applications : filtering, numerical computation, wavelets / Claude Gasquet, Patrick Witomski.
p. cm. – (Texts in applied mathematics; 30)
Includes bibliographical references and index.
ISBN 0-387-98485-2 (hardcover)
1. Fourier analysis. I. Witomski, Patrick. II. Title.
III. Series.
QA403.5.G37 1998
515′.2433–dc21 98-4682

Printed on acid-free paper.

Production managed by A.D. Orrantia; manufacturing supervised by Jacqui Ashri.
Camera-ready copy prepared from the authors' LaTeX files.
Printed and bound by Edwards Brothers, Inc., Ann Arbor, MI.
Printed in the United States of America.

9 8 7 6 5 4 3 2 1

ISBN 0-387-98485-2 Springer-Verlag New York Berlin Heidelberg SPIN 10658148

Translator's Preface

This book combines material from two sources: *Analyse de Fourier et applications: Filtrage, Calcul numérique, Ondelettes* by Claude Gasquet and Patrick Witomski (Masson, Paris, second printing, 1995) and *Analyse de Fourier et applications: Exercices corrigés* by Robert Delmasso and Patrick Witomski (Masson, Paris, 1996). The translation of the first book forms the core of this Springer edition; to this have been added all of the exercises from the second book. The exercises appear at the end of the lessons to which they apply. The solutions to the exercises were not included because of space constraints.

When Springer offered me the opportunity to translate the book by Gasquet and Witomski, I readily accepted because I liked both the book's content and its style. I particularly liked the structure in 42 lessons and 12 chapters, and I agree with the authors that each lesson is a "chewable piece," which can be assimilated relatively easily. Believing that the structure is important, I have maintained as much as possible the "look and feel" of the original French book, including the page format and numbering system. I believe that this page structure facilitates study, understanding, and assimilation. With regard to content, again I agree with the authors: Mathematics students who have worked through the material will be well prepared to pursue work in many directions and to explore the proofs of results that have been assumed, such as the development of measure theory and the representation theorems for distributions. Physics and engineering students, who perhaps have a different outlook and motivation, will be well equipped to manipulate Fourier transforms and distributions correctly and to apply correctly results such as the Poisson summation formula.

Translating is perhaps the closest scrutiny a book receives. The process of working through the mathematics and checking in-text references always uncovers typos, and a number of these have been corrected. On the other hand, I have surely introduced a few. I have also added material: I have occasionally added details to a proof where I felt a few more words of explanation were appropriate. In the case of Proposition 31.1.3 (which is a key result), Exercise 31.12 was added to complete the proof. I have also completed the proofs in Lesson 42 and added some comments. Several new references on wavelets have been included in the bibliography, a few of them with annotations. All of these modifications have been made with the knowledge and concurrence of Patrick Witomski.

Although the book was written as a textbook, it is also a useful reference book for theoretical and practical results on Fourier transforms and distributions. There are several places where the Fourier transforms of specific functions and distributions are summarized, and there are also summaries of general results. These summaries have been indexed for easy reference.

The French edition was typeset in Plain TEX and printed by Louis–Jean in Gap, France. Monsieur Albert at Louis–Jean kindly sent me a copy of the TEX source for the French edition, thus allowing many of the equations and arrays to be copied. This simplified the typesetting and helped to avoid introducing errors. My sincere thanks to M. Albert. Similarly, thanks go to Anastis Antoniadis (IMAG, Grenoble) for providing the LATEX source for the exercises, which was elegantly prepared by his wife. I had the good fortune to have had the work edited by David Kramer, a mathematician and freelance editor. He not only did a masterful job of straightening out the punctuation and other language-based lapses, but he also added many typesetting suggestions, which, I believe, manifestly improved the appearance of the book. I also thank David for catching a few of the typos that I introduced; those that remain are my responsibility and embarrassment.

Robert Ryan
Paris, July 14, 1998

Preface to the French Edition

This is a book of applied mathematics whose main topics are Fourier analysis, filtering, and signal processing.

The development proceeds from the mathematics to its applications, while trying to make a connection between the two perspectives. On one hand, specialists in signal processing constantly use mathematical concepts, often formally and with considerable intuition based on experience. On the other hand, mathematicians place more priority on the rigorous development of the mathematical concepts and tools.

Our objective is to give mathematics students some understanding of the uses of the fundamental notions of analysis they are learning and to provide the physicists and engineers with a theoretical framework in which the "well known" formulas are justified.

With this in mind, the book presents a development of the fundamentals of analysis, numerical computation, and modeling at levels that extend from the junior year through the first year of graduate school. One aim is to stimulate students' interest in the coherence among the following three domains:

- Fourier analysis;
- signal processing;
- numerical computation.

On completion, students will have a general background that allows them to pursue more specialized work in many directions.

The general concept

We have chosen a modular presentation in lessons of an average size that can be easily assimilated ... or passed over. The density and the level of the material vary from lesson to lesson. We have purposefully modulated the pace and the concentration of the book, since as lecturers know, this is necessary to capture and maintain the attention of their audience. Each lesson is devoted to a specific topic, which facilitates reading "à la carte." The lessons are grouped into twelve chapters in a way that allows one to navigate easily within the book.

A progressive approach

The program we have adopted is progressive; it is written on levels that range from the third year of college through the first year of graduate school.

JUNIOR LEVEL

Lessons 1 through 7 are accessible to third-year students. They introduce, at a practical level, Fourier series and the basic ideas of filtering. Here one finds some simple examples that will be re-examined and studied in more depth later in the book. The Lebesgue integral is introduced for convenience, but in superficial way. On the other hand, emphasis is placed on the geometric aspects of mean quadratic approximation, in contrast to the point of view of pointwise representation. The notion of frequency is illustrated in Lesson 7 using musical scales.

SENIOR LEVEL

The reader will find a presentation and overview of the Lebesgue integral in Chapter IV, where the objective is to master the practical use of the integral. The lesson on measure theory has been simplified. This chapter, however, serves as a good guide for a more thorough study of measure and integration. Chapter VI contains concentrated applications of integration techniques that lead to the Fourier transform and convolution of functions. One can also include at this level the algorithmic aspects of the discrete Fourier transform via the fast Fourier transform (Chapter III), the concepts of filtering and linear differential equations (Chapter VII), an easy version of Shannon's theorem, and an introduction to distributions (Chapter VIII).

MASTER LEVEL

According to our experience, the rest of the book, which is a good half of it, demands more maturity. Here one finds precise results about the fundamental relation $\widehat{f * g} = \widehat{f} \cdot \widehat{g}$, the Young inequalities (Chapter VI), and various aspects of Poisson's formula related to sampling (Chapter XI). Finally, time-frequency analysis based on Gabor's transform and wavelet analysis (Chapter XII) call upon all of the tools developed in the first eleven chapters and lead to recent applications in signal processing.

The content of this book is not claimed to be exhaustive. We have, for example, simply treated the z-transform without speaking of the Laplace transform. We chose not to deal with signals of several variables in spite of the fact that they are clearly important for image processing.

Possible uses of time

This book is an extension of a course given for engineering students during their second year at E.N.S.I.M.A.G.[1] and at I.U.P.[2]. We have been confronted, as are all teachers, with class schedules that constrain the time available for instruction. The 40 hours available to us per semester at E.N.S.I.M.A.G. or at I.U.P., which is divided equally between lectures and work in sections, provides enough time to present the essential material.

Nevertheless, the material is very rich and requires a certain level of maturity on the part of the students. We are thus led to assume in our lectures some of the results that are proved in the book. This is facilitated by the partition of the book into lessons, and it is not incompatible with a good mathematics education. The time thus saved is more usefully invested in practicing proofs and the use of the available tools. The material is written at a level that leads to a facility in manipulating distributions, to a rigorous formulation of the fundamental formula $\widehat{f * g} = \widehat{f} * \widehat{g}$ under various assumptions, to an exploration of the formulas of Poisson and Shannon, and finally, to precise ideas about the wavelet decomposition of a signal.

Our presentation contrasts with those that simply introduce certain formulas such as

$$\int_{-\infty}^{+\infty} e^{-2i\pi(\lambda-a)}\,dt = \delta(\lambda - a)$$

out of thin air, where one ignores all of the fundamental background for a very short-term advantage.

Different possible courses

One can work through the book linearly, or it is possible to enter at other places as suggested below:

Juniors
Chapters I, II, and III.

Seniors and Masters in Mathematics
Chapters IV, V, VI, VIII, and IX.

Seniors and Masters in Physics
Chapters VII, X, XI, and XII.

This book comes from many years of teaching students at E.N.S.I.M.A.G. and I.U.P. and pre-doctoral students. In fact, it was for pre-doctoral instruction that a course in applied mathematics oriented toward signal processing

[1]Ecole Nationale Supérieure d'Informatique et de Mathématiques Appliquées de Grenoble (Institut National Polytechnique de Grenoble)

[2]Institut Universitaire Professionnalisé de Mathématiques Appliquées et Industrielles (Université Joseph Fourier Grenoble I)

was established by Raoul Robert. His initiative in this subject, which was not his area of research, has played a decisive role, and the current explosion of numerical work based on wavelets shows that his vision was correct. Our thanks go equally to Pierre Baras for the numerous animated discussions we have had. Their ideas and comments have been a valuable aid and irreplaceable inspiration for us.

The second printing of this book is an opportunity to make several remarks. We have chosen not to include any new developments. We have listed at the end of the book several references on wavelets, which show that this area has exploded during these last years. But for the student or the teacher to whom we address the book, the path to follow remains the same, and the basics must be even more solidly established to understand these new areas of applications. It seems to us that our original objective continues to be appropriate today.

We have made the necessary corrections to the original text, and a book of exercises with solutions will soon be available to complete the project.

Claude Gasquet
Patrick Witomski
Grenoble, June 30, 1994

Contents

Chapter VI Convolution and the Fourier Transform of Functions 153

Chapter I

Signals and Systems

Lesson 1

Signals and Systems

1.1 General considerations

The purpose of signal theory is to study signals and the systems that transmit them. The notion of *signal* is extensive. The observation of some phenomenon yields certain quantities that depend on time (on space, on frequency, or on something else!). These quantities, which are assumed to be measurable, will be called signals. They correspond in mathematics to the notion of function (of one or more variables of time, space, etc.), and thus signals are modeled by functions. We will see later that the notion of distribution provides a model for signals that is both more general and more satisfactory than that of function.

Examples of signals:

- Intensity of an electric current
- Potential difference between two points in a circuit
- Position of an object, located with respect to time, $M = M(t)$, or with respect to space, $M = M(x, y, z)$
- Gray levels of the points of an image $g(i, j)$
- Components of a field $\mathbf{V}(x, y, z)$
- A sound

There are different ways to think about a signal:

(i) It can be modeled deterministically or statistically. The deterministic point of view will be the only one used here.

(ii) The variable can be continuous; one is then said to have an *analog signal* $x = x(t)$. If the variable is discrete, one is said to have a *discrete signal* $x = (x_n)_{n \in \mathbb{Z}}$. A discrete signal will most often result from *sampling* (also called *discretizing*) an analog signal. (See Figure 1.1.)

(iii) Finally, we will consider the values $x = x(t)$ of a signal to be exact real or complex numbers. However, for computer processing, it is necessary to store these numbers in some finite form, for example, as multiples of an elementary quantity q. This approximation of the exact values is called

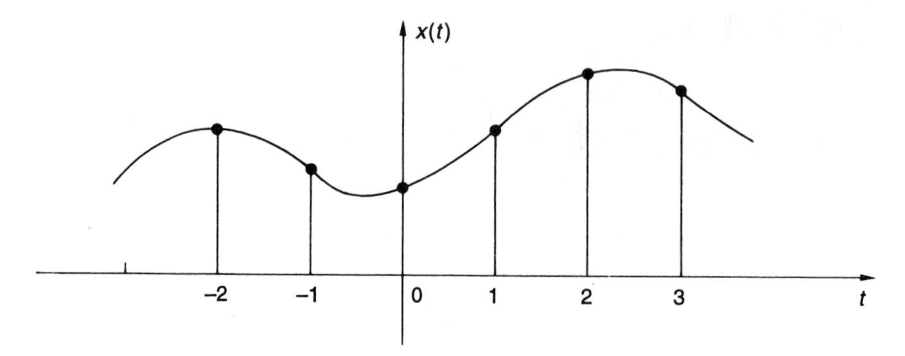

FIGURE 1.1. Sampling an analog signal.

quantization. We will not examine the effects of this operation. A discrete, quantized signal is called a *digital signal*.

Any entity, or apparatus, where one can distinguish input signals and output signals will be called a (transmission) *system* (Figure 1.2). The input and output signals are not necessarily of the same kind (see, for example, Section 1.3.7).

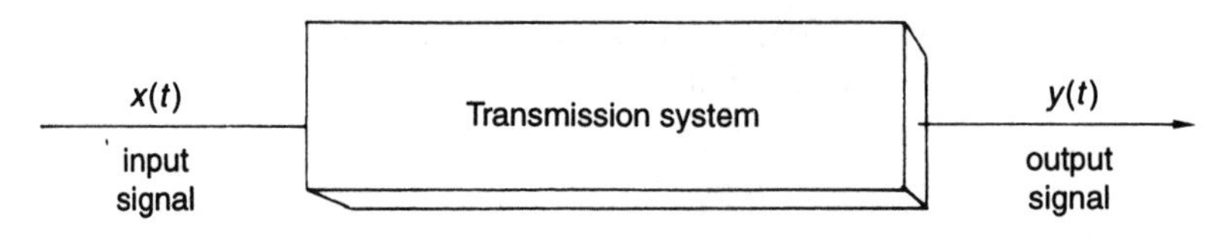

FIGURE 1.2. Diagram of a system.

When there are several input or output signals, the functions $x(t)$ and $y(t)$ are vectors. We will limit our discussion to the scalar case, where there is a single input signal and a single output signal.

In signal theory, one is not necessarily interested in the system's components, but rather in the way it transforms the input signal into the output signal. It is a "black box." It will be modeled by an operator acting on functions, and we write

$$y = Ax,$$

where $x \in X$, the set of input signals, and $y \in Y$, the set of output signals.

Examples of systems:

- An electric circuit
- An amplifier
- The telephone
- The Internet

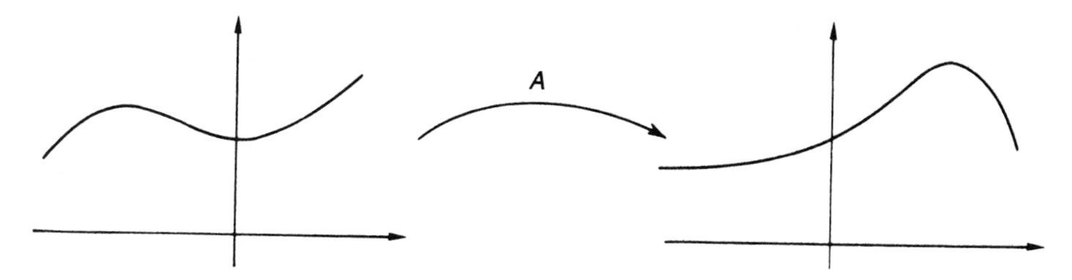

FIGURE 1.3. Analog system.

One distinguishes:

- *Analog systems* that transform an analog signal into another analog signal (Figure 1.3)
- *Discrete systems* that transform a discrete signal into another discrete signal (Figure 1.4)

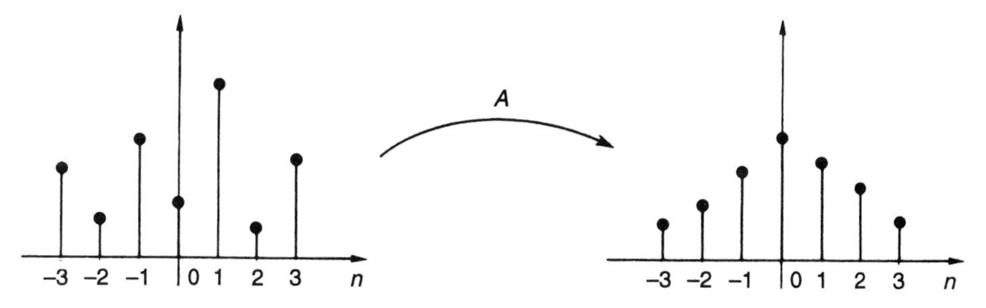

FIGURE 1.4. Discrete system.

One can go from a discrete signal to an analog signal, or conversely, using *converters* that are called *hybrid systems*:

- An analog-to-digital converter, like a sampler, for example;
- A digital-to-analog converter, which produces an analog signal from a digital signal. We mention as an example the *clamper*, or clamping circuit (Figure 1.5). This device yields the last value of the digital signal until the point when the next value arrives.

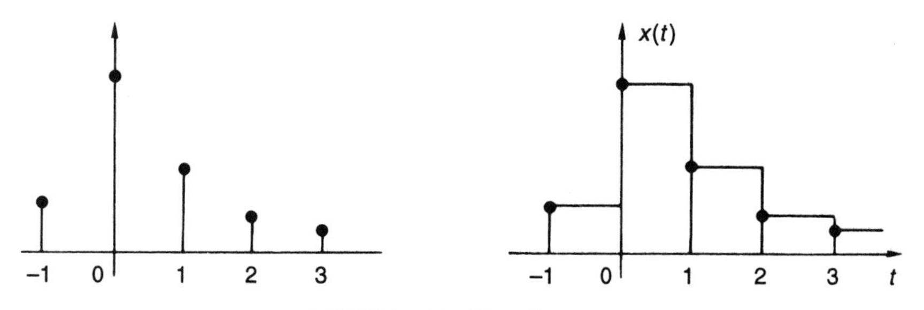

FIGURE 1.5. The clamper.

1.2 Some elementary signals

1.2.1 The Heaviside function

The Heaviside function is the signal, denoted throughout the book by $u(t)$, defined by

$$u(t) = \begin{cases} 0 & \text{if} \quad t < 0, \\ 1 & \text{if} \quad t > 0. \end{cases}$$

(See Figure 1.6.) The value at $t = 0$ can be specified or not. This value is not important for integration. The Heaviside signal models the instantaneous establishment of a steady state.

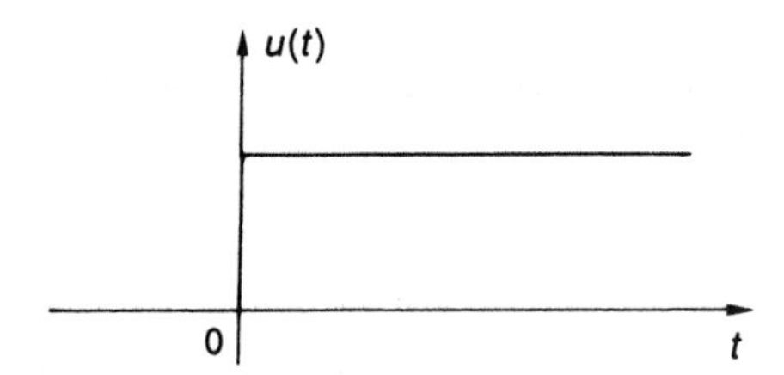

FIGURE 1.6. The Heaviside function.

1.2.2 A rectangular window

The (centered) rectangular signal $r(t)$ (Figure 1.7) is defined, for $a > 0$, by

$$r(t) = \begin{cases} 1 & \text{if} \quad |t| < a, \\ 0 & \text{if} \quad |t| > a. \end{cases}$$

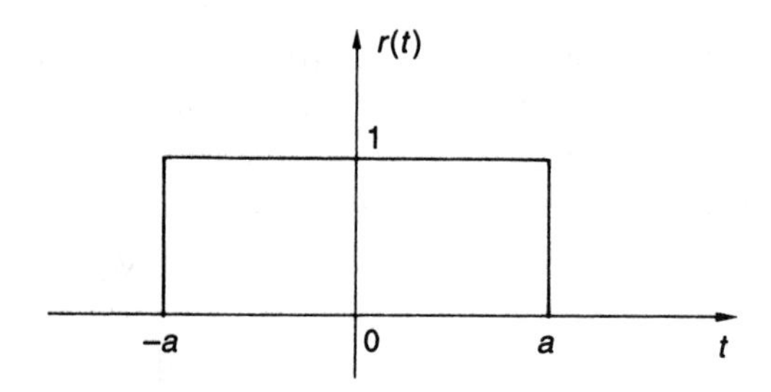

FIGURE 1.7. A rectangular window.

1.2.3 A pure sinusoidal, or monochromatic, signal

A sinusoidal signal is of the form

$$x(t) = \alpha \cos(\omega t + \varphi),$$

where the parameters have the following interpretations:

$$\begin{aligned}
|\alpha| &= \max|x(t)| \text{ is the amplitude of the signal;}\\
\omega &= \text{the angular rate;}\\
a &= 2\pi/\omega \text{ is the (smallest) period;}\\
\lambda &= 1/a \text{ is the frequency;}\\
\varphi &= \text{the initial phase.}
\end{aligned}$$

Signal values are, in principle, real numbers, and the frequency is a positive number. However, for reasons of convenience (Fresnel representation, derivation, multiplication, ...) a complex-valued function

$$z(t) = \alpha e^{i(\omega t+\varphi)}$$

is often used, and one has

$$x(t) = \mathrm{Re}(z(t)) = \frac{1}{2}(z(t) + \overline{z}(t)),$$

where $\overline{z}$ is the complex conjugate of z. Writing the signal this way involves negative frequencies, which make no sense physically. Nevertheless, this is a useful convention. It is always understood that the frequencies of opposite sign will be combined to reproduce the real signal.

A signal of the form $z(t) = ce^{2i\pi\lambda t}$, where $c = |c|e^{i\varphi}$ and where $\lambda \in \mathbb{R}$ — but where c is can be complex and can thus include a phase φ — is often represented by plotting the modulus and argument of c in frequency space, that is, as a function of the frequency λ. This is illustrated in Figures 7.1 and 7.2 for more general functions of the form

$$z(t) = \sum_{n=-\infty}^{\infty} c_n e^{2i\pi\lambda_n t}.$$

1.3 Examples of systems

1.3.1 Ideal amplifier

$$y(t) = kx(t), \quad \text{where } k \text{ is a fixed constant.}$$

1.3.2 Delay line

$$y(t) = x(t-a), \quad \text{where } a \text{ is a real constant.}$$

1.3.3 Differentiator

$$y(t) = x'(t), \quad \text{where } x' \text{ is the derivative of } x.$$

1.3.4 A discrete system

One taps the output y_k, subjects it to a unit time delay, multiplies it by a, and adds it to the input x_k. This gives the recursion equation

$$y_k = x_k + a y_{k-1}, \quad k \in \mathbb{Z}. \tag{1.1}$$

Such a system is typically represented by the diagram in Figure 1.8.

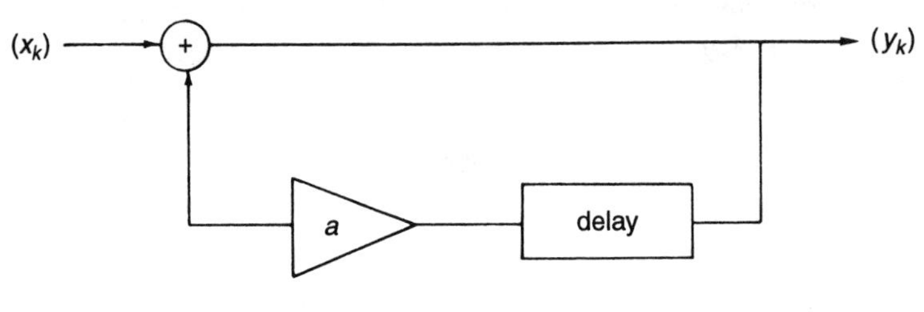

FIGURE 1.8.

1.3.5 An RC circuit

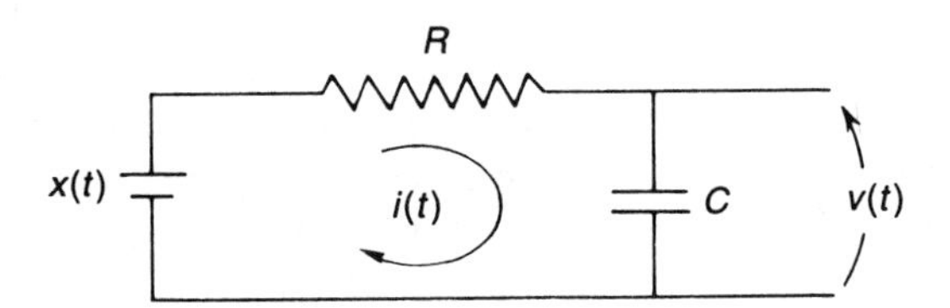

FIGURE 1.9. An RC circuit.

The input to the circuit shown in Figure 1.9 is the voltage $x(t)$; the output is the voltage $v(t)$ across the capacitor. Thus,

$$A : x \mapsto v.$$

The potential difference across a capacitor with charge Q is $v = Q/C$, so by Ohm's law

$$Ri(t) + v(t) = x(t).$$

By writing $i(t) = Q'(t)$, this becomes

$$RCv'(t) + v(t) = x(t). \tag{1.2}$$

This system is governed by a first-order linear differential equation with constant coefficients. In general, the solution v will depend on a parameter that is computed using an additional condition, for example, the initial condition $v(0) = 0$.

1.3.6 A mechanical example

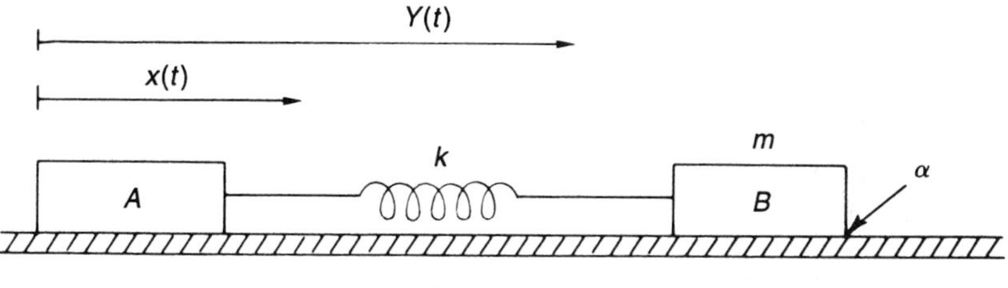

FIGURE 1.10.

The mechanical system shown in Figure 1.10 consists of two bodies A and B that slide in one direction on a fixed surface. The mass B has a coefficient of sliding friction α. A and B are connected by a spring with restoring constant k. A is driven by the controlled movement $x(t)$ (the input), and this causes the motion of B, which is measured by the distance $y(t)$ (the output).

If B has mass m and acceleration $\gamma(t)$, then by Newton's law, $m\gamma(t)$ is equal at each instant t to the sum of the forces acting on B. These are the restoring force, $-k(y(t) - x(t))$, and the friction force, $-\alpha y'(t)$. Combining these forces gives the equation

$$my''(t) + \alpha y'(t) + ky(t) = kx(t), \tag{1.3}$$

and the output is completely determined if two initial conditions are known, for example, $y(0) = 0$ and $y'(0) = 0$.

1.3.7 A system of resistors

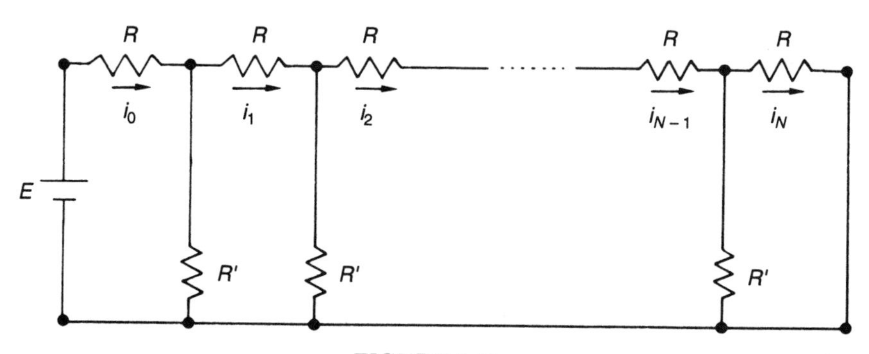

FIGURE 1.11.

Consider the electrical circuit in Figure 1.11. The input is the constant voltage E, and the currents $i_0, i_1, \ldots, i_N$ constitute the output:

$$A : E \mapsto (i_0, i_1, \ldots, i_N).$$

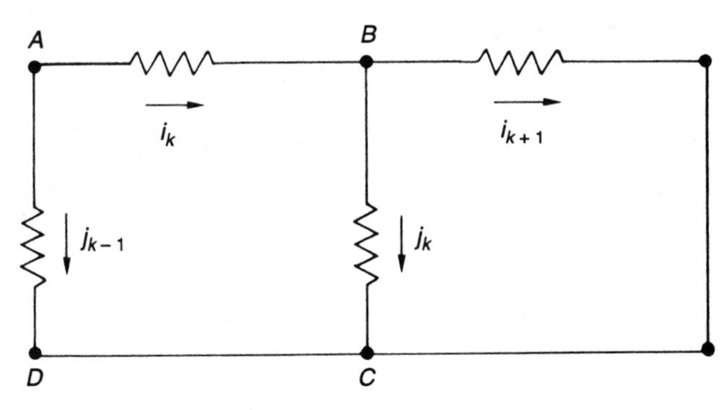

FIGURE 1.12.

Summing the currents at the kth node gives

$$i_k = i_{k+1} + j_k.$$

On the other hand, by summing the voltage drops around the loop $DABC$ we have (see Figure 1.12)

$$0 = v_A - v_B + v_B - v_C + v_C - v_A,$$
$$0 = Ri_k + R'j_k - R'j_{k-1},$$
$$0 = Ri_k + R'(i_k - i_{k+1}) - R'(i_{k-1} - i_k).$$

Finally, taking $i_{N+1} = 0$ shows that for $k = 1, \dots, N$,

$$E = (R + R')i_0 - R'i_1,$$
$$0 = R'i_{k-1} - (2R' + R)i_k + R'i_{k+1},$$
$$0 = i_{N+1}.$$

This is a second-order linear recursion system with boundary conditions.

Lesson 2

Filters and Transfer Functions

Systems have properties, at least sometimes. We are going to review several of the more standard properties of systems.

2.1 Algebraic properties of systems

The set of input signals X and the set of output signals Y are assumed to be vector spaces (real or complex). A system A can have several properties:

2.1.1 Linearity

Consider the system

$$A : X \to Y.$$

A is said to be linear if

$$A(x + u) = A(x) + A(u)$$

and

$$A(\lambda x) = \lambda A(x)$$

for all $x, u \in X$ and all $\lambda \in \mathbb{R}$ (or $\mathbb{C}$ if X is complex). This is also called the *principle of superposition*. The systems in Section 1.3 are all linear, which is easily verified by examining the governing equations (1.1) through (1.4).

2.1.2 Causality

A is said to be *realizable* (or *causal*) if the equality of any two input signals up to time $t = t_0$ implies the equality of the two output signals at least to time t_0:

$$x_1(t) = x_2(t) \text{ for } t < t_0 \quad \Longrightarrow \quad Ax_1(t) = Ax_2(t) \text{ for } t < t_0.$$

This property is completely natural for a physical system in which the variable is time. It says that the response at time t depends only on what has happened before t. In particular, the system does not respond before there is an input. Thus causality is a necessary condition for the system to be physically realizable.

2.1.3 Invariance

A is said to be *invariant*, or *stationary*, if a translation in time of the input leads to the same translation of the output; that is,

$$x(t) \to y(t) \quad \Longrightarrow \quad x(t-a) \to y(t-a).$$

Let τ_a be the delay operator defined by

$$\tau_a x(t) = x(t-a).$$

If the system A is invariant, then

$$A(\tau_a x) = \tau_a(Ax)$$

for all $x \in X$ and $a \in \mathbb{R}$. Thus, for all $a \in \mathbb{R}$,

$$A\tau_a = \tau_a A,$$

which says that A commutes with all translations. For discrete systems, one considers only a that are multiples of the sampling interval.

2.2 Continuity of a system

The system

$$A : X \to Y$$

is said to be *continuous* if the sequence Ax_n $(= y_n)$ tends to Ax $(= y)$ when the sequence x_n tends to x. *This concept assumes that there exists some notion of sequential limit for signals in both X and Y.*

Continuity is a natural hypothesis; it expresses that idea that if two input signals are close, then the output signals are also close.

2.2.1 The analog case

When the signals are functions, the notion of limit is often defined in terms of a *norm* $\|\cdot\|$ defined on each of the vector spaces X and Y. In this case

$$x_n \to x \quad \text{means that} \quad \|x_n - x\| \to 0.$$

These are the three most frequently used norms:

(i) The norm for *uniform convergence*:

$$\|x\|_\infty = \sup_{t \in I} |x(t)|.$$

(ii) The norm for *mean convergence*:

$$\|x\|_1 = \int_I |x(t)|\, dt.$$

(iii) The norm for *convergence "in energy"* (*mean quadratic convergence*):

$$\|x\|_2 = \Big(\int_I |x(t)|^2\, dt \Big)^{1/2}.$$

In all cases, I is the interval of interest.

This last norm has the advantage over the other two of being derived from a *scalar product*,

$$(x, y) = \int_I x(t)\overline{y}(t)\, dt,$$

where $\overline{y}(t)$ denotes the complex conjugate of $y(t)$. Thus, $\|x\|_2 = \sqrt{(x, x)}$. Such a structure allows one to introduce the notion of orthogonality between two signals. This generalizes the concept of orthogonality in $\mathbb{R}^n$ and is expressed by the relation

$$(x, y) = 0.$$

One often uses a less restrictive notion of continuity, namely, *continuity in the sense of distributions*. This concept will be studied Chapter VIII.

2.2.2 The discrete case

When the signals are discrete, one can use the analogous norms:

$$\|x\|_\infty = \sup_{n \in \mathbb{Z}} |x_n|; \quad \|x\|_1 = \sum_{n=-\infty}^{+\infty} |x_n|; \quad \|x\|_2 = \Big(\sum_{n=-\infty}^{+\infty} |x_n|^2 \Big)^{1/2}.$$

The *simple convergence* of a sequence of signals,

$$x_n = (x_{nk})_{k \in \mathbb{Z}},$$

is also used. By this we mean that the sequence x_n tends to x if the limit exists for each of the components:

$$x_n \to x \quad \Longleftrightarrow \quad x_{nk} \to x_k \text{ for each } k \in \mathbb{Z} \text{ as } n \to +\infty.$$

Example: A differentiator is not a continuous system in the uniform convergence norm. Indeed, if we take $x_n = (1/n)\sin(nt)$, then $x_n \to 0$ uniformly in t. But $y_n(t) = x'_n(t) = \cos(nt)$ does not tend to zero as $n \to +\infty$. On the other hand, we will show that the integrator

$$y(t) = \int_{-\infty}^{t} e^{-(t-s)} x(s)\, ds$$

is continuous with respect to uniform convergence.

2.3 The filter and its transfer function

The term *filter* refers both to a physical system having certain properties and to its mathematical model defined in terms of the following objects:

(i) two vector spaces X and Y of input and output signals, respectively, that are endowed with a notion of convergence;

(ii) a linear operator $A : X \mapsto Y$ that is continuous and translation-invariant.

We will say informally that *a filter is a continuous, translation-invariant linear system.*

Such a system satisfies the *principle of superposition*, which is another name for linearity. Thus,

$$A\Big(\sum_{n=0}^{k} a_n x_n\Big) = \sum_{n=0}^{k} a_n A x_n,$$

and by continuity, one can pass to the limit when the infinite sums converge:

$$A\Big(\sum_{n=0}^{+\infty} a_n x_n\Big) = \sum_{n=0}^{+\infty} a_n A x_n.$$

Later we will see that a periodic signal (under rather general conditions) can be written as an infinite sum of *monochromatic* signals in the form

$$x(t) = \sum_{n=-\infty}^{+\infty} c_n e^{2i\pi\lambda nt}.$$

Hence, at the output of a filter we will have

$$y = Ax = \sum_{n=-\infty}^{+\infty} c_n A(e_\lambda^n), \quad \text{where} \quad e_\lambda(t) = e^{2i\pi\lambda t}.$$

It is thus sufficient to know the outputs for each of the inputs $(e_\lambda^n)_{n\in\mathbb{Z}}$ to know the image of an arbitrary periodic signal. Furthermore, it is easy to determine the image f_λ of the signal e_λ, assuming that the latter belongs to the space of input signals for the filter. Indeed, for all values of t and u,

$$e_\lambda(t+u) = e_\lambda(t)e_\lambda(u) = \tau_{-t}e_\lambda(u),$$

where t is considered to be a parameter and u to be the variable. The image of this signal is $f_\lambda(t+u)$. As a result, we see that

$$f_\lambda(t+u) = A(e_\lambda(t)e_\lambda)(u) = e_\lambda(t)f_\lambda(u)$$

for all $u \in \mathbb{R}$. Hence, for $u = 0$,

$$f_\lambda(t) = e_\lambda(t)f_\lambda(0),$$

which we write as

$$A(e_\lambda) = H(\lambda)e_\lambda, \quad \text{where} \quad H(\lambda) = f_\lambda(0).$$

This result can be expressed as follows:

2.3.1 Proposition *Assume that e_λ is an admissible input function for the filter A, which is otherwise arbitrary. Then e_λ is an eigenfunction of the filter A. That is to say, there exists a scalar function $H(\lambda)$ such that for $\lambda \in \mathbb{R}$,*

$$A(e_\lambda) = H(\lambda)e_\lambda.$$

The function $H : \mathbb{R} \mapsto \mathbb{C}$ is called the *transfer function* of the filter A. There will be many occasions in the rest of the book where we will see the essential role that this function plays in the action of a filter on the spectrum of an input signal.

2.4 A standard analog filter: the RC cell

We will illustrate the general ideas presented above with the RC circuit shown in Section 1.3.5.

2.4.1 System response

Writing the unknown function v as $v(t) = w(t)e^{-\frac{t}{RC}}$ reduces the initial equation

$$RCv'(t) + v(t) = x(t)$$

to

$$w'(t) = \frac{1}{RC}e^{\frac{t}{RC}}x(t).$$

Assuming that the input signal $x(t)$ is such that the second member is integrable on every interval $(-\infty, t)$, we have

$$w(t) = \frac{1}{RC} \int_{-\infty}^{t} e^{\frac{s}{RC}} x(s)\, ds + K$$

and

$$v(t) = \frac{1}{RC} \int_{-\infty}^{t} e^{-\frac{t-s}{RC}} x(s)\, ds + K e^{-\frac{t}{RC}}.$$

The constant K is determined by an auxiliary condition. For example, if we assume that the response to the zero input is zero, we see that $K = 0$.

One can *define* the response of the system A to the input x to be

$$v(t) = Ax(t) = \frac{1}{RC} \int_{-\infty}^{t} e^{-\frac{t-s}{RC}} x(s)\, ds. \tag{2.1}$$

It is clear from this expression that A is linear, realizable, and invariant. It is also continuous, for example, in the uniform norm, since

$$|Ax(t)| \leq \|x\|_\infty \frac{1}{RC} \int_{-\infty}^{t} e^{-\frac{t-s}{RC}}\, ds = \|x\|_\infty,$$

and thus

$$\|Ax\|_\infty \leq \|x\|_\infty.$$

This shows that the RC cell is a filter.

2.4.2 An expression for the output

If we write

$$h(t) = \frac{1}{RC} e^{-\frac{t}{RC}} u(t),$$

where u is the Heaviside function, then we can express (2.1) as

$$Ax(t) = \int_{-\infty}^{+\infty} h(t-s) x(s)\, ds = (h * x)(t). \tag{2.2}$$

This operation is, by definition, the *convolution* of the two signals h and x. It is denoted by $h * x$, and we have

$$Ax = h * x.$$

In this situation, one is said to have a *convolution system*. The function h, called the *impulse response* of the system, characterizes the filter because knowing h implies that the output of the filter is known for any input x. Throughout the book, we will use h to denote the impulse response of a system. A companion notion, the response of a system to the unit step function $u(t)$, will be defined in Lesson 24.

2.4.3 The transfer function of the RC filter

The response to the input $x(t) = e_\lambda(t)$ is $v(t) = H(\lambda)e_\lambda(t)$. Substitution in equation (1.2) gives

$$(2i\pi\lambda RC + 1)H(\lambda)e_\lambda(t) = e_\lambda(t),$$

and we have

$$H(\lambda) = \frac{1}{1 + 2i\pi\lambda RC}.$$

We see that signals for which $|\lambda|$ is small, the low-frequency signals, are transmitted by the filter almost as if it were the identity mapping (see Figure 2.1). On the other hand, the high-frequency signals, for which $|\lambda|$ is large, are almost completely attenuated. This explains why this filter is called a *low-pass filter*. The action of the filter on different frequencies is clearly apparent from the graph of the function

$$|H(\lambda)|^2 = \frac{1}{1 + 4\pi^2\lambda^2R^2C^2},$$

which is called the *energy spectrum of the filter.* The function $|H(\lambda)|$ is called the *spectral amplitude.*

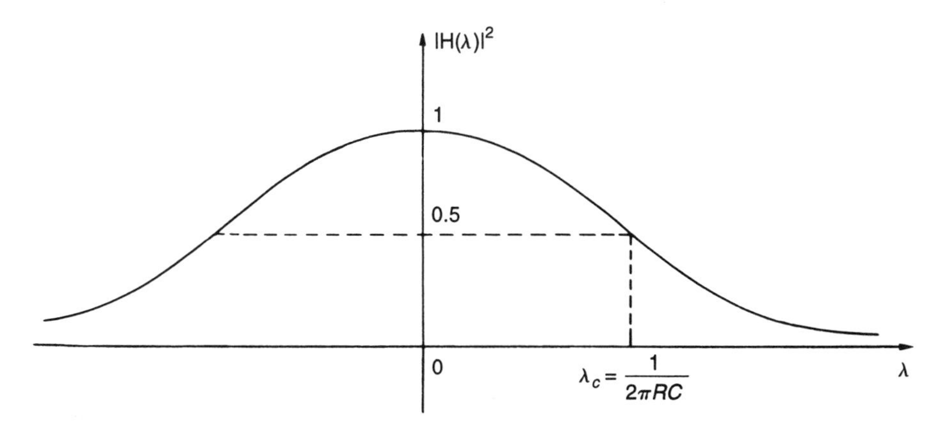

FIGURE 2.1. Energy spectrum of the low-pass RC filter.

The frequency $\lambda_C = 1/(2\pi RC)$, beyond which the amplitudes of the input frequencies are reduced by more than the factor $1/\sqrt{2}$, is considered to be the cutoff frequency.

We will return to the RC filter in Lesson 25. In fact, the analysis of this filter and of the systems described by generalizing the equation that governs the RC filter will be the main application of the mathematical tools that are developed in the book.

2.5 A first-order discrete filter

The discrete analogue of the last case is the example in Section 1.3.4, where for $a \neq 0$,

$$y_k = ay_{k-1} + x_k, \quad k \in \mathbb{Z}. \tag{2.3}$$

The analysis of this system follows that of the last example. We try to express the output explicitly as a function of the input. Thus we change the unknown by letting $y_k = a^k v_k$, which transforms (2.3) into

$$v_k - v_{k-1} = a^{-k} x_k. \tag{2.4}$$

By successive additions, we see that

$$v_k - v_{k-p} = \sum_{l=k-p+1}^{k} a^{-l} x_l, \quad p \in \mathbb{N}^*. \tag{2.5}$$

Suppose that the input signal (x_k) is such that the series

$$\sum_{n=-\infty}^{0} a^{-n} x_n \tag{2.6}$$

is absolutely convergent. Then from (2.5), the sequence $(v_k)_{k \leq 0}$ is a Cauchy sequence and thus converges to some limit b as $k \to -\infty$. Letting $p \to +\infty$ in (2.5) shows that for each fixed $k \in \mathbb{Z}$

$$v_k = b + \sum_{l=-\infty}^{k} a^{-l} x_l$$

and

$$y_k = ba^k + \sum_{n=-\infty}^{k} a^{k-n} x_n.$$

Conversely, for each complex constant b, the sequence (y_k) of this form is a solution of (2.3). It is logical that y_k is not completely determined, since we have not specified an initial condition. Now assume that the response to a null input is null; then necessarily $b = 0$. If we define (h_n) by

$$h_n = \begin{cases} a^n & \text{if} \quad n \geq 0, \\ 0 & \text{if} \quad n < 0, \end{cases}$$

the output (y_k) can be written as

$$y_k = \sum_{n=-\infty}^{+\infty} h_{k-n} x_n, \quad \text{or} \quad y = h * x,$$

which is called the *discrete convolution* of the two signals $x = (x_n)$ and $h = (h_n)$. This system is linear and invariant. One can easily verify that the condition $h_n = 0$ if $n < 0$ implies that it is realizable. The system is continuous in the uniform norm whenever $|a| < 1$, and we have

$$\|y\|_\infty \leq \|h\|_1 \cdot \|x\|_\infty.$$

The signal h is called the *impulse response* of the system. It is the response to a unit impulse at time $t = 0$ defined by

$$x = e = (e_k), \quad \text{where} \quad e_k = \begin{cases} 1 & \text{if} \quad k = 0, \\ 0 & \text{if} \quad k \neq 0. \end{cases}$$

As in the analog case, we examine the response to an exponential input $x = (z^k)$ where z is a fixed complex number with $|z| > |a|$ so that the hypothesis concerning the convergence of (2.6) is satisfied. We obtain the relation

$$y_k = a^k \sum_{n=-\infty}^{k} \left(\frac{z}{a}\right)^n = \frac{z}{z-a} x_k.$$

Thus the output is proportional to the input. The exponential signals are eigenfunctions of the filter, as expected, and the eigenvalue

$$H(z) = \frac{z}{z-a}$$

is again called the *transfer function* of the discrete filter. It is a function of the complex variable z that is defined and analytic in the domain $|z| > |a|$. We return to this function, from another point of view, in Lesson 40.

Chapter II

Periodic Signals

Lesson 3

Trigonometric Signals

We have seen that the pure sinusoidal signals are eigenfunctions for all filters. They are also the simplest periodic signals. These two facts explain their importance. We will see in the next two lessons that they enter into the structure of all periodic signals.

3.1 Trigonometric polynomials

A function f is said to be *periodic* with period a, $a > 0$, if for all $t \in \mathbb{R}$,

$$f(t+a) = f(t).$$

(Note that a is not necessarily the smallest period.) Since $e_n(t) = e^{2i\pi n \frac{t}{a}}$ has period a for each integer n, the same is true for functions p of the form

$$p(t) = \sum_{n \in I} c_n e^{2i\pi n \frac{t}{a}},$$

where I is any fixed, finite set of integers and the c_n are arbitrary complex numbers. By adding zero terms if necessary, we may assume that

$$p(t) = \sum_{n=-N}^{+N} c_n e^{2i\pi n \frac{t}{a}}. \tag{3.1}$$

This function is called a *trigonometric polynomial* of degree less than or equal to N. These functions model the superposition of a finite number of monochromatic signals; the real part of such a function can be represented graphically as a function of time as in Figure 3.1.

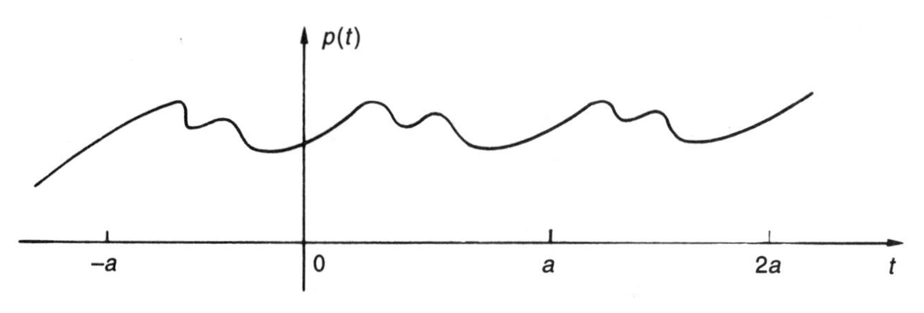

FIGURE 3.1.

3.2 Representation in sines and cosines

Expression (3.1) can be transformed to express $p(t)$ as a linear combination of sines and cosines. Thus,

$$p(t) = c_0 + \sum_{n=1}^{N} \left(c_n e^{2i\pi n\frac{t}{a}} + c_{-n} e^{-2i\pi n\frac{t}{a}}\right),$$

and by expanding the exponentials, this becomes

$$p(t) = \frac{1}{2}a_0 + \sum_{n=1}^{N} \left(a_n \cos\left(2\pi n\frac{t}{a}\right) + b_n \sin\left(2i\pi n\frac{t}{a}\right)\right), \tag{3.2}$$

where, for $n \geq 0$,

$$\begin{aligned} a_n &= c_n + c_{-n}, \\ b_n &= i(c_n - c_{-n}). \end{aligned} \tag{3.3}$$

The inverse formulas are

$$\begin{aligned} c_n &= \frac{1}{2}(a_n - ib_n), \\ c_{-n} &= \frac{1}{2}(a_n + ib_n). \end{aligned} \tag{3.4}$$

3.3 Orthogonality

A simple computation shows that the following important relation holds for the functions $e_n(t)$:

$$\int_0^a e_n(t)\bar{e}_m(t)\,dt = \begin{cases} a & \text{if} \quad n = m, \\ 0 & \text{if} \quad n \neq m. \end{cases} \tag{3.5}$$

We let T_N denote the set of all trigonometric polynomials p of degree less than or equal to N. T_N is obtained by letting the c_n in formula (3.1) vary over all possible values. If we endow this vector space, which has finite dimension $\leq 2N+1$, with the scalar product

$$(p, q) = \int_0^a p(t)\overline{q}(t)\,dt,$$

the relation (3.5) expresses the fact that the functions e_n, $n \in \mathbb{Z}$, are orthogonal:

$$(e_n, e_m) = 0 \text{ if } n \neq m, \text{ and } \|e_n\|_2 = \sqrt{a}.$$

It follows that the vectors e_n are independent and that the dimension of T_N is exactly $2N+1$. If p is of the form (3.1), we have

$$(p, e_n) = c_n \|e_n\|_2^2 = a c_n,$$

and

$$c_n = \frac{1}{a}\int_0^a p(t) e^{-2i\pi n \frac{t}{a}}\,dt. \tag{3.6}$$

This is called Fourier's formula; it gives the coefficients c_n explicitly in terms of the function p. One easily obtains the following formulas for the coefficients a_n and b_n, $n \geq 0$:

$$\begin{aligned} a_n &= \frac{2}{a}\int_0^a p(t)\cos\left(2\pi n\frac{t}{a}\right)dt, \\ b_n &= \frac{2}{a}\int_0^a p(t)\sin\left(2\pi n\frac{t}{a}\right)dt. \end{aligned} \tag{3.7}$$

Theoretically, this has solved the problem of spectral analysis for a trigonometric signal: knowing the values of p, calculate the coefficients c_n in (3.1). Later we will see how to do this important calculation efficiently.

Since p is periodic, the integral (3.6) can be taken over any interval of length a. By taking it to be $(-a/2, a/2)$, we immediately have the following properties:

$$\begin{aligned} &p \text{ even} &\Leftrightarrow\quad & c_{-n} = c_n, & n \in \mathbb{Z} \quad &\Leftrightarrow\quad b_n = 0,\ n \in \mathbb{N}; \\ &p \text{ odd} &\Leftrightarrow\quad & c_{-n} = -c_n, & n \in \mathbb{Z} \quad &\Leftrightarrow\quad a_n = 0,\ n \in \mathbb{N}. \end{aligned}$$

Finally, computing the quadratic norm of p from (3.1) gives

$$\|p\|_2^2 = \sum_{n=-N}^{N}\sum_{m=-N}^{N} c_n \overline{c}_m (e_n, e_m),$$

and this, combined with (3.5), yields *Parseval's equality* for trigonometric polynomials:

$$\sum_{m=-N}^{N} |c_n|^2 = \frac{1}{a}\int_0^a |p(t)|^2\,dt. \tag{3.8}$$

3.4 Exercises

Exercise 3.1 If $f : \mathbb{R} \to \mathbb{R}$ is a periodic function with period a, integrable on bounded intervals, show that the integral $\int_x^{x+a} f(t)\,dt$ does not depend on x.

Lesson 4

Periodic Signals and Fourier Series

The question is this: If $f : \mathbb{R} \to \mathbb{C}$ is an arbitrary function with period a, can we find a decomposition of f of the form

$$f(t) = \sum c_n e^{2i\pi n \frac{t}{a}} \ ? \tag{4.1}$$

The immediate answer is "no" if one considers only finite sums. The sum on the right is infinitely differentiable, while there is no reason for f to be. For example, f could be the periodic window function in Figure 4.1.

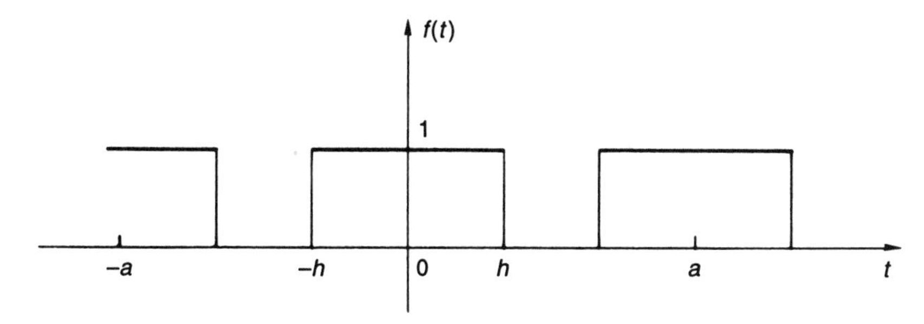

FIGURE 4.1. Periodic window.

4.1 The space $L^2_{\mathrm{p}}(0, a)$

In a famous paper dated 1807, Joseph Fourier asserted that the answer to this question was "yes," provided that infinite sums are allowed. He arrived at his results by a very circuitous route using the "tools at hand," which is to say, the mathematical techniques available at that time. Recall that at the beginning of the nineteenth century, not only was the notion of convergence rather vague, but the definition of function itself was open to controversy. For example, the following question was debated: Is a function defined on two consecutive intervals by two different formulas still a function? (For more about this subject, we recommend the interesting popular book [DH82].)

Today we approach this problem with two centuries of experience, during which the tools of mathematical analysis have been considerably developed and refined. In particular, we now understand the parts of the problem that are simple and those that are difficult. In this section, we will approach the problem from a geometric point of view.

Note first that a periodic function defined on $\mathbb{R}$ with period a is completely determined by its values on any interval $[x, x+a)$ of length a.

In addition to periodicity, we need to assume that the functions considered are such that the integral

$$\int_0^a |f(t)|^2 \, dt$$

exists and is finite. We will not dwell on issues of integration at this point; they will be discussed in Lessons 11 through 15. We introduce the notation

$$L^2_{\mathrm{p}}(0,a) = \Big\{ f : \mathbb{R} \to \mathbb{C} \;\Big|\; f \text{ has period } a \text{ and } \int_0^a |f(t)|^2 \, dt < +\infty \Big\}.$$

The index "p" is to remind us that the functions are periodic.

This set, endowed with the usual addition for functions and scalar multiplication, is a complex vector space. (At this point, it is not obvious that the sum of two such functions is a member of the set (see Proposition 16.1.4). We define a scalar product (or in the complex case, a *Hermitian form*) on this set by

$$(f, g) = \int_0^a f(t)\overline{g}(t) \, dt.$$

The associated norm is given by

$$\|f\|_2 = \sqrt{(f,f)} = \Big(\int_0^a |f(t)|^2 \, dt \Big)^{1/2}.$$

It is important to note that the norm of f can be zero even though the function $f \in L^2_{\mathrm{p}}(0,a)$ is not zero at every point. For example, f could be zero at all but a finite number of points. Thus, to have a true norm, it is necessary to identify such a function with the function that is identically zero, which we sometimes call the *null function*. Generally, we must identify any two functions f and g for which $\int_0^a |f(t) - g(t)| \, dt = 0$ (see Section 13.3). In this case, we say that the two functions are equal *almost everywhere*, and we write $f = g$ a.e. At this point, it is sufficient to remember that

$$\int_0^a |f(t)| \, dt = 0 \quad \Longleftrightarrow \quad f = 0 \text{ a.e. on } (0, a).$$

4.2 The idea of approximation

If equality (4.1) cannot hold exactly for a finite sum, one can try to have it hold "as well as possible." More precisely, we can try to answer the following question: Given an integer N, is it possible to find coefficients x_n such that

$$\|f - \sum_{n=-N}^{N} x_n e_n\|_2 \text{ attains a minimum?}$$

Geometrically, this amounts to finding an element f_N in the subspace T_N of $L^2_{\mathrm{p}}(0,a)$ that has minimum distance from f. When such an element f_N exists, we say that it is the *best approximation* of f in T_N (see Figure 4.2).

FIGURE 4.2. Orthogonal projection on a subspace.

To solve this approximation problem, we first try to evaluate the distance between f and an arbitrary trigonometric polynomial in T_N,

$$p(t) = \sum_{n=-N}^{N} x_n e_n.$$

Thus,

$$\|f - p\|_2^2 = \|f\|_2^2 - 2\mathrm{Re}(f,p) + \|p\|_2^2.$$

We know from (3.8) that

$$\|p\|_2^2 = a \sum_{n=-N}^{N} |x_n|^2 \quad \text{and} \quad (f,p) = \sum_{n=-N}^{N} \overline{x}_n (f, e_n).$$

By writing

$$c_n = c_n(f) = \frac{1}{a}(f, e_n),$$

we have

$$\|f - p\|_2^2 = \|f\|_2^2 + a \sum_{n=-N}^{N} (|c_n - x_n|^2 - |c_n|^2). \tag{4.2}$$

From (4.2) it is perfectly clear that the minimum is attained when $x_n = c_n$, and only for this value.

In summary, the best approximation f_N exists and is unique, and it is given by

$$f_N(t) = \sum_{n=-N}^{N} c_n e_n(t).$$

4.2.1 Theorem *There exists a unique trigonometric polynomial f_N in T_N such that*

$$\|f - f_N\|_2 = \min_{p \in T_N} \|f - p\|_2.$$

This polynomial is given by

$$f_N(t) = \sum_{n=-N}^{N} c_n e^{2i\pi n \frac{t}{a}}, \tag{4.3}$$

where

$$c_n = \frac{1}{a} \int_0^a f(t) e^{-2i\pi n \frac{t}{a}} \, dt. \tag{4.4}$$

4.2.2 Bessel's inequality

For $x_n = c_n$, equality (4.2) becomes

$$a \sum_{n=-N}^{N} |c_n|^2 + \|f - f_N\|_2^2 = \|f\|_2^2. \tag{4.5}$$

An immediate consequence is the inequality

$$\sum_{n=-N}^{N} |c_n|^2 \le \frac{1}{a} \int_0^a |f(t)|^2 \, dt, \quad N \in \mathbb{N},$$

which is traditionally known as *Bessel's inequality.*

4.2.3 Corollary *For any $f \in L^2_{\mathrm{p}}(0, a)$, we have the inequality*

$$\sum_{n=-\infty}^{+\infty} |c_n|^2 < +\infty,$$

and hence $c_n(f) \to 0$ as $|n| \to +\infty$.

4.3 Convergence of the approximation

One can ask what happens to f_N as $N \to +\infty$. Here is an example. Take $a = 2\pi$ and define

$$f(t) = \begin{cases} +1 & \text{if} \quad 0 \le t < \pi, \\ -1 & \text{if} \quad \pi \le t < 2\pi. \end{cases}$$

By writing the exponentials in terms of sines, we have the following approximations for $N = 1, 3, 5$:

$$\begin{aligned} f_1(t) &= \frac{4}{\pi} \sin t; \\ f_3(t) &= \frac{4}{\pi}(\sin t + \frac{1}{3} \sin 3t); \\ f_5(t) &= \frac{4}{\pi}(\sin t + \frac{1}{3} \sin 3t + \frac{1}{5} \sin 5t). \end{aligned}$$

These functions are shown graphically in Figures 4.3–4.5.

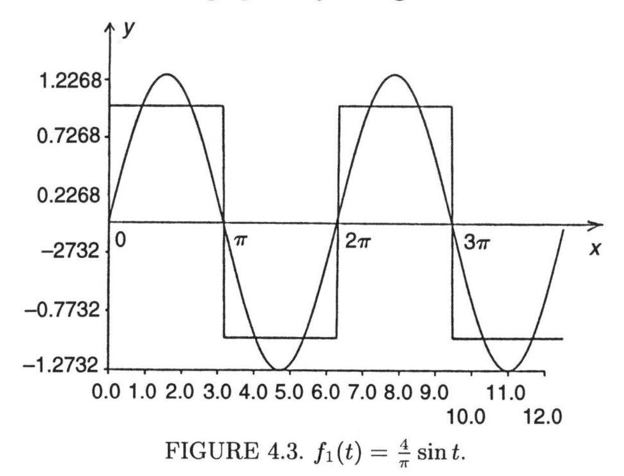

FIGURE 4.3. $f_1(t) = \frac{4}{\pi} \sin t$.

From this example it appears that f_N tends to f as N increases. In fact, we have following important general result.

4.3.1 Theorem *If $f \in L^2_{\mathrm{p}}(0, a)$, then the best approximation of f in T_N, which is given by*

$$f_N = \sum_{n=-N}^{N} c_n e^{2i\pi n \frac{t}{a}}$$

with the c_n defined by (4.4), tends to f in $L^2_{\mathrm{p}}(0, a)$ as $N \to +\infty$. Expressed otherwise,

$$\int_0^a |f(t) - f_N(t)|^2 \, dt \to 0 \text{ as } N \to +\infty.$$

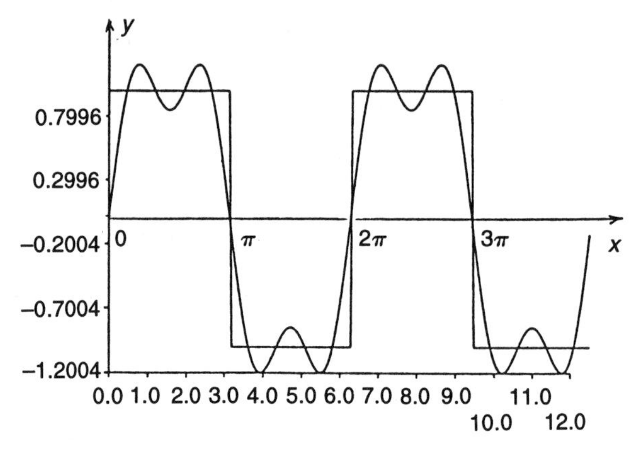

FIGURE 4.4. $f_3(t) = \frac{4}{\pi}(\sin t + \frac{1}{3}\sin 3t)$.

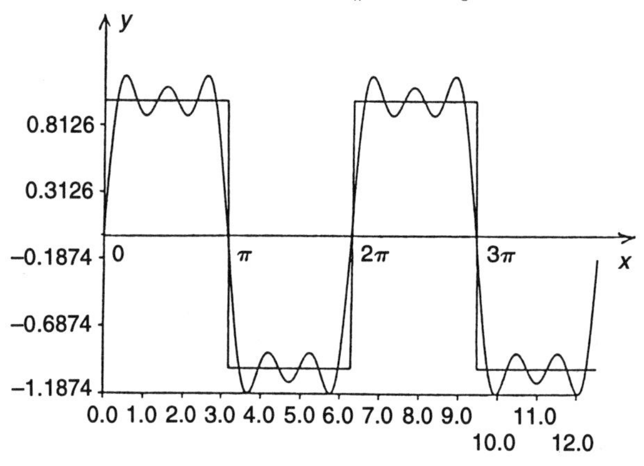

FIGURE 4.5. $f_5(t) = \frac{4}{\pi}(\sin t + \frac{1}{3}\sin 3t + \frac{1}{5}\sin 5t)$.

The proof of this theorem requires more background than is available in these early lessons. We will prove it in Lesson 16 as an illustration of results about Lebesgue integration.

It is this theorem that gives meaning to the formula

$$f(t) = \sum_{n=-\infty}^{+\infty} c_n e^{2i\pi n\frac{t}{a}}, \tag{4.6}$$

or, as in (3.2), to the expression

$$f(t) = \frac{a_0}{2} + \sum_{n=1}^{+\infty}\left(a_n \cos\left(2\pi n\frac{t}{a}\right) + b_n \sin\left(2i\pi n\frac{t}{a}\right)\right). \tag{4.7}$$

Two remarks are indicated at this point:

(a) Formulas (4.6) and (4.7) are equalities in the $L^2_{\mathrm{p}}(0,a)$ norm. In particular, they do not mean that for each value of t, the complex number $f(t)$

is equal to the sum of the series on the right. In the last example, the sum of the series is zero for $t = \pi$, whereas $f(\pi) = -1$. Similarly, if f is modified at a point t_0, the Fourier series is unchanged for $t = t_0$. This touches on the problem of pointwise representation that will be studied in the next lesson.

(b) A more scholarly way to express Theorem 4.3.1 is to say that the family of functions $(e_n)_{n\in\mathbb{Z}}$ is a *topological basis* for the space $L^2_{\mathrm{p}}(0,a)$ and that the series

$$\sum_{n=-\infty}^{+\infty} c_n e_n$$

is summable to f in this space. The meaning of this will be explained in Lesson 16, where we develop the theory of Hilbert spaces.

The c_n are called the *Fourier coefficients* of the periodic function f.

4.3.2 Parseval's equality

Equation (4.5) and Theorem 4.3.1 imply that

$$\sum_{n=-\infty}^{+\infty} |c_n|^2 = \frac{1}{a}\int_0^a |f(t)|^2\,dt. \tag{4.8}$$

This is called Parseval's equality. In fact, in view of (4.5), it is equivalent to the statement of Theorem 4.3.1. Another way to express (4.8) is to say that the energy of a periodic signal is equal to the sum of the energies of its harmonic components. If the development in sines and cosines (4.7) is used, then it follows from (3.4) that

$$\frac{1}{4}|a_0|^2 + \frac{1}{2}\sum_{n=1}^{+\infty}(|a_n|^2 + |b_n|^2) = \frac{1}{a}\int_0^a |f(t)|^2\,dt. \tag{4.9}$$

4.3.3 Uniqueness of the Fourier coefficients

Are two functions that have the same Fourier coefficients equal? This is the question of uniqueness. One might expect to have at least equality almost everywhere, since

$$f = g \text{ a.e.} \quad\Longrightarrow\quad c_n(f) = c_n(g) \text{ for all } n \in \mathbb{Z}.$$

By linearity, an affirmative answer to the uniqueness question reduces to the following property:

$$f \in L^2_{\mathrm{p}}(0,a) \text{ and } c_n(f) = 0 \text{ for all } n \in \mathbb{Z} \quad\Longrightarrow\quad f = 0 \text{ a.e.} \tag{4.10}$$

But this is an immediate consequence of (4.8), knowing that

$$\int_0^a |f(t)|^2\, dt = 0 \quad \Longrightarrow \quad f = 0 \text{ a.e. on } (0, a).$$

In fact, in the context of the space $L^2_{\mathrm{p}}(0, a)$ there is an equivalence between Parseval's formula (and thus Theorem 4.3.1) and the uniqueness of the Fourier coefficients. One of the implications is immediate, as we have just seen; the converse is more difficult to establish. It will be done in Lesson 16. A direct proof (which does not use Theorem 4.3.1) of the uniqueness of the Fourier coefficients for a piecewise continuous function is the content of Exercise 4.7.

4.4 Fourier coefficients of real, odd, and even functions

At the end of Lesson 3, we showed that certain properties of a trigonometric polynomial p are reflected as conditions on its coefficients, and conversely. It is easy to prove the same relations for a periodic function f. These relations are important in practice because they can be used to reduce the number of numerical operations in certain calculations. We note specifically the following relations:

(a) f real $\Leftrightarrow$ $c_{-n} = \bar{c}_n,\ n \in \mathbb{Z}$ $\Leftrightarrow$ a_n and b_n real, $n \in \mathbb{N}$

(b) f even $\Leftrightarrow$ $c_{-n} = c_n,\ n \in \mathbb{Z}$ $\Leftrightarrow$ $b_n = 0,\ n \in \mathbb{N}$

(c) f odd $\Leftrightarrow$ $c_{-n} = -c_n,\ n \in \mathbb{Z}$ $\Leftrightarrow$ $a_n = 0,\ n \in \mathbb{N}$

(d) f real, even $\Leftrightarrow$ the sequence (c_n) is real and even

(e) f real, odd $\Leftrightarrow$ the sequence (c_n) is pure imaginary and odd

The properties of f on the left are to be understood in the sense of almost everywhere. For example, if $c_{-n} = c_n$ for all n, then the two functions $f(t)$ and $f(-t)$ have the same Fourier series development and are equal almost everywhere by (4.10). Thus one must understand by "f even" the property

$$f(-t) = f(t) \quad \text{for a.e. } t,$$

where "a.e. t" means "almost every t" or almost everywhere.

4.5 Formulary

$$f(t) = \sum_{n=-\infty}^{+\infty} c_n e^{2i\pi n \frac{t}{a}}$$

$$f(t) = \frac{a_0}{2} + \sum_{n=1}^{+\infty} \left(a_n \cos\left(2\pi n \frac{t}{a}\right) + \left(b_n \sin\left(2\pi n \frac{t}{a}\right)\right)\right)$$

$$c_n = \frac{1}{a}\int_0^a f(t) e^{-2i\pi n \frac{t}{a}}\, dt \qquad \begin{cases} a_n = \dfrac{2}{a}\displaystyle\int_0^a f(t)\cos\left(2\pi n \frac{t}{a}\right) dt \\ b_n = \dfrac{2}{a}\displaystyle\int_0^a f(t)\sin\left(2\pi n \frac{t}{a}\right) dt \end{cases} \quad (n \geq 0)$$

$$\begin{cases} c_n = \dfrac{1}{2}(a_n - ib_n) \\ c_{-n} = \dfrac{1}{2}(a_n + ib_n) \end{cases} (n \geq 0) \qquad \begin{cases} a_n = c_n + c_{-n} \\ b_n = i(c_n - c_{-n}) \end{cases} (n \geq 0)$$

$$\sum_{n=-\infty}^{+\infty} |c_n|^2 = \frac{1}{a}\int_0^a |f(t)|^2\, dt = \frac{1}{4}|a_0|^2 + \frac{1}{2}\sum_{n=1}^{+\infty}(|a_n|^2 + |b_n|^2)$$

4.6 Exercises

Exercise 4.1 Calculate the Fourier series expansions of the following functions and verify the symmetric properties of the coefficients:

(a) f has period 2 and $f(t) = |t|$ if $|t| < 1$.

(b) f has period a and $f(t) = \dfrac{t}{a}$ if $0 \leq t < a$.

(c) $f(t) = |\sin t|$.

(d) $f(t) = \sin^3 t$.

Exercise 4.2 If the Fourier coefficients of $f(t)$ are c_n, what are the Fourier coefficients of the translated function $f(t - t_0)$? Deduce from Exercise 4.1 the Fourier series expansion of $f(t) = |\cos t|$.

Exercise 4.3 Prove relation (4.9). (*Hint*: Use (4.8) and (3.4).)

Exercise 4.4 Write Parseval's equality for each function in Exercise 4.1.

Exercise 4.5 Assume that $f \in L^2_{\mathrm{P}}(0, a)$ and c_n are its Fourier coefficients. Then f is also in $L^2_{\mathrm{P}}(0, 2a)$ with Fourier coefficients c'_n. How are the coefficients c_n and c'_n related? Verify that the two Fourier series are identical.

Exercise 4.6 Find the Fourier series expansion of the function f with period $a = 2$ defined on $[-1, +1)$ for $z \in \mathbb{C}\backslash\mathbb{Z}$ by

$$f(t) = e^{i\pi zt}.$$

Deduce the relation

$$\frac{\pi^2}{\sin^2 \pi x} = \sum_{n=-\infty}^{\infty} \frac{1}{(x-n)^2}$$

for all $x \in \mathbb{R}\backslash\mathbb{Z}$ from Parseval's equality.

Exercise 4.7 (Uniqueness of the Fourier coefficients for a periodic, piecewise continuous function.) Assume for simplicity that the piecewise continuous function f has period 1. We wish to show that if all the Fourier coefficients are zero, that is, if

$$(\mathrm{H}_1) \qquad c_n(f) = \int_0^1 f(t)e^{-2i\pi nt}\,dt = 0, \quad n \in \mathbb{Z},$$

then f vanishes everywhere except perhaps at points where f is discontinuous.

Let $s \in [0, 1]$ be a point where f is continuous and suppose that $f(s) \neq 0$. One can assume (by a translation and multiplication by -1, if necessary) that $s = 0$ and $f(0) > 0$.

(1) Show that (H_1) implies

$$(\mathrm{H}_2) \qquad \int_{-\frac{1}{2}}^{\frac{1}{2}} f(t)p(t)\,dt = 0$$

for all trigonometric polynomials p with period 1.

(2) Take $\alpha \in (0, 1/4]$ such that

$$|t| \leq \alpha \Rightarrow f(t) \geq \frac{1}{2}f(0). \quad \text{(Why does such an } \alpha \text{ exist?)}$$

Define

$$\sigma_N(t) = \frac{1}{N}\sum_{k=0}^{N-1}\sum_{n=-k}^{k} e^{2i\pi nt}.$$

(a) Calculate $\sigma_N(t)$ and show that

$$\int_{-\frac{1}{2}}^{\frac{1}{2}} \sigma_N(t)\,dt = 1.$$

(b) Show that

$$\lim_{N\to\infty} \int_{\alpha\leq|t|\leq\frac{1}{2}} \sigma_N(t)dt = 0.$$

(c) Deduce from this that

$$\int_{-\alpha}^{\alpha} \sigma_N(t) f(t)\, dt \geq \frac{f(0)}{4}$$

for sufficiently large N.

(3) Show that

$$\int_{-\frac{1}{2}}^{-\frac{1}{2}} \sigma_N(t) f(t)\, dt > 0$$

for sufficiently large N and deduce the result.

Lesson 5

Pointwise Representation

A function used for numerical computation is necessarily evaluated at only a finite number of points. It is therefore important to determine whether formulas (4.6) and (4.7) can express equality at a given point t. This is the problem of *pointwise representation.* We begin by extending the notion of Fourier series beyond the space $L^2_{\mathrm{p}}(0,a)$.

5.1 The Riemann–Lebesgue theorem

The Lebesgue integral has the remarkable property that a function is integrable on an interval I if and only if its modulus is integrable on I:

$$f \text{ is Lebesgue-integrable on } I \quad \Longleftrightarrow \quad \int_I |f(x)|\,dt < \infty. \tag{5.1}$$

This is false for the Riemann integral; take, for example, the function equal to 1 on the rationals and -1 on the irrationals.

It follows immediately that the Fourier coefficients of a periodic function

$$c_n(f) = \frac{1}{a}\int_0^a f(t)e^{-2\pi n\frac{t}{a}}\,dt$$

exist if and only if f is integrable on $(0,a)$. We introduce the notation

$$L^1_{\mathrm{p}}(0,a) = \left\{ f : \mathbb{R} \to \mathbb{C} \;\middle|\; f \text{ has period } a \text{ and } \int_0^a |f(t)|\,dt < +\infty \right\}.$$

Note that $L^2_{\mathrm{p}}(0,a) \subset L^1_{\mathrm{p}}(0,a)$, so saying f is in $L^1_{\mathrm{p}}(0,a)$ is less restrictive than saying f is in $L^2_{\mathrm{p}}(0,a)$. For any f in $L^1_{\mathrm{p}}(0,a)$, we can consider the Fourier series

$$\sum_{n=-\infty}^{+\infty} c_n(f)e^{2\pi n\frac{t}{a}}.$$

We do not know whether this series converges, and if it does converge, we do not know the value of its limit.

That $c_n(f) \to 0$ as $n \to +\infty$ is an important necessary, but not sufficient, condition for convergence. This limit was established for $f \in L^2_{\mathrm{p}}(0,a)$ using Bessel's inequality (4.2.3), which makes sense only for functions in $L^2_{\mathrm{p}}(0,a)$. However, the property $c_n(f) \to 0$ remains true for f in $L^1_{\mathrm{p}}(0,a)$.

5.1.1 Theorem (Riemann–Lebesgue) *Let (a,b) be a bounded interval and assume that f is integrable on (a,b). Then the integral*

$$I_n = \int_a^b f(x)e^{2i\pi nx}\,dx$$

tends to 0 *as* $|n| \to +\infty$.

Proof. This is easy to establish when f is continuously differentiable on $[a,b]$. Integration by parts shows that

$$I_n = \frac{1}{2i\pi n}\left[f(x)e^{2i\pi nx}\right]_a^b - \frac{1}{2i\pi n}\int_a^b f'(x)e^{2i\pi nx}\,dx,$$

which yields the estimate

$$|I_n| \leq \frac{1}{2\pi|n|}\Big(|f(a)| + |f(b)| + \int_a^b |f'(x)|\,dx\Big).$$

The right hand-side, and thus I_n, tends to 0 as when $|n| \to +\infty$.

We now use a density argument that is based on the following provisional assumption: The functions that are continuously differentiable on $[a,b]$ are dense in $L^1_{\mathrm{p}}(a,b)$. This means that given $\varepsilon > 0$, there exists a $g_\varepsilon \in C^1([a,b])$ such that

$$\int_a^b |f(x) - g_\varepsilon(x)|\,dx \leq \frac{\varepsilon}{2},$$

which implies that

$$|I_n| \leq \int_a^b |f(x) - g_\varepsilon(x)|\,dx + \left|\int_a^b g_\varepsilon(x)e^{2i\pi nx}\,dx\right|.$$

From the previous argument, there exists $N > 0$ such that the last integral is dominated by $\varepsilon/2$ if $|n| \geq N$. Thus, $|n| \geq N$ implies $|I_n| \leq \varepsilon$, and this proves the theorem. □

5.2 Pointwise convergence?

We have already noted following Theorem 4.3.1 that the mean quadratic convergence of the Fourier series f_N to f gives no information about the convergence of f_N at a given point. For this, one needs more refined assumptions about the function. In practice, these hypotheses will generally hold. However, we emphasize that these additional hypotheses are essential, since several "natural" results that one might expect to hold for $f \in L^1_p(0,a)$ are indeed false. We cite three of these:

(i) $f_N \to f$ in the $L^1_p(0,a)$ norm.

(ii) $f_N(t) \to f(t)$ for almost all t.

(iii) If f is also continuous on $\mathbb{R}$, then $f_N(t) \to f(t)$ for all $t \in \mathbb{R}$.

It has even been shown (see [KF74]) that there exists an f in $L^1_p(0,a)$ such that $f_N(t)$ diverges for all t as $N \to +\infty$! These examples represent difficult problems that have played a central role during the last century in research on the theory of functions.

5.2.1 Piecewise continuous functions on $[a,b]$

Let $f : [a,b] \to \mathbb{C}$ be a complex-valued function. We say that f is *piecewise continuous* on $[a,b]$ if it is continuous on $[a,b]$ except at a finite (possibly zero) number of points and if both the right and left limits exist and are finite at these exceptional points. These limits are defined by

$$f(t+) = \lim_{h \to 0^+} f(t+h),$$
$$f(t-) = \lim_{h \to 0^-} f(t+h).$$

At the end points a and b we require that only the one-sided limits exist. We denote this function space by $C_{\text{pw}}[a,b]$, where "pw" stands for "piecewise." Then

$$f \in C_{\text{pw}}[a,b] \quad \Longrightarrow \quad \begin{cases} f \text{ is bounded on } [a,b], \\ f \text{ is integrable on } [a,b]. \end{cases}$$

The integral of f involves only the integration of continuous functions, since

$$\int_a^b f(t)\,dt = \sum_{k=0}^{n} \int_{a_k}^{a_{k+1}} f(t)\,dt,$$

where $a_0 = a$, $a_{n+1} = b$, and where $a_1, \ldots, a_n$ are the points of discontinuity of f in (a,b). Each integral in the sum is understood to be the integral of the continuous extension of f to the interval $[a_k, a_{k+1}]$. Note that as far as the integral is concerned, f does not need to be defined at the points a_k.

EXAMPLE: Define f on $[-1,1]$ by

$$f(t) = \begin{cases} t^2+1 & \text{if} \quad -1 \le t < 0, \\ -t+2 & \text{if} \quad 0 \le t \le 1. \end{cases}$$

Then f has a derivative

$$f'(t) = \begin{cases} 2t & \text{if} \quad -1 \le t < 0, \\ -1 & \text{if} \quad 0 < t \le 1, \end{cases}$$

and f' is in $C_{\text{pw}}[-1, +1]$; however, f' is not defined at $t = 0$.

5.2.2 Functions of bounded variation on $[a, b]$

We say that a function f is of *bounded variation* on $[a, b]$, denoted by $f \in BV[a, b]$, if there exists an M such that

$$\sum_{k=0}^{n} |f(t_{k+1}) - f(t_k)| \le M$$

for all subdivisions $a = t_0 < t_1 < \cdots < t_{n+1} = b$, where $n \in \mathbb{N}$ is arbitrary. For f to be of bounded variation, it is necessary and sufficient that its real and imaginary parts be of bounded variation. Any real function that is monotonic on $[a, b]$ is of bounded variation on $[a, b]$. In fact, we have the following characterization for real functions:

$$f \in BV[a, b] \quad \Longleftrightarrow \quad \begin{cases} \text{There exist } g \text{ and } h\text{, monotonic} \\ \text{on } [a, b]\text{, such that } f = g - h. \end{cases}$$

The implication from right to left is immediate. For the other direction, see, for example, [KF74]. We deduce from this last equivalence that

$$f \in BV[a, b] \quad \Longrightarrow \quad \begin{cases} f \text{ is Riemann integrable on } [a, b]\text{, and} \\ f(t-) \text{ and } f(t+) \text{ exist for all } t \in (a, b). \end{cases} \tag{5.2}$$

5.2.3 An expression for the remainder $f_N(t_0) - f(t_0)$

Let f be in $L^1_{\text{p}}(0, a)$ and assume that at a point t_0 the limits $f(t_0+)$ and $f(t_0-)$ exist. From (4.3) and (4.4) we have

$$f_N(t_0) = \frac{1}{a} \int_{-\frac{a}{2}}^{\frac{a}{2}} \Big(\sum_{n=-N}^{N} e^{2i\pi n \frac{t_0 - x}{a}} \Big) f(x) \, dx.$$

An easy computation shows that

$$\sum_{n=-N}^{N} e^{2i\pi nt} = \frac{\sin \pi(2N+1)t}{\sin \pi t}. \tag{5.3}$$

Thus we obtain

$$f_N(t_0) = \frac{1}{a}\int_{-\frac{a}{2}}^{\frac{a}{2}} \frac{\sin \pi(2N+1)\dfrac{t_0 - x}{a}}{\sin \pi \dfrac{t_0 - x}{a}} f(x)\, dx,$$

which, by a change of variable, becomes

$$f_N(t_0) = \frac{1}{a}\int_{-\frac{a}{2}-t_0}^{\frac{a}{2}-t_0} \frac{\sin \pi(2N+1)\dfrac{x}{a}}{\sin \pi \dfrac{x}{a}} f(x+t_0)\, dx.$$

Since the integrand has period a, the integral can be taken over $[-a/2, a/2]$. By making the variable change $x \mapsto -x$ in the integral over $[-a/2, 0]$, we have the expression

$$f_N(t_0) = \frac{1}{a}\int_0^{\frac{a}{2}} [f(t_0+x) + f(t_0-x)] \frac{\sin \pi(2N+1)\dfrac{x}{a}}{\sin \pi \dfrac{x}{a}}\, dx.$$

From (4.3) and (4.4), we see that $f = 1$ implies $f_N = 1$ for all N. Thus, for all $N \geq 0$,

$$\frac{1}{a}\int_0^{\frac{a}{2}} \frac{\sin \pi(2N+1)\dfrac{x}{a}}{\sin \pi \dfrac{x}{a}}\, dx = \frac{1}{2}.$$

If we write $y_0 = \frac{1}{2}[f(t_0+) + f(t_0-)]$, we obtain

$$f_N(t_0) - y_0 = \\ \frac{1}{a}\int_0^{\frac{a}{2}} [f(t_0+x) - f(t_0+) + f(t_0-x) - f(t_0-)] \frac{\sin \pi(2N+1)\dfrac{x}{a}}{\sin \pi \dfrac{x}{a}}\, dx. \tag{5.4}$$

5.2.4 Theorem (Dirichlet's theorem) *Let f be in $L^1_p(0,a)$. If the limits $f(t_0+)$ and $f(t_0-)$ exist at a point t_0, and if the left- and right-hand derivatives also exist at t_0, then*

$$f_N(t_0) \to \frac{1}{2}[f(t_0+) + f(t_0-)]$$

as $N \to +\infty$. If, in addition, f is continuous at t_0, then $f_N(t_0) \to f(t_0)$.

Proof. The result will follow from (5.4). The assumption that the left and right derivatives exist means that the quantities

$$\frac{1}{x}\big(f(t_0+x) - f(t_0+)\big) \quad \text{and} \quad \frac{1}{x}\big(f(t_0-x) - f(t_0-)\big)$$

tend to finite limits as $x \to 0^-$. Hence, the same is true for

$$\varphi(x) = \frac{f(t_0+x) - f(t_0+) + f(t_0-x) - f(t_0-)}{\sin \pi \dfrac{x}{a}}.$$

Thus, there exist $\alpha > 0$ and $M > 0$ such that $|\varphi(x)| \leq M$ for all $x \in (0, \alpha]$. Since $f \in L^1_{\mathrm{p}}(0,a)$, φ is integrable on $[\alpha, a/2]$, and

$$|\varphi(x)| \leq M + |\varphi(x)|\chi_{[\alpha,a/2]}(x)$$

for all $x \in (0, a/2]$. The function on the right is integrable because it is the sum of two integrable functions. From (5.1) we know that

$$\left.\begin{array}{l} |g(x)| \leq h(x) \text{ for all } x \in I, \\ h \text{ Lebesgue integrable on } I \end{array}\right\} \implies g \text{ Lebesgue integrable on } I. \quad (5.5)$$

This implies that φ is integrable on $(0, a/2)$, and by the Riemann–Lebesgue theorem (Theorem 5.1.1), the integral

$$f_N(t_0) - y_0 = \frac{1}{a}\int_0^{\frac{a}{2}} \varphi(x) \sin\left((2N+1)\frac{x}{a}\right) dx$$

tends to 0 as $N \to +\infty$. □

This result shows that the convergence of the Fourier series of f at a point t_0 depends only on the behavior of f in a neighborhood of t_0.

Here is a global result that we will not prove.

5.2.5 Theorem *Assume that $f : \mathbb{R} \to \mathbb{C}$ is a periodic function of bounded variation with period a. Then we have the following results:*

(i) *For all $t_0 \in \mathbb{R}$, $f_N(t_0) \to \frac{1}{2}\big(f(t_0+) + f(t_0-)\big)$ as $N \to +\infty$.*

(ii) *If, in addition, f is continuous on a closed and bounded interval $[\alpha, \beta]$, then f_N converges uniformly to f on $[\alpha, \beta]$.*

REMARK: The fact that $f_N(t_0)$ has a limit for each fixed t_0 as $N \to +\infty$ does not imply that either of the series

$$\sum_{n=0}^{+\infty} c_n e^{2i\pi n \frac{t}{a}} \quad \text{or} \quad \sum_{n=-\infty}^{0} c_n e^{2i\pi n \frac{t}{a}}$$

is convergent. On the other hand, this does imply the convergence of the series (4.7) of sines and cosines because it is obtained by symmetrically regrouping the terms of f_N.

5.3 Uniform convergence of Fourier series

5.3.1 Theorem *Assume that f has period a and is continuous on $\mathbb{R}$; that f is differentiable on $[0, a]$, except possibly at a finite number of points; and that f' is piecewise continuous.*

(i) *The Fourier series of f' is obtained by differentiating the Fourier series of f term by term.*

(ii) *The Fourier coefficients of f satisfy*

$$\sum_{n=-\infty}^{+\infty} |c_n(f)| < +\infty.$$

(iii) *The Fourier series of f converges uniformly to f on $\mathbb{R}$.*

Proof. The hypotheses have been fashioned so that the expression for $c_n(f)$ can be integrated by parts and so that f' is in $L^2_{\mathrm{p}}(0, a)$, and hence in $L^1_{\mathrm{p}}(0, a)$. Integration by parts shows that

$$c_n(f) = \frac{1}{2i\pi n} \int_0^a f'(t) e^{-2i\pi n \frac{t}{a}}\, dt,$$

since $f(0+) = f(a-)$, and hence the Fourier coefficients $c_n(f')$ of f' are given by

$$c_n(f') = \frac{2i\pi n}{a} c_n(f),$$

which proves (i). We deduce (ii) directly from the inequality

$$|c_n(f)| = \frac{a}{2\pi |n|} |c_n(f')| \leq \frac{a}{4\pi} \left(\frac{1}{n^2} + |c_n(f')|^2 \right).$$

To prove (iii), note that $\sum_{n=-\infty}^{+\infty} |c_n(f)| < +\infty$ implies that (f_N) is a Cauchy sequence in the uniform norm on $[0, a]$ (and hence on $\mathbb{R}$), so f_N converges uniformly on $\mathbb{R}$ to some continuous function g. Since uniform convergence implies convergence in the $L^2_{\mathrm{p}}(0, a)$ norm, it follows from 4.3.1 and the uniqueness of the limit that $f = g$ almost everywhere. But both f and g are continuous, so $f = g$ everywhere. □

The last part of this proof established the following corollary.

5.3.2 Corollary *If $f \in L^2_{\mathrm{p}}(0, a)$ and if its Fourier coefficients satisfy*

$$\sum_{n=-\infty}^{+\infty} |c_n(f)| < +\infty,$$

then f is equal to a continuous function $\tilde{f}$ almost everywhere and the Fourier series of f converges uniformly to $\tilde{f}$ on $\mathbb{R}$.

5.3.3 Conclusions

We use the terms "regular" and "smooth" informally to mean that a function has a number (undetermined) of derivatives. Thus the more regular, or the smoother, a function f is, the faster the coefficients $c_n(f)$ tend to 0. This can be seen by repeatedly integrating by parts. This is summarized in the following display, where the regularity of f is increasing and where $C_{\mathrm{p}}^k[0, a]$ denotes the space of functions $f \in C^k(\mathbb{R})$ that are a-periodic.

$$\begin{array}{llll} \text{(a)} & f \in L_{\mathrm{p}}^1(0,a) & \Longrightarrow & c_n(f) \to 0 \\ \text{(b)} & f \in L_{\mathrm{p}}^2(0,a) & \Longrightarrow & \sum_{n=-\infty}^{+\infty} |c_n(f)|^2 < +\infty \\ \text{(c)} & f \in C_{\mathrm{p}}^1[0,a] & \Longrightarrow & \sum_{n=-\infty}^{+\infty} |c_n(f)| < +\infty \\ \text{(d)} & f \in C_{\mathrm{p}}^2[0,a] & \Longrightarrow & |c_n(f)| \le K/n^2 \\ \text{(e)} & f \in C_{\mathrm{p}}^\infty[0,a] & \Longrightarrow & |n^k c_n(f)| \to 0,\ k \in \mathbb{N}, \\ & & & \text{as } |n| \to +\infty \end{array}$$

The property

$$|n^k c_n(f)| \to 0 \text{ for all } k \in \mathbb{N} \text{ as } |n| \to +\infty$$

will be abbreviated by the expression *the sequence $c_n(f)$ is rapidly decreasing* or variations such as *$c_n(f)$ decreases rapidly.* Note that these expressions, although widely used, are a slight abuse of the language: They in no way imply that the sequence $|c_n(f)|$ is monotonic.

We now show that implication (e) is in fact an equivalence.

5.3.4 Proposition *Assume that $f \in L_{\mathrm{p}}^2(0,a)$. Then the following two properties are equivalent.*

(i) *The Fourier coefficients of f are rapidly decreasing.*

(ii) *The function f is infinitely differentiable.*

Proof. We have seen that (ii) $\Rightarrow$ (i). Conversely, if the $c_n(f)$ are rapidly decreasing, then in particular, $n^2|c_n(f)| \to 0$. Hence $\sum_{n=-\infty}^{+\infty} |c_n(f)| < +\infty$, and the sequence of functions

$$f_N(t) = \sum_{n=-N}^{N} c_n e^{2i\pi n \frac{t}{a}}$$

converges uniformly to f (Corollary 5.3.2). In the same way,

$$f_N'(t) = \sum_{n=-N}^{N} \frac{2i\pi n}{a} c_n e^{2i\pi n \frac{t}{a}}$$

converges uniformly to a continuous function g. Then it is a classical result that f is differentiable and that $f' = g$. By iterating this argument, f is shown to have derivatives up to any finite order. □

EXAMPLE: The Fourier coefficients of $f(t) = \dfrac{\sin t}{2 + \cos t}$ decrease rapidly.

5.4 Exercises

Exercise 5.1 Prove that a piecewise continuous function on $[a, b]$ is bounded. Note that a and b must be finite! Find a counterexamples if $a = -\infty$ or $b = +\infty$.

Exercise 5.2 Show that a function in $C^1[a, b]$ is of bounded variation on $[a, b]$. (As in Exercise 5.1, the bounds a and b must be finite.)

***Exercise 5.3** Define f by

$$f(x) = \begin{cases} x \sin \dfrac{1}{x} & \text{if} \quad x \neq 0, \\ 0 & \text{if} \quad x = 0. \end{cases}$$

Show that f is continuous on $[0, 1]$ and differentiable $(0, 1]$ but that it is not of bounded variation on $[0, 1]$.

Exercise 5.4 Prove formula (5.3).

Exercise 5.5 Develop the Fourier series of the function f, with period $a = 2$, defined on $[-1, 1)$ by

$$f(t) = \cos \pi z t, \quad z \in \mathbb{C} \backslash \mathbb{Z}.$$

From this deduce the equalities

$$\pi \cot \pi z = \frac{1}{z} + 2z \sum_{n=1}^{\infty} \frac{1}{z^2 - n^2},$$

$$\frac{\pi}{\sin \pi z} = \frac{1}{z} + 2z \sum_{n=1}^{\infty} \frac{(-1)^n}{z^2 - n^2}.$$

Exercise 5.6 Show that

$$\sum_{n=1}^{\infty} \frac{\sin nx}{n} = \frac{\pi}{2} - \frac{x}{2}, \quad x \in (0, 2\pi).$$

Derive from this the value of

$$f(x) = \sum_{\substack{n=-\infty \\ n \neq 0}}^{\infty} \frac{1}{n} e^{2i\pi n \frac{x}{a}}, \quad 0 < x < a.$$

Exercise 5.7 Let f be the 2π-periodic function defined on $(0, 2\pi)$ by

$$f(x) = \ln\left(2\sin\frac{x}{2}\right).$$

(a) Verify that f is even.

(b) Show that $f \in L^1_p(0, 2\pi)$.

(c) Is f of bounded variation on $(0, 2\pi)$? Is it $L^2_p(0, 2\pi)$?

(d) We wish to determine the Fourier series expansion of f :

$$f(x) = \frac{a_0}{2} + \sum_{n=1}^{\infty} a_n \cos nx.$$

Compute a_n, $n \geq 1$, by noticing that the integral

$$I_n = \int_0^{\pi} \cot\frac{x}{2} \sin nx \, dx$$

does not depend on n (compute $I_n - I_{n-1}$).

(e) Determine the value of a_0 and prove that

$$\sum_{n=1}^{\infty} \frac{1}{n} \cos nx = -\ln\left(2\sin\frac{x}{2}\right), \quad x \in (0, 2\pi).$$

***Exercise 5.8** Let x be a real parameter and let f be defined by

$$f(t) = e^{xe^{it}}.$$

(a) What is the period of f? Show that

$$c_n(f) = \begin{cases} 0 & \text{if } n < 0, \\ \dfrac{x^n}{n!} & \text{if } n \geq 0. \end{cases}$$

(b) Deduce that

$$\int_0^{2\pi} e^{2x\cos t} \, dt = 2\pi \sum_{n=0}^{\infty} \frac{x^{2n}}{(n!)^2}.$$

(c) Define

$$I_n = \int_0^{\pi} (\cos t)^n \, dt.$$

Show that

$$\int_0^{\pi} e^{2x\cos t} \, dt = \sum_{n=0}^{\infty} \frac{2^n}{n!} I_n x^n,$$

and from this determine I_n.

Note that $I_{2p+1} = 0$. Was this foreseeable?

Exercise 5.9 Consider the sequence of polynomials B_k defined by

$$B_0(x) = 1,$$
$$B_k'(x) = kB_{k-1}(x) \text{ and } \int_0^1 B_k(x)\,dx = 0, \quad k \geq 1.$$

(a) Compute B_1, B_2, B_3 and draw their graphs in the same coordinate system.

(b) Show that each of these graphs is mapped into itself by reflection about one or two axes. Is this generally true for the polynomials B_k? Express this property algebraically for B_k.

(c) Let f_k be the function with period 1 that coincides with B_k on $[0, 1)$. Show that for $k \geq 3$,

$$f_k \in C^1(\mathbb{R}) \quad \text{and} \quad f_k' = kf_{k-1}$$

and that f_2 satisfies the hypotheses of Theorem 5.3.1.

(d) Show that

$$f_k \in C^{k-2}(\mathbb{R}), \quad k \geq 2.$$

(e) Compute the Fourier series of f_1 and use it to determine the Fourier series of f_k for all $k \geq 2$.

Exercise 5.10 Let f be the 2π-periodic function defined on $[-\pi, \pi)$ by

$$f(x) = \cosh(ax), \quad a > 0.$$

(a) Show that the Fourier series of f converges uniformly to f.

(b) Compute the expansion of f in a series of cosines.

(c) Conclude from this that

$$\sum_{n=1}^{\infty} \frac{1}{a^2 + n^2} = \frac{\pi}{2a}\left[\coth(\pi a) - \frac{1}{\pi a}\right], \quad a \in \mathbb{R}\backslash\{0\}.$$

(d) Justify the term-by-term differentiation of the series for f and show that

$$\sinh(ax) = \frac{2\sinh(a\pi)}{\pi} \sum_{n=1}^{\infty} (-1)^{n+1} \frac{n}{n^2 + a^2} \sin nx, \quad x \in (-\pi, \pi).$$

Exercise 5.11

(a) Show that if $f \in C_{\mathrm{p}}^2[0, a]$, then $|c_n(f)| \leq \dfrac{K}{n^2}$.

(b) Show that $f \in C_{\mathrm{p}}^{\infty}[0, a]$ implies $\lim_{|n|\to\infty} |n^k c_n(f)| = 0$ for all $k \in \mathbb{N}$.

Exercise 5.12 Take $f \in L_{\mathrm{p}}^1(0, a)$ and let f_k be a sequence in $L_{\mathrm{p}}^1(0, a)$ such that

$$\lim_{k\to\infty} \int_0^a |f(t) - f_k(t)|\,dt = 0.$$

Show that for fixed n, $\lim_{k\to\infty} c_n(f_k) = c_n(f)$.

Exercise 5.13 (Fourier series of a product) Suppose f and g are in $L^2_{\mathrm{p}}(0,a)$. We wish to compute the Fourier series of the product fg.

(a) Verify that $fg \in L^1_{\mathrm{p}}(0,a)$.

(b) Write

$$f_N(t) = \sum_{n=-N}^{N} c_n(f) e^{2i\pi n \frac{t}{a}},$$

$$g_N(t) = \sum_{n=-N}^{N} c_n(g) e^{2i\pi n \frac{t}{a}}.$$

Show that

$$c_n(f_N g_N) = \sum_{k=-N}^{N} c_{n-k}(f) c_k(g).$$

(c) Prove that $f_N g_N$ tends to fg in $L^1_{\mathrm{p}}(0,a)$ and use Exercise 5.12 to show that

$$c_n(fg) = \sum_{k=-\infty}^{\infty} c_{n-k}(f) c_k(g), \quad n \in \mathbb{Z},$$

where the series on the right is absolutely convergent.

Lesson 6

Expanding a Function in an Orthogonal Basis

6.1 Fourier series expansions on a bounded interval

Let (a, b) be a bounded interval and let f be a complex function defined on (a, b). A priori, this function does not have a Fourier series expansion, since this notion has been defined only for functions that are defined and periodic on all of $\mathbb{R}$. However, f can be extended periodically, with period $b - a$, to all of $\mathbb{R}$ as in Figure 6.1.

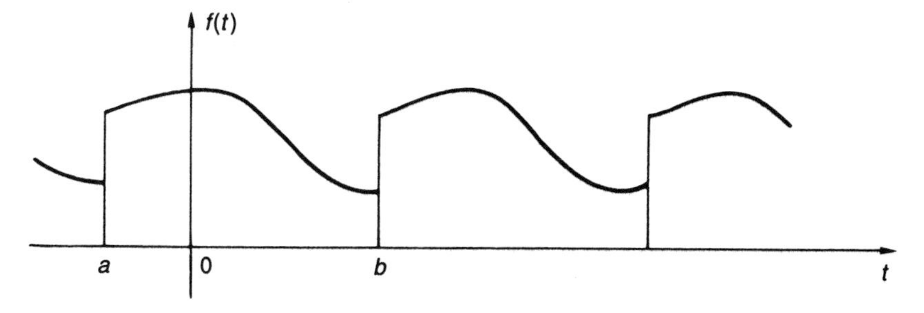

FIGURE 6.1.

The extended function $\tilde{f}$ has a Fourier series expansion if f is in $L^2(a, b)$, and this is called the expansion of f on the interval (a, b). Here are two other ways to obtain a Fourier expansion of f. For simplicity, we assume that f is defined on $(0, a)$.

6.1.1 Expansion of f in a series of sines

Define the function $\tilde{f}$ by $\tilde{f}(t) = f(t)$ for $t \in (0, a)$,$\tilde{f}(t) = -f(-t)$ for $t \in (-a, 0)$, and extend $\tilde{f}$ periodically to $\mathbb{R}$ with period $2a$. (See Figure 6.2.) The Fourier series expansion of $\tilde{f}$ is then of the form (4.7) and contains only sine terms.

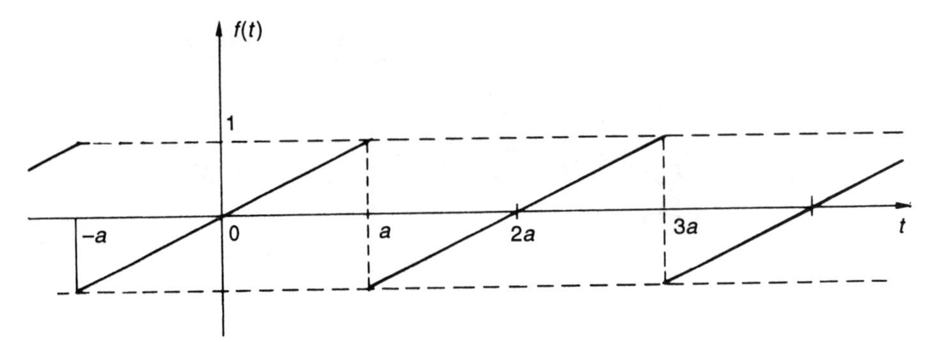

FIGURE 6.2. Example where $f(t) = t/a$.

6.1.2 Expansion of f in a series of cosines

The construction of the cosine series is similar, except that here $\tilde{f}$ is the even extension of f defined by $\tilde{f}(t) = f(-t)$ for $t \in (-a, 0)$ (see Figure 6.3). This time the series expansion contains only cosines.

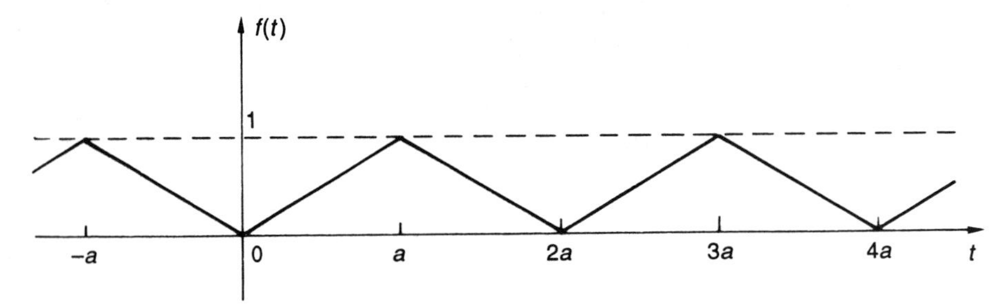

FIGURE 6.3. Example where $f(t) = t/a$.

If nothing else is said, the expansion of f on the interval $(0, a)$ leads to the expansion of the a-periodic function illustrated in Figure 6.4.

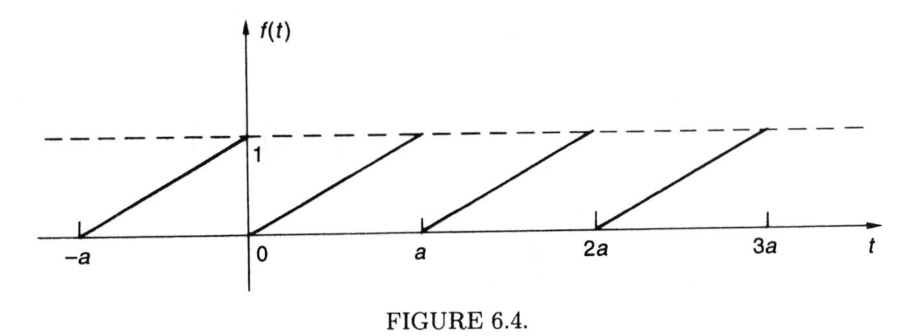

FIGURE 6.4.

AN IMPORTANT CONSEQUENCE: Although the Fourier series expansion of a periodic function is unique (Section 4.3.3), the Fourier series expansion of a function f defined on a bounded interval is not unique. It depends fundamentally on the way one extends f periodically to the whole line.

6.2 Expansion of a function in an orthogonal basis

6.2.1 General method

What has been done here and in Section 4.2 for periodic functions using complex exponentials as the basis can be done for functions f in $L^2(a,b)$ using other basis functions. The simplicity of the formulas and their similarity to those of Section 4.2 are consequences of the fact that the basis functions are orthogonal.

Assume that there is a family of functions $\{\phi_n\}_{n\in\mathbb{N}}$ that are orthogonal in the space $L^2(a,b)$, which we assume to be endowed with either the usual scalar product or perhaps a scalar product weighted with a positive function $w(t)$. Then it is always possible to project f orthogonally onto the subspace V_N generated by $\{\phi_0, \phi_1, \dots, \phi_N\}$ and thereby obtain the best approximation f_N of f in this subspace. We find that

$$f_N = \sum_{n=0}^{N} c_n \phi_n \quad \text{with} \quad c_n = \frac{(f, \phi_n)}{\|\phi_n\|_2^2},$$

and

$$\|f - f_N\|_2 = \min_{p\in V_N} \|f - p\|_2.$$

If, for all $f \in L^2(a,b)$, f_N tends to f in the norm of $L^2(a,b)$ as N tends to infinity, the family $\{\phi_n\}_{n\in\mathbb{N}}$ is said to be a *topological basis* for $L^2(a,b)$. One also says that this family is a *complete system* or a *total family* (see Section 16.3). In this case, we have the Parseval relation

$$\sum_{n=0}^{+\infty} |c_n|^2 \|\phi_n\|_2^2 = \int_a^b |f(t)|^2\, dt,$$

and we write, in the sense of the $L^2(a,b)$ norm,

$$f = \sum_{n=0}^{+\infty} c_n \phi_n.$$

One can also use functions defined on an unbounded interval, for example $(a,b) = \mathbb{R}$ or $(a,b) = \mathbb{R}_+$, where $\mathbb{R}_+ = \{x \in \mathbb{R} \mid x \geq 0\}$. A problem in these cases is that the polynomials are no longer in $L^2(a,b)$. This can be solved, however, by multiplying the polynomials by an appropriate weighting function that tends to 0 sufficiently fast at infinity. One then obtains a family of functions in $L^2(a,b)$ that can be used to expand $f \in L^2(a,b)$.

6.2.2 Examples

(a) LEGENDRE POLYNOMIALS

Take $(a, b) = (-1, 1)$ and consider only real-valued functions. An orthogonal basis is obtained by "orthogonalizing" the set $\{1, t, t^2, \dots\}$ of linearly independent polynomials with respect to the inner product

$$(f, g) = \int_{-1}^{1} f(t)g(t)\, dt.$$

This process yields a family of orthogonal polynomials that are, up to a factor depending only on n, the Legendre polynomials:

$$P_0 = 1,\ P_1(t) = t,\ P_2 = \frac{3}{2}t^2 - \frac{1}{2},\ P_3(t) = \frac{5}{2}t^3 - \frac{3}{2}t, \dots .$$

In general,

$$P_n(t) = \frac{1}{n!2^n} \frac{d^n}{dt^n}(t^2 - 1)^n.$$

These polynomials are orthogonal, and

$$\|P_n\|_2^2 = \frac{2}{2n+1}.$$

Thus, for $f \in L^2(-1, 1)$,

$$f(t) = \sum_{n=0}^{+\infty} c_n P_n(t) \quad \text{with} \quad c_n = \frac{2n+1}{2}(f, P_n).$$

(b) CHEBYSHEV POLYNOMIALS

Suppose that $f \in L^2(-1, 1)$ and associate with f the function

$$F(x) = f(\cos x),$$

which is even and 2π-periodic. The scalar product for $L^2(0, \pi)$ is

$$(F, G) = \int_0^{\pi} F(x)G(x)\, dx = \int_0^{\pi} f(\cos x)g(\cos x)\, dx;$$

by letting $t = \cos x$, this becomes

$$(F, G) = \int_{-1}^{1} \frac{f(t)g(t)}{\sqrt{1 - t^2}}\, dt.$$

This formula defines the new scalar product on $L^2(-1, 1)$

$$(f, g)_w = \int_{-1}^{1} \frac{f(t)g(t)}{\sqrt{1 - t^2}}\, dt$$

involving the weight

$$w(t) = \frac{1}{\sqrt{1-t^2}}.$$

The functions $F_n(x) = \cos nx$ form a basis for the subspace of even functions in $L^2(-\pi, \pi)$. The corresponding functions

$$T_n(t) = \cos(n \arccos t)$$

form an orthogonal basis in $L^2(-1, 1)$ with respect to the scalar product $(\cdot, \cdot)_w$. These polynomials T_n are called the Chebyshev polynomials, and they play an important role in the theory of approximation (see Section 10.4).

(c) HERMITE POLYNOMIALS IN $L^2(\mathbb{R})$

One uses the scalar product

$$(f, g)_H = \int_{-\infty}^{+\infty} f(t)g(t)e^{-t^2}\,dt,$$

which leads to an orthogonal family $\{\phi_0, \phi_1, \dots\}$ of functions of the form

$$\phi_n(t) = H_n(t)e^{-t^2/2},$$

where H_n is a polynomial of degree n called a Hermite polynomial. The Hermite polynomials are orthogonal with respect to the scalar product $(\cdot, \cdot)_H$. One can show that the family $\{\phi_n\}$ forms a topological basis for the Hilbert space $L^2(\mathbb{R})$ [KF74].

(d) LAGUERRE POLYNOMIALS IN $L^2(0, +\infty)$

The basis $\{1, t, t^2, \dots\}$ is orthogonalized with respect to the scalar product

$$(f, g)_L = \int_{0}^{+\infty} f(t)g(t)e^{-t}\,dt.$$

One obtains a topological basis for the space $L^2(0, +\infty)$ [KF74]. This basis consists of the functions $L_n e^{-t/2}$, where the L_n are the Laguerre polynomials, which are orthogonal with respect to the scalar product $(\cdot, \cdot)_L$.

6.2.3 Comments and references

Many facts about orthogonal polynomials and about their use can be found in books on numerical analysis or the theory of approximation. See, for example, [Lau72] and [Sze59].

The rather sophisticated proofs that these families form topological bases can be found in the book by Kolmogorov and Fomine [KF74]. These examples serve to justify, among other things, the theoretical study of Hilbert spaces and their usual topological bases. This will be done in Lesson 16.

6.3 Exercises

Exercise 6.1

(a) Expand the functions in Figures 6.2–6.4 in Fourier series.

(b) Determine the rates at which their Fourier coefficients c_n converge to 0.

(c) Write Parseval's equality for these three cases.

(d) Express the pointwise convergence in the three cases for $t = a/2$ and $t = a$.

In which cases do (c) and (d) produce interesting identities?

Exercise 6.2

(a) Use the Legendre polynomials P_0, P_1, P_2, P_3 to compute the best approximations f_i, $i = 0, 1, 2, 3$, to $f(t) = |t|$ on $[-1, +1]$ in the sense of the usual $L^2(-1, 1)$ norm.

(b) Represent f, f_1, f_2 on the same graph.

Exercise 6.1 Compute the Hermite polynomials H_0, H_1, and H_2.

Exercise 6.2 Let f be defined on $[0, 1]$ by $f(x) = x(1 - x)$.

(a) We wish to consider the expansion of f in a series of sines. Sketch the graph of the periodic (period 2) extension g of f. Is the sine series expansion of g uniformly convergent? Can it be differentiated term by term?

(b) Compute the expansion of g.

(c) Deduce from (b) that

$$\sum_{n=0}^{\infty} \frac{(-1)^n}{(2n+1)^3} = \frac{\pi^3}{32} .$$

(d) Compute $\int_0^1 f(x)\, dx$ and deduce that

$$\sum_{n=0}^{\infty} \frac{1}{(2n+1)^4} = \frac{\pi^4}{96} .$$

(e) Compute the expansion in cosines of

$$f'(x) = 1 - 2x, \quad x \in [0, 1],$$

and the expansion of

$$f''(x) = -2, \quad x \in (0, 1).$$

(f) Deduce from (e) that

$$\sum_{n=0}^{\infty} \frac{1}{(2n+1)^2} = \frac{\pi^2}{8} \quad \text{and} \quad \sum_{n=0}^{\infty} \frac{(-1)^n}{2n+1} = \frac{\pi}{4} .$$

(g) Expand f in a series of cosines and address the same questions as above.

Lesson 7

Frequencies, Spectra, and Scales

7.1 Frequencies and spectra

7.1.1 The notion of the spectrum of a periodic signal

If f is a periodic signal with period a, and if it has the Fourier series expansion

$$f(t) = \sum_{n=-\infty}^{+\infty} c_n e^{2i\pi n \frac{t}{a}}, \tag{7.1}$$

then the *spectrum* of f is defined to be the set of pairs $(n/a, c_n)$, $n \in \mathbb{Z}$.

7.1.2 Amplitude spectrum and phase spectrum

In Section 1.2.3, we mentioned the frequency representation of a pure sinusoidal signal in terms of amplitude and phase. By superposition, we have a representation of the amplitude and phase spectra as a function of frequency for all periodic signals by representing the various values with arrows parallel to the y-axis.

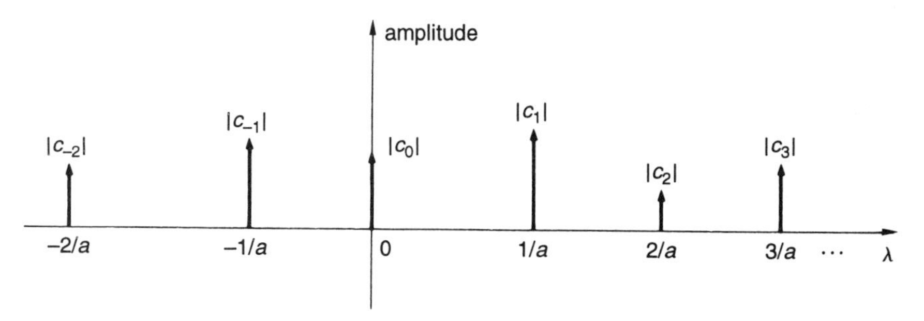

FIGURE 7.1. Amplitude spectrum.

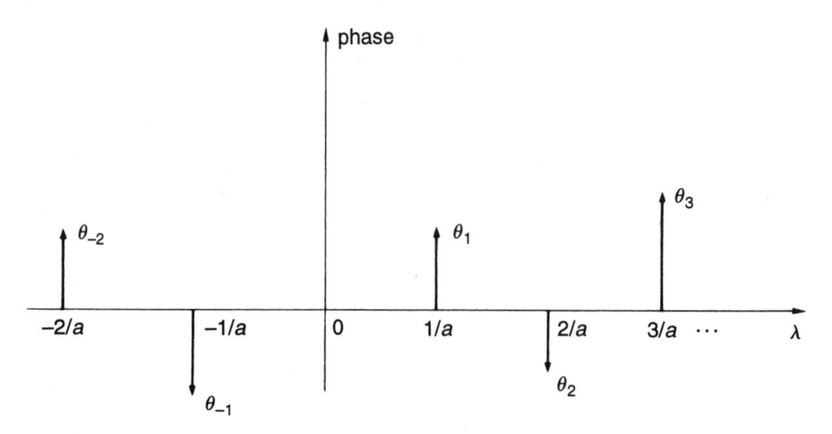

FIGURE 7.2. Phase spectrum (of a real signal).

The amplitude spectrum (Figure 7.1) consists of *spectral lines* regularly spaced at the frequencies n/a. For $|n| = 1$, the two lines correspond to the *fundamental frequency.* The other lines are called *harmonics* of the signal. The phase spectrum (Figure 7.2) is the set of pairs $(n/a, \theta_n)$, $n \in \mathbb{Z}$, where $c_n = |c_n|e^{i\theta_n}$, $\theta_n \in [-\pi, \pi)$.

7.1.3 Action of a filter on a periodic signal

We saw in Section 2.3 that the action of a filter on a sinusoidal signal with frequency λ was simply to multiply the signal by $H(\lambda)$. For a periodic signal f given by (7.1) the output from the filter with transfer function H will be

$$g(t) = (Af)(t) = \sum_{n=-\infty}^{+\infty} c_n H\left(\frac{n}{a}\right) e^{2i\pi n \frac{t}{a}}.$$

The positions of the spectral lines are not changed; only their relative values change as they are multiplied by $H(n/a)$. Thus, this process is properly called *frequency filtering.*

7.1.4 Orders of magnitude

Vibrating or oscillating phenomena are widespread in nature. In fact, some would claim that everything reduces to vibrations! Listed below are some common oscillatory phenomena encountered in the physical sciences along with their frequencies (1 hertz (Hz)= 1 cycle per second):

- Household current: 60 Hz
- Quartz in a watch: 10^5 Hz
- Radar wave: 10^{10} Hz
- Vibration of a caesium atom: 10^{14} Hz

- Electromagnetic waves:

very long: (telegraph)	$1.5 \cdot 10^4$ to $6 \cdot 10^4$ Hz
long: (radio)	$6 \cdot 10^4$ to $3 \cdot 10^5$ Hz
medium: (radio)	$3 \cdot 10^5$ to $3 \cdot 10^6$ Hz
short: (radio)	$3 \cdot 10^6$ to $3 \cdot 10^7$ Hz
meters: (television)	$3 \cdot 10^7$ to $3 \cdot 10^8$ Hz
centimeters: (radar)	$3 \cdot 10^8$ to 10^{11} Hz
visible light:	$3.7 \cdot 10^{14}$ to $7.5 \cdot 10^{14}$ Hz

The human ear can, in the best cases, detect sounds whose frequencies range from 20 to 20,000 Hz.

7.2 Variations on the scale

Sound is measured, as a function of time, as variations in air pressure either by the ear or other sensors. A pure tone with a fixed frequency f (or a note) is sensed by the ear as a periodic variation in air pressure where the maxima occur every $1/f$ seconds.

7.2.1 The octave

An *octave* is the interval between two notes, one with frequency f and the other with frequency $2f$. This definition may seem arbitrary, but it is clearly based on a spectral decomposition: An octave is the interval that separates the fundamental frequency and its first harmonic.

When one hears middle C, say, one also hears, hidden immediately behind, its first harmonic, which is the C in the next (higher) octave. These two notes are closely related, and one has the impression that they are from the same family. And indeed, we give them the same name, C. A note on the scale is thus determined modulo multiplication by a power of 2; it is the exponent of 2 that determines its octave.

EXAMPLE: Standard pitch assigns 440 Hz to A_3. This standard leads to the following A-ladder (in hertz):

27.5	55	110	220	440	880 ...
A_{-1}	A_0	A_1	A_2	A_3	A_4 ...

7.2.2 The harmonic scale

When we hear a C with frequency f, we also hear the harmonics $2f$, $3f$, etc., and it is in this progression that we encounter G and E:

f	$2f$	$3f$	$4f$	$5f$	$6f$
C	C	G	C	E	G

Thus the common cord (C, G, E) is found within the C-scale. Bringing these notes back to the same octave, we have the cord (C, E, G, C):

$$\begin{array}{cccc} f & \frac{5}{4}f & \frac{3}{2}f & 2f \\ \text{C} & \text{E} & \text{G} & \text{C} \end{array}$$

If we similarly analyze a G, we find the cord (G, D, B):

$$\begin{array}{ccccc} \frac{3}{2}f & 3f & \frac{9}{2}f & 6f & \frac{15}{2}f \\ \text{G} & \text{G} & \text{D} & \text{G} & \text{B} \end{array}$$

Brought back to the octave $(f, 2f)$, we obtain the following:

$$\begin{array}{cccccc} f & \frac{9}{8}f & \frac{5}{4}f & \frac{3}{2}f & \frac{15}{8}f & 2f \\ \text{C} & \text{D} & \text{E} & \text{G} & \text{B} & \text{C} \end{array}$$

Starting with E leads to no new note, at least in the first three harmonics:

$$\begin{array}{cccc} \frac{5}{4}f & \frac{5}{2}f & \frac{15}{4}f & 5f \\ \text{E} & \text{E} & \text{B} & \text{E} \end{array}$$

The frequencies of all of these notes are simple fractions of the fundamental frequency f. Furthermore, all of the denominators are powers of 2. The first simple fraction with denominator 3 leads to the discovery of the F with frequency $\frac{4}{3}f$, from which the cord (F, C, A) is constructed:

$$\begin{array}{ccccc} \frac{4}{3}f & \frac{8}{3}f & 4f & \frac{16}{3}f & \frac{20}{3}f \\ \text{F} & \text{F} & \text{C} & \text{F} & \text{A} \end{array}$$

We now have all 7 notes of the *harmonic scale*:

$$\begin{array}{cccccccc} f & \frac{9}{8}f & \frac{5}{4}f & \frac{4}{3}f & \frac{3}{2}f & \frac{5}{3}f & \frac{15}{8}f & 2f \\ \text{C} & \text{D} & \text{E} & \text{F} & \text{G} & \text{A} & \text{B} & \text{C} \end{array}$$

This scale is also call the *physicists' scale.*

7.2.3 Tones and semitones

The sound of two tones whose frequencies have a fixed ratio is perceived by the ear as a fixed interval between two notes. Thus fixed intervals are expressed by frequencies in geometric progression.

EXAMPLE: For the interval called a *fifth*, the ratio is 3 to 2:

$$\begin{array}{ccc} f & \frac{3}{2}f & \frac{9}{4}f \\ \text{C} & \text{G} & \text{D} \end{array}$$

If one wishes to relate these ratios between frequencies to the length of the intervals, the thing to do is to take the logarithms (for example to base 2) of these ratios. This shows that there are intervals with three different lengths: the major interval, the minor interval, and the semitone or half-step. The following table illustrates this idea:

(C,D)	$\log 9/8$	$= 0.170$	Major interval
(D,E)	$\log 10/9$	$= 0.152$	Minor interval
(E,F)	$\log 16/15$	$= 0.093$	Half-step
(F,G)	$\log 9/8$	$= 0.170$	Major interval
(G,A)	$\log 10/9$	$= 0.152$	Minor interval
(A,B)	$\log 9/8$	$= 0.170$	Major interval
(B,C)	$\log 16/15$	$= 0.093$	Half-step

The difference between a major interval and a minor interval is called the comma. It is the minimum interval perceived by the ear.

$$1 \text{ comma} = \log 81/80 = 0.018.$$

This is roughly a ninth of the major second: $0.170/9 = 0.0188$. Although a very approximate definition, it is generally taught in courses on music theory.

7.2.4 The tempered scale

The two kinds of intervals, major and minor, have the disadvantage that they do not partition the octave into 12 equal half-steps, nor do they have one fixed value for sharps and flats. This can be remedied directly by dividing the octave into 12 equal intervals. This process defines the tempered scale:

f	$a^2 f$	$a^4 f$	$a^5 f$	$a^7 f$	$a^9 f$	$a^{11} f$	$2f$
C	D	E	F	G	A	B	C

where

$$a = 2^{1/12}.$$

The semitone, or half-step, is thus defined by the interval

$$(f, 2^{1/12} f).$$

It is easier to see the nuances between the harmonic scale (H.S.) and the tempered scale (T.S.) by looking at the decimal values:

	C	D	E	F	G	A	B	C
H.S.	f	$1.125f$	$1.250f$	$1.333f$	$1.5f$	$1.667f$	$1.875f$	$2f$
T.S.	f	$1.122f$	$1.260f$	$1.335f$	$1.498f$	$1.682f$	$1.888f$	$2f$

The tempered scale has no need for the comma. It is this scale that is, in principle, used for the piano.

7.3 Exercises

Exercise 7.1 The Pythagorean scale is built on the "fifth," which is defined by two vibrating strings whose lengths are in the ratio 3 to 2:

F	C	G	D	A	E	B
$\frac{2}{3}f$	f	$\frac{3}{2}f$	$\left(\frac{3}{2}\right)^2 f$	$\left(\frac{3}{2}\right)^3 f$	$\left(\frac{3}{2}\right)^4 f$	$\left(\frac{3}{2}\right)^5 f$

(a) Bring these frequencies back to the same octave and compare this scale with those described in Section 7.2.2 and 7.2.4.

(b) Note that this scale has only one major tone T and a very "tight" half-tone m. Describe the succession of tones and half-tones.

Exercise 7.2 Let f be a real periodic signal. Show that its amplitude spectrum is even. Investigate the properties of its phase spectrum.

Chapter III

The Discrete Fourier Transform and Numerical Computations

Lesson 8

The Discrete Fourier Transform

8.1 Computing the Fourier coefficients

We will work with the following assumptions: We know the period a of the function f as well as N of its values that are regularly spaced over one period:

$$f\left(k\frac{a}{N}\right) = y_k, \quad k = 0, 1, 2, \ldots, N-1.$$

The signal $f(t)$ is thus assumed to have been sampled at regularly spaced times separated by a/N units. Using this information, we wish to approximate the Fourier coefficients of f. We also assume that the Fourier series of f converges pointwise to f and that at points of discontinuity

$$f(t) = \frac{1}{2}\big(f(t+) + f(t-)\big).$$

Given N data points, it is logical to try to compute N Fourier coefficients c_n. Since these coefficients tend to zero as n tends to infinity, we choose to compute c_n for $n = -N/2, \ldots, N/2 - 1$ (or a centered interval if N is odd). These considerations lead one to compute an approximation of the integral

$$c_n = \frac{1}{a}\int_0^a f(t)e^{-2i\pi n\frac{t}{a}}\,dt. \tag{8.1}$$

FIRST METHOD: Integrating (8.1) by the trapezoid formula gives the approximate value

$$c'_n = \frac{1}{N}\sum_{k=0}^{N-1} y_k e^{-2i\pi n\frac{k}{N}},$$

or

$$c'_n = \frac{1}{N}\sum_{k=0}^{N-1} y_k \omega_N^{-nk} \quad \text{with} \quad \omega_N = e^{2i\pi\frac{1}{N}}. \tag{8.2}$$

SECOND METHOD: One can also compute the Fourier coefficients, denoted by c_n^N, of the trigonometric polynomial

$$p(t) = \sum_{n=-\frac{N}{2}}^{\frac{N}{2}-1} c_n^N e^{2i\pi n\frac{t}{a}} \tag{8.3}$$

that interpolates f at the points $k(a/N)$, $k = 0, 1, 2, \dots, N-1$. Note that with the notation c_n^N, N is fixed; it is not a running index like n. c_n^N, like c_n', is destined to be an approximation of c_n. One is thus led to solve the linear system of order N,

$$\sum_{n=-\frac{N}{2}}^{\frac{N}{2}-1} c_n^N \omega_N^{nk} = y_k, \quad k = 0, 1, 2, \dots, N-1. \tag{8.4}$$

For convenience, we bring all of the indices n into the interval $[0, N-1]$ by translating the negative indices to the right by N. This is possible because the functions involved are N-periodic. Thus,

$$\sum_{n=-\frac{N}{2}}^{-1} c_n^N \omega_N^{nk} = \sum_{p=\frac{N}{2}}^{N-1} c_{p-N}^N \omega_N^{k(p-N)} = \sum_{n=\frac{N}{2}}^{N-1} c_{n-N}^N \omega_N^{nk}.$$

By defining

$$Y_n = \begin{cases} c_n^N & \text{if} \quad 0 \le n \le \frac{N}{2} - 1, \\ c_{n-N}^N & \text{if} \quad \frac{N}{2} \le n \le N-1, \end{cases}$$

the system (8.4) can be written as

$$\sum_{n=0}^{N-1} Y_n \omega_N^{nk} = y_k, \quad k = 0, 1, 2, \dots, N-1.$$

This system has the advantage that it can be solved explicitly. Let p be an integer between 0 and $N-1$ and compute the sum

$$\sum_{k=0}^{N-1} y_k \omega_N^{-kp} = \sum_{k=0}^{N-1}\sum_{n=0}^{N-1} Y_n \omega_N^{k(n-p)} = \sum_{n=0}^{N-1} Y_n \sum_{k=0}^{N-1} \omega_N^{k(n-p)}.$$

The last sum is a geometric series:

$$\sum_{k=0}^{N-1} \omega_N^{k(n-p)} = \begin{cases} 0 & \text{if} \quad p \ne n, \\ N & \text{if} \quad p = n. \end{cases}$$

This show that

$$\sum_{k=0}^{N-1} y_k \omega_N^{-kp} = NY_p,$$

and hence the unknowns Y_n are given by

$$Y_n = \frac{1}{N} \sum_{k=0}^{N-1} y_k \omega_N^{-nk}, \quad n = 0, 1, 2, \ldots, N-1.$$

We see that this is the same as formula (8.2)! After a change of indices, we discover that

$$c_n^N = c_n', \quad -\frac{N}{2} \leq n < \frac{N}{2}.$$

8.1.1 Conclusions

Integrating (8.1) using the trapezoid method yields N approximate Fourier coefficients c_n^N that are equal to the Fourier coefficients of the trigonometric polynomial (8.3) that interpolates f at the points $t_k = k(a/N)$. We have the equivalent formulas

$$\begin{aligned} y_k &= \sum_{n=0}^{N-1} Y_n \omega_N^{nk}, \qquad k = 0, 1, 2, \ldots, N-1, \\ Y_n &= \frac{1}{N} \sum_{k=0}^{N-1} y_k \omega_N^{-nk}, \quad n = 0, 1, 2, \ldots, N-1, \end{aligned} \tag{8.5}$$

and the approximate Fourier coefficients are

$$c_n \approx c_n^N = \begin{cases} Y_N & \text{if} \quad 0 \leq n < \dfrac{N}{2}, \\ Y_{n+N} & \text{if} \quad -\dfrac{N}{2} \leq n < 0. \end{cases}$$

8.1.2 Definition The second formula in (8.5) defines a transformation $\mathscr{F}_N$ from $\mathbb{C}^N$ into itself

$$(y_k) \overset{\mathscr{F}_N}{\longmapsto} (Y_n)$$

that is called the discrete Fourier transform of order N.

$\mathscr{F}_N$ is linear and bijective:

$$y = \mathscr{F}_N Y \iff Y = \mathscr{F}_N^{-1} y.$$

If Ω_N is the matrix representation of $\mathscr{F}_N$, then

$$\Omega_N = (\omega_N^{nk}) = \begin{bmatrix} 1 & 1 & 1 & \dots & 1 \\ 1 & \omega_N & \omega_N^2 & \dots & \omega_N^{N-1} \\ 1 & \omega_N^2 & \omega_N^4 & \dots & \omega_N^{2(N-1)} \\ \vdots & & & & \\ 1 & \omega_N^{N-1} & \omega_N^{2(N-1)} & \dots & \omega_N^{(N-1)^2} \end{bmatrix},$$

and

$$\Omega_N^{-1} = \frac{1}{N}\overline{\Omega}_N.$$

8.1.3 Remarks

(a) Be careful to note that the computation of the Y_n yields the Fourier coefficients c_n^N in the following order ($N = 8$):

Y_0	Y_1	Y_2	Y_3	Y_4	Y_5	Y_6	Y_7
c_0^8	c_1^8	c_2^8	c_3^8	c_{-4}^8	c_{-3}^8	c_{-2}^8	c_{-1}^8

(b) It is convenient, particularly for manipulating the formulas, to extend the vector $y \in \mathbb{C}^N$ to a periodic sequence with period N, which just comes back to its original definition, since $y_k = f(ka/N)$, where f has period a. This allows us to write y_k with $k \in \mathbb{Z}$. We will use this convention in the rest of the book. Formula (8.5) shows that the Y_n are also periodic with period N. We will use the same convention and define Y_n for $n \in \mathbb{Z}$.

These conventions mean that the approximate Fourier coefficients c_n^N also form a periodic sequence. Here it is important to remember that c_n^N is an approximation (see Figure 8.1) of c_n only for

$$-\frac{N}{2} \leq n < \frac{N}{2}.$$

(There is no reason to believe otherwise, but an easy way to remember this is to keep in mind that $c_n \to 0$ as $n \to +\infty$.)

With these conventions, $Y_n = c_n^N$ for all $n \in \mathbb{Z}$ (although the c_n^N are still out of order), and the sums in (8.5) can be taken over any set of N consecutive integers.

8.2 Some properties of the discrete Fourier transform

All the sequences in this section will be complex and periodic with the same period N. The sequence (y_n) is said to be *even* if $y_{-n} = y_n$ for all $n \in \mathbb{Z}$. It is said to be *odd* if $y_{-n} = -y_n$ for all $n \in \mathbb{Z}$.

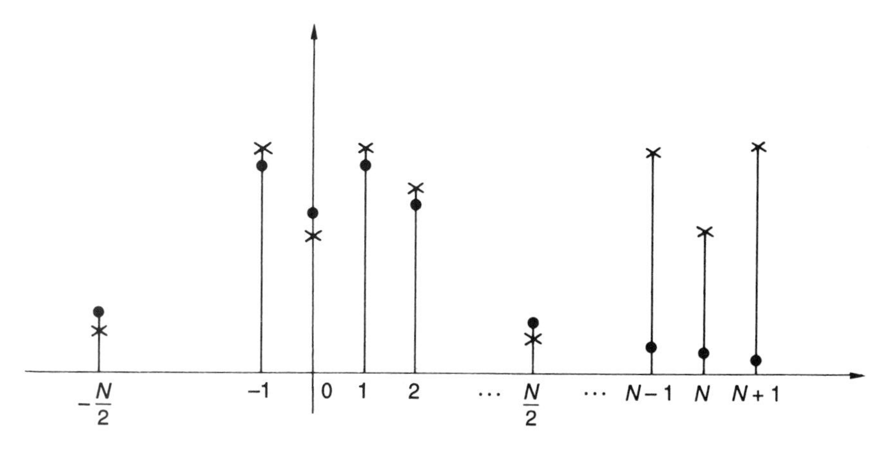

FIGURE 8.1. Fourier coefficients exact (•) and computed (×).

We note without further comment that the properties described in this section are also true for the inverse transform

$$(Y_n) \overset{\mathscr{F}_N^{-1}}{\longmapsto} (y_k).$$

8.2.1 Proposition *If* $(y_k) \overset{\mathscr{F}_N}{\longmapsto} (Y_n)$, *then*

(i) $(y_{-k}) \overset{\mathscr{F}_N}{\longmapsto} (Y_{-n})$;

(ii) $(\overline{y}_k) \overset{\mathscr{F}_N}{\longmapsto} (\overline{Y}_{-n})$;

(iii) $(\overline{y}_{-k}) \overset{\mathscr{F}_N}{\longmapsto} (\overline{Y}_n)$.

Proof. For example, for (i), write $\mathscr{F}_N(y_{-k}) = (Y_n')$. Then

$$Y_n' = \frac{1}{N}\sum_{k=0}^{N-1} y_{-k}\omega_N^{-nk} = \frac{1}{N}\sum_{k=-N+1}^{0} y_k\omega_N^{nk} = Y_{-n}. \qquad \square$$

8.2.2 Proposition *If* $(y_k) \overset{\mathscr{F}_N}{\longmapsto} (Y_n)$, *the following relations hold:*

(i) (y_k) *is even (odd)* $\Leftrightarrow$ (Y_n) *is even (odd);*

(ii) (y_k) *is real* $\Leftrightarrow$ $Y_{-n} = \overline{Y}_n$ *for all* $n \in \mathbb{Z}$;

(iii) (y_k) *is real and even* $\Leftrightarrow$ (Y_n) *is real and even;*

(iv) (y_k) *is real and odd* $\Leftrightarrow$ (Y_n) *is imaginary and odd.*

Proof. These follow directly from Proposition 8.2.1. □

We now have a result that explains the important relation between the discrete Fourier transform and the periodic discrete convolution.

8.2.3 Theorem *Let* (x_k) *and* (y_k) *be two complex sequences with period* N *and let* (X_k) *and* (Y_k) *denote their discrete Fourier transforms.*

(i) *The sequence defined by circular convolution,*

$$z_k = \sum_{q=0}^{N-1} x_q y_{k-q}, \quad k \in \mathbb{Z},$$

has as its transform

$$(z_k) \overset{\mathscr{F}_N}{\longmapsto} (Z_n = N X_n Y_n).$$

(ii) *The transform of the pointwise product of the sequences* (x_k) *and* (y_k) *is*

$$(p_k = x_k y_k) \overset{\mathscr{F}_N}{\longmapsto} (P_n),$$

where

$$P_n = \sum_{q=0}^{N-1} X_q Y_{n-q}.$$

Proof. By definition,

$$Z_n = \frac{1}{N} \sum_{k=0}^{N-1} \sum_{q=0}^{N-1} x_q y_{k-q} \omega_N^{-nk},$$

and interchanging the order of summation shows that

$$Z_n = \frac{1}{N} \sum_{q=0}^{N-1} x_q \omega_N^{-nq} \sum_{k=0}^{N-1} y_{k-q} \omega_N^{-n(k-q)} = N X_n Y_n.$$

If we do the "same" computation for the inverse transform applied to the vector (P_n), we see that

$$(P_n) \overset{\mathscr{F}_N^{-1}}{\longmapsto} (x_k y_k),$$

which proves (ii). □

We will see later that a similar property holds in the continuous case.

8.2.4 Proposition *If* $(y_k) \overset{\mathscr{F}_N}{\longmapsto} (Y_n)$,

$$\sum_{k=0}^{N-1} |y_k|^2 = N \sum_{n=0}^{N-1} |Y_n|^2.$$

The proof is left as an exercise.

This result means that the discrete Fourier transform is, up to a factor N, an isometry on the Euclidean space $\mathbb{C}^N$ into itself. Furthermore, a quadratic error of ε in the data (y_k) appears as a quadratic error ε/N in the result. This means that the computation is stable.

8.3 The Fourier transform of real data

The discrete Fourier transform applies to complex-valued vectors, hence to the case where (y_k) is real-valued. However, in this case it is possible to reduce the cost of computation by half by treating two sets of real data with a single complex transformation. We wish to compute the transforms of two real vectors (x_k) and (y_k):

$$(x_k) \xmapsto{\mathscr{F}_N} (X_n),$$
$$(y_k) \xmapsto{\mathscr{F}_N} (Y_n).$$

We know from Proposition 8.2.2 that

$$X_{N-n} = \overline{X}_n \quad \text{and} \quad Y_{N-n} = \overline{Y}_n.$$

Let $z_k = x_k + iy_k$ and denote the transform of (z_k) by (Z_n):

$$(z_k) \xmapsto{\mathscr{F}_N} (Z_n).$$

By linearity,

$$Z_n = X_n + iY_n.$$

But note that X_n and Y_n are not necessarily real! With this notation,

$$X_n = \frac{1}{2}(Z_n + \overline{Z}_{N-n}),$$
$$Y_n = \frac{1}{2i}(Z_n - \overline{Z}_{N-n}).$$

It is only necessary to compute these values for $n = 0, 1, \dots, N/2$, since the values for n between $N/2$ and $N-1$ will have already appeared as their conjugates. Thus it is sufficient to compute the transform of (z_k) to obtain the transforms of (x_k) and (y_k).

It is also possible to compute the Fourier transform of a single real vector with N components by using a transform of length $N/2$ (see [CLW70]).

8.4 A relation between the exact and approximate Fourier coefficients

Assume for simplicity that the periodic function f is expressed by

$$f(t) = \sum_{n=-\infty}^{+\infty} c_n e^{2i\pi n \frac{t}{a}} \tag{8.6}$$

and that the series is absolutely convergent:

$$\sum_{n=-\infty}^{+\infty} |c_n| < +\infty.$$

It is sufficient, for example, that f satisfy the hypotheses of Theorem 5.3.1. Since (8.6) is absolutely convergent, we can, for each t, rearrange the order of summation. In particular, we can first sum all of the terms whose indices have a fixed residue modulo N and then sum these N terms. Thus, by taking $t = k(a/N)$,

$$f\left(k\frac{a}{N}\right) = y_k = \sum_{m=-\infty}^{+\infty} c_m \omega_N^{mk} = \sum_{n=0}^{N-1} \left(\sum_{q=-\infty}^{+\infty} c_{n+qN} \right) \omega_N^{nk},$$

where we have written $m = n + qN$ with $n \in \{0, 1, \ldots, N-1\}$ and $q \in \mathbb{Z}$.

We deduce from this, using the inverse transform, that

$$c_n^N = \sum_{q=-\infty}^{+\infty} c_{n+qN}. \tag{8.7}$$

A surprising relation! It expresses the approximate coefficients in terms of the exact coefficients, and we obtain an expression for the approximation error:

$$c_n^N - c_n = \sum_{q \neq 0} c_{n+qN}.$$

From this we see that for a fixed N, the faster c_n tends to zero as n tends to infinity, the better will be the approximation

$$c_n \approx c_n^N \quad \text{for} \quad -\frac{N}{2} \leq n \leq \frac{N}{2} - 1.$$

Consequently, the smoother f is, the better the approximation (compare with Section 5.3.3). On the other hand, the approximation for a discontinuous function can be rather bad. The determination of the rate at which the c_n tend to zero and, if possible, of a bound on the sums

$$\left| \cdots + c_{-2N+n} + c_{-N+n} + c_{N+n} + c_{2N+n} + \cdots \right|$$

are serious issues for the numerical analysis of this approximation problem.

The sum in (8.7) is, of course, finite if f is a trigonometric polynomial. Take, for example, the general trigonometric polynomial of degree 6:

$$f(t) = \sum_{n=-6}^{+6} c_n e^{2i\pi n \frac{t}{a}}.$$

Then we have for $N = 4$:

$$c_0^4 = c_{-4} + c_0 + c_4,$$
$$c_1^4 = c_{-3} + c_1 + c_5,$$
$$c_2^4 = c_{-2} + c_2 + c_6,$$
$$c_3^4 = c_{-1} + c_3, \text{ etc.}$$

For $N = 8$:

$$c_0^8 = c_0,$$
$$c_1^8 = c_1,$$
$$c_2^8 = c_{-6} + c_2, \text{ etc.}$$

For $N \geq 13$:

$$c_n^N = c_n.$$

This simple example illustrates the following general result, which follows directly from (8.7):

For a trigonometric polynomial of degree P, the values of the approximate coefficients computed with the discrete Fourier transform are exact whenever $N \geq 2P + 1$.

8.5 Exercises

Exercise 8.1 Consider two consecutive discrete Fourier transforms:

$$(y_k) \overset{\mathcal{F}_N}{\longmapsto} (Y_n) \quad \text{and} \quad (Y_n) \overset{\mathcal{F}_N}{\longmapsto} (z_q).$$

Compute z_q as a function of y_k.

Exercise 8.2 Let (x_k) and (y_k) be two complex periodic sequences (with period N) such that

$$x_{N-k} = \bar{x}_k \quad \text{and} \quad y_{N-k} = \bar{y}_k$$

for all $k \in \mathbb{Z}$. Show that the discrete Fourier transforms (X_n) and (Y_n) are real and that they can be computed with a single transform of order N.

Exercise 8.3 Compute the successive powers of the matrix Ω_N.

Exercise 8.4 Prove Proposition 8.2.4 by computing $(\overline{\Omega}_N \overline{Y})^t\, (\Omega_N Y)$.

Exercise 8.5 Calculate the discrete Fourier transform of the vector $x_k = k$, $k = 0, 1, \ldots, N - 1$.

Lesson 9

A Famous, Lightning-Fast Algorithm

Computing the vector $(Y_0, Y_1, \ldots, Y_{N-1})$ using formula (8.5) requires

$$(N-1)^2 \text{ complex multiplications,}$$
$$N(N-1) \text{ complex additions,}$$

assuming that the values of ω_N^j, the sines and cosines of the given angles, have already been computed and stored.

A typical value of N is of the order of 1000, which implies about a million operations of each kind. Considering the frequency of this computation, it was natural to seek to lower the cost. In 1965, two American scientists, J. W. Cooley and J. W. Tukey, developed a much more efficient algorithm that has since been known as the *fast Fourier transform* (FFT). This algorithm takes into consideration the special form of the transformation matrix, which is constructed from the roots of unity. From the beginning, the FFT, including its many extensions, has enjoyed enormous success. In fact, it is safe to say that it has been the backbone of signal and image processing in the last half of the twentieth century. Furthermore, it has been the inspiration for numerous investigations in algebra independently of its intensive use in signal processing. It was indeed a marvelous discovery. The fast Fourier transform marked an important step in the theory of the *complexity of algorithm.* This field of research is concerned with determining and minimizing the cost of a given computation or class of computations, where the cost is measured by the number of numerical operations. For example, we will see that the cost of the FFT is of the order $N \log N$.

9.1 The Cooley–Tukey algorithm

Assume that N is even, $N = 2m$, and rearrange the terms of (8.5) into two groups—those with even indices and those with odd indices. Then

$$Y_k = \frac{1}{2}\,(P_k + \omega_N^{-k} I_k),$$

where The P_k and I_k are given by the formulas

$$P_k = \frac{1}{m}\left(y_0 + y_2\omega_N^{-2k} + \cdots + y_{N-2}\omega_N^{-(N-2)k}\right),$$
$$I_k = \frac{1}{m}\left(y_1 + y_3\omega_N^{-2k} + \cdots + y_{N-1}\omega_N^{-(N-2)k}\right).$$

Note that we have the relations

$$P_{k+m} = P_k, \quad I_{k+m} = I_k, \quad \omega_N^{-(k+m)} = -\omega_N^{-k}$$

for $k = 0, 1, \ldots, m-1$. These identities provide the key to the algorithm, whose essential idea is this:

Step 1: Compute P_k and $\omega_N^{-k} I_k$;

Step 2: Form $Y_k = \frac{1}{2}(P_k + \omega_N^{-k} I_k)$;

Step 3: Deduce $Y_{k+m} = \frac{1}{2}(P_k - \omega_N^{-k} I_k)$.

These computations are done successively only for $k = 0, 1, \ldots, m-1$. This scheme is illustrated schematically in Figure 9.1, where the arrows indicate dependent relations in the calculations.

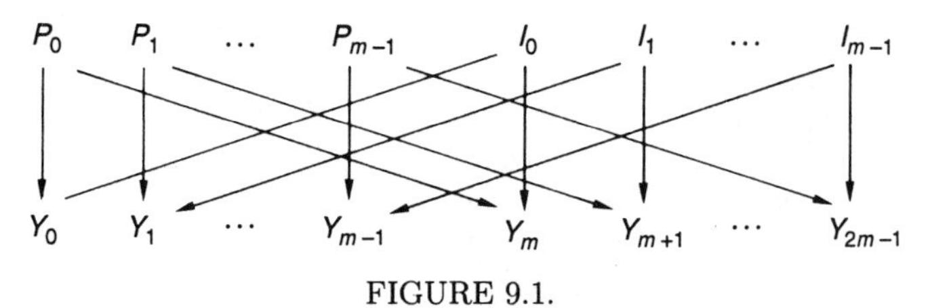

FIGURE 9.1.

The cost of Step 1 is $2(m-1)^2 + m - 1$ (or roughly $N^2/2$) multiplications. Steps 2 and 3 cost nothing in complex multiplications. Thus one obtains the same result for about half the work. (Note we have neglected the divisions by 2 and m. In practice, m is a power of 2, and these are binomial shifts.)

One could consider that this saving is sufficient and stop here. But most readers probably have noticed that P_k and I_k are themselves two independent discrete Fourier transforms of order $m = N/2$. In any case, it takes only a moment to be convinced that

$$(y_0, y_2, \ldots, y_{2m-2}) \overset{\mathscr{F}_{N/2}}{\longmapsto} (P_0, P_1, \ldots, P_{m-1}),$$
$$(y_1, y_3, \ldots, y_{2m-1}) \overset{\mathscr{F}_{N/2}}{\longmapsto} (I_0, I_1, \ldots, I_{m-1}).$$

An obvious strategy is to repeat this clever decomposition, provided that m is even. The best case is where N is a power of 2, $N = 2^p$. We can then iterate the process until we arrive at discrete Fourier transforms of order 2. These are particularly simple computations, since they are of the form

$$Y = (y + z)/2,$$
$$Z = (y - z)/2.$$

We illustrate the algorithm for $N = 8$. The first step is to rearrange the sequence $(y_1, y_2, \ldots, y_8)$ into two sequences of length 4, the first having the odd indices and the second the even indices. The process is repeated, and we obtain the four vectors of length 2 shown in Figure 9.2. The computation begins with the vectors of length 2. As in Figure 9.1, the arrows indicate

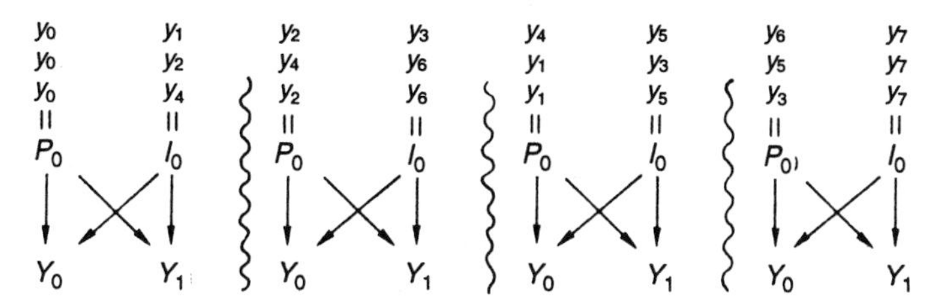

FIGURE 9.2. Rearrangement of the data.

the dependencies of the Y-vectors on the data. To simplify the notation, we have written P_0, I_0, Y_0, Y_1 four times, but they are clearly not the same values. The wiggly lines separate independent computations. Going from one level (vectors of length m) to the next (vectors of length $2m$) is done using the formulas

$$Y_k = \frac{1}{2}(P_k + \omega_N^{-k} I_k),$$
$$Y_{k+m} = \frac{1}{2}(P_k - \omega_N^{-k} I_k),$$

for $k = 0, 1, \ldots, m-1$. Figure 9.3 illustrates the complete algorithm for $N = 2^3$. The wiggly lines separate the independent computations. In an actual program, a single vector is used. This is ultimately the output vector $(Y_0, Y_1, \ldots, Y_{N-1})$; it is the result of successively transforming the vector obtained by appropriately rearranging the original data.

9.2 Evaluating the cost of the algorithm

The only arithmetic operations that appear in the FFT are multiplications and additions of complex numbers. (We neglect the successive divisions by 2; these reduce to a single division by $N = 2^p$, at the outset, for example.) We denote the cost of r complex multiplications and of s complex additions by $[r;\ s]$.

For $N = 2^p$, let M_p be the number of multiplications used in the algorithm and let A_p be the number of additions. Formulas (9.1) are used to evaluate the cost for $N = 2^p$ in terms of the cost for $N = 2^{p-1}$:

Cost of computing the P_k: $[M_{p-1};\ A_{p-1}]$;
Cost of computing the I_k: $[M_{p-1};\ A_{p-1}]$;

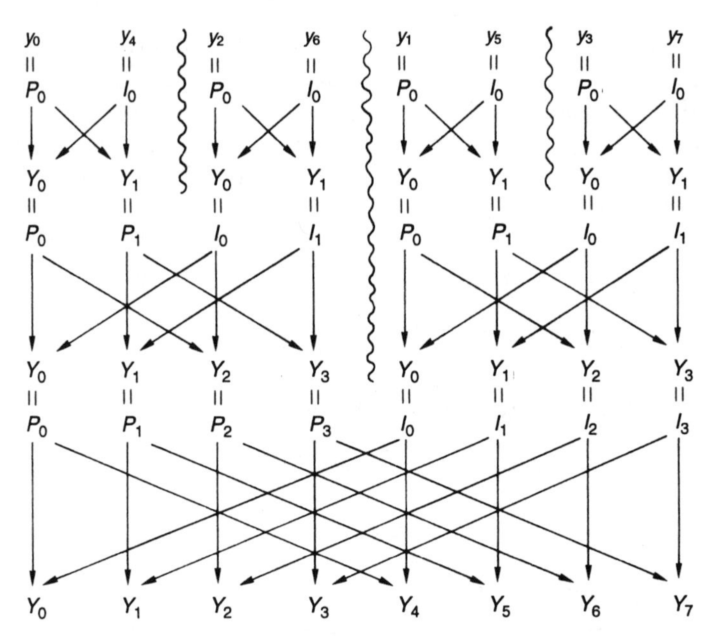

FIGURE 9.3. The FFT algorithm of order 8.

Multiplications by ω_N^{-k} $(k \geq 1)$: $[2^{p-1} - 1; 0]$;
Additions: $[0; 2^p]$.

From these relations we have

$$M_1 = 0, \qquad A_1 = 2,$$
$$M_p = 2M_{p-1} + 2^{p-1} - 1, \qquad A_p = 2A_{p-1} + 2^p.$$

A computation, which is left as an exercise, provides an explicit expressions for M_p and A_p, namely,

$$M_p = (p-2)2^{p-1} + 1,$$
$$A_p = p2^p.$$

We see from this that the global cost, as a function of N, is

$$\left[\frac{1}{2}N(\log_2 N - 2) + 1;\ N \log_2 N\right]. \tag{9.1}$$

Table 9.1 compares the FFT with the "old" method. It shows the savings for the two operations as a function of N. For $N = 1024$, we see that the FFT divides the cost by 250: a fantastic gain.

9.3 The mirror permutation

If we wish to obtain the values $Y_0, Y_1, \ldots, Y_{N-1}$ in this order, it is clear from Figure 9.3 that we must begin with a vector (y_n), $n = p(k)$, where

	Multiplications			Additions		
N	Old Method	FFT	Ratio	Old Method	FFT	Ratio
2	0	0		2	2	1
4	0	0		12	8	1.5
8	49	5	10	56	24	2.3
16	225	17	13	240	64	3.8
32	961	49	20	992	160	6.2
64	3,969	129	31	4,032	384	10
128	16,129	321	50	16,256	896	18
256	65,025	769	85	65,280	2,048	32
512	261,121	1,793	145	261,632	4,608	57
1,024	1,046,529	4,097	255	1,047,552	10,240	102

TABLE 9.1.

p is a permutation of the indices $k = 0, 1, \ldots, N-1$. This permutation of the data at the outset is an important issue, particularly for programming the algorithm. There are a number of ways to do this, and it is a problem that generally stimulates much imagination from students. The only restriction is not to introduce so many operations that the gain realized by the algorithm is compromised.

For these consecutive even–odd permutations, one feels that the representation of the indices in base 2 must come into play. Take the case $N = 8$ and notice what happens:

$$\begin{array}{rcl|rcl}
0 = 000 & \cdot\cdot & & \cdot\cdot & 0 = 000 \\
1 = 001 & \cdot\cdot & & \cdot\cdot & 4 = 100 \\
2 = 010 & \cdot\cdot & & \cdot\cdot & 2 = 010 \\
3 = 011 & \cdot\cdot & & \cdot\cdot & 6 = 110 \\
4 = 100 & \cdot\cdot & & \cdot\cdot & 1 = 001 \\
5 = 101 & \cdot\cdot & & \cdot\cdot & 5 = 101 \\
6 = 110 & \cdot\cdot & & \cdot\cdot & 3 = 011 \\
7 = 111 & \cdot\cdot & & \cdot\cdot & 7 = 111
\end{array}$$

Each number has been written in binary form using three places, which is possible, since we stop at 7 ($N = 2^3$). We notice a surprising property: The required permutation of an index is given by reversing the order of its binary representation. It is as if they were reflected in a mirror. One can verify that this holds for $N = 16$, and it is an excellent exercise to show that it is true in general.

This "mirror" permutation leads to a method for programming the initial permutation. For this, it is necessary to work with the binary representations of the indices. These, however, are not directly accessible in high-level languages like PASCAL; consequently, this is not the best method.

9.4 A recursive program

Here, to finish the chapter, is a program (written in a simplified pseudo-language) for computing the FFT of a vector y. It is taken from [Lip81]. We include it because it is astonishingly simple to program and because it follows step by step the approach we have taken. The particularity of this procedure is that it is *recursive*, which means that calls are made within the program to the program itself.

```
Procedure FFT(n,w,y,Y);
    begin
        if n=1 then Y[0]:=y[0] else
        begin
            m:=n div 2;
            for k:=0 to m−1 do
            begin
                b[k]:=y[2*k];
                c[k]:=y[2*k+1]
            end; w2=w*w;
            TFR(m,w2,b,B);
            TFR(m,w2,c,C);
            wk:=1;
            for k:=0 to m−1 do
            begin
                X:=B[k]; T:=wk*C[k];
                Y[k]:=(X+T)/2;
                Y[k+m]:=(X−T)/2;
                wk=wk*w
            end
        end
    end.
```

We note that compilers deal with these recursions more or less well, particularly on microcomputers. While the program itself is concisely written, which is very attractive to the programmer, its execution, by contrast, requires a great deal of processing and a large amount of memory: At each call to FFT, the procedure is completely recopied with new parameters.

Finally, it is not obvious that this procedure does indeed compute the desired FFT. For example, the second call to FFT is executed only after many other such calls.

9.5 Exercises

Exercise 9.1 Consider the discrete Fourier transform of order N defined by the formulas (8.5)

$$Y_n = \frac{1}{N}\sum_{k=0}^{N-1} y_k \omega_N^{-nk}, \quad n = 0, 1, \ldots, N-1.$$

Write the discrete Fourier transform in its matrix form $Y = S_N y$. What is the matrix associated with the inverse transform?

****Exercise 9.2** Let u and v be two complex periodic sequences with period N. Consider the periodic convolution $w = u * v$ defined by (10.1):

$$w_n = \sum_{q=0}^{N-1} u_{n-q} v_q, \quad n = 0, 1, \ldots, N-1.$$

(a) Write this convolution in matrix form,

$$w = C(u)v \quad \text{where} \quad w = (w_0, w_1, \ldots, w_{N-1})^t, \; v = (v_0, v_1, \ldots, v_{N-1})^t,$$

and $C(u)$ is an $N \times N$ circulant matrix.

(b) Let U and V denote, respectively, the discrete Fourier transforms of u and v. Express W, the discrete Fourier transform of w, in matrix form as a function of S_N, $C(u)$, and V.

(c) Show that $W = D(U)V$, where $D(U)$ is a diagonal matrix $(D(U))_{ii} = NU_i$, $i = 0, 1, \ldots, N-1$. (Write $C(u)$ as a linear combination of permutation matrices.)

Exercise 9.3 (binary mirror permutation) For $M \in \mathbb{N}^*$, let $N = 2^M$ and define $E_M = \{j \in \mathbb{N} \mid 0 \le j \le 2^M - 1\}$. Each integer $j \in E_M$ can be represented as a binary number of length M:

$$j = \sum_{k=0}^{M-1} \alpha_k 2^k, \quad \text{or} \quad j = \alpha_{M-1}\alpha_{M-2}\cdots\alpha_1\alpha_0,$$

where α_k is 0 or 1. With each $j \in E_M$ we associate $j^* \in E_M$ obtained by reversing the order of the α_k :

$$j^* = \alpha_0\alpha_1\cdots\alpha_{M-2}\alpha_{M-1}, \quad \text{or} \quad j^* = \sum_{k=0}^{M-1} \alpha_{M-k-1} 2^k.$$

This defines a permutation on E_M.

(a) Write the matrix P_N associated with this permutation for $N = 8$.

(b) Verify that P_N is symmetric and that $P_N^2 = I$, the identity matrix.

(c) How does this generalize for arbitrary M?

****Exercise 9.4 (matrix version of the FFT)** The exercise is for the case $M = 3$, that is, $N = 2^3$. It generalizes to an arbitrary M. (A description of the general algorithm is given in [Ebe70].)

We will use the results of Exercises 9.1 and 9.3 with modified notation: In place of S_N, P_N, Ω_N we write S_3, P_3, Ω_3 $(N = 2^3)$.

We will use the mirror permutation matrix P_3 to compute $Y = S_3 y$; in fact, we compute $Y = P_3(P_3 S_3)y$. The work proceeds in two phases:

- Compute $Y^* = (P_3 S_3)y$.
- Compute $Y = P_3 Y^*$.

(1) Phase 1: Compute Y^*.
Let $T_3 = P_3 S_3$. Define $E_3(S_3)$ and $E_3(T_3)$ to be the 8×8 matrices formed from the exponents (nk mod 8) of the terms $(\omega_N^{-1})^{nk}$ appearing in S_3 and T_3 respectively.

(a) Express $E_3(S_3)$ and $E_3(T_3)$ explicitly.

(b) Show that T_3 can be written as

$$T_3 = \frac{1}{2}\begin{bmatrix} T_2 & T_2 \\ L_2 T_2 & -L_2 T_2 \end{bmatrix}$$

with

$$L_2 = \begin{bmatrix} 1 & 0 & 0 & 0 \\ 0 & e^{-2i\pi\frac{1}{2^3}} & 0 & 0 \\ 0 & 0 & e^{-2i\pi\frac{2}{2^3}} & 0 \\ 0 & 0 & 0 & e^{-2i\pi\frac{3}{2^3}} \end{bmatrix}.$$

(c) Now consider the algorithm for computing $Y^* = T_3 y$.

First step:

- Write $R^0 = T_3$. Verify that R^0 can be decomposed as a product of 3 matrices:

$$R^0 = R^1 \Delta^0 \Sigma^0,$$

where

$$R^1 = \begin{bmatrix} T_2 & 0 \\ 0 & T_2 \end{bmatrix}, \quad \Delta^0 = \begin{bmatrix} I_2 & 0 \\ 0 & L_2 \end{bmatrix}, \quad \text{and } \Sigma^0 = \frac{1}{2}\begin{bmatrix} I_2 & I_2 \\ I_2 & -I_2 \end{bmatrix},$$

I_2 being the 4×4 identity matrix.

- Write $v^0 = y$ and cut v^0, a vector of length 8, into two vectors, v_0^0 and v_1^0, of length 4:

$$v^0 = \begin{bmatrix} v_0^0 \\ v_1^0 \end{bmatrix}.$$

Compute $v^1 = \Delta^0 \Sigma^0 v^0$ as a function of v_0^0, v_1^0, and L_2. How many complex multiplications are needed to compute v_1?

Second step:

This step is to compute $Y^* = R^1 v^1$.

- Show that R^1 can be written as the product of 3 matrices:

$$R^1 = R^2 \Delta^1 \Sigma^1,$$

where

$$R^2 = \begin{bmatrix} T_1 & 0 & 0 & 0 \\ 0 & T_1 & 0 & 0 \\ 0 & 0 & T_1 & 0 \\ 0 & 0 & 0 & T_1 \end{bmatrix}, \quad \Delta^1 = \begin{bmatrix} I_1 & 0 & 0 & 0 \\ 0 & L_1 & 0 & 0 \\ 0 & 0 & I_1 & 0 \\ 0 & 0 & 0 & L_1 \end{bmatrix},$$

and

$$\Sigma^1 = \frac{1}{2} \begin{bmatrix} I_1 & I_1 & 0 & 0 \\ I_1 & -I_1 & 0 & 0 \\ 0 & 0 & I_1 & I_1 \\ 0 & 0 & I_1 & -I_1 \end{bmatrix}.$$

I_1 is the (2×2) identity matrix, $T_1 = P_1 S_1$ is a 2×2 matrix, and

$$L_1 = \begin{bmatrix} 1 & 0 \\ 0 & e^{-2i\pi \frac{1}{2^2}} \end{bmatrix}.$$

- Cut the vector v^1 into 4 vectors v_0^1 , v_1^1, v_2^1, and v_3^1 of length 2. Compute $v^2 = \Delta^1 \Sigma^1 v^1$ as a function of v_i^1, $i = 0, \ldots, 3$.
How many complex multiplications are needed to compute v^2?

Third step:

This is the last step when $M = 3$. Here we compute $Y^* = R^2 v^2$.

- Show that R^2 can be decomposed as a product of 3 matrices:

$$R^2 = R^3 \Delta^2 \Sigma^2,$$

where R^3 is the (8×8) identity matrix, $\Delta^2 = R^3$, and

$$\Sigma^2 = \frac{1}{2} \begin{bmatrix} 1 & 1 & 0 & 0 & 0 & 0 & 0 & 0 \\ 1 & -1 & 0 & 0 & 0 & 0 & 0 & 0 \\ 0 & 0 & 1 & 1 & 0 & 0 & 0 & 0 \\ 0 & 0 & 1 & -1 & 0 & 0 & 0 & 0 \\ 0 & 0 & 0 & 0 & 1 & 1 & 0 & 0 \\ 0 & 0 & 0 & 0 & 1 & -1 & 0 & 0 \\ 0 & 0 & 0 & 0 & 0 & 0 & 1 & 1 \\ 0 & 0 & 0 & 0 & 0 & 0 & 1 & -1 \end{bmatrix}.$$

- From this, deduce the value of Y^*.
- What is the total number of complex multiplications needed to compute Y^*?

(2) Phase 2: Rearrange the components of Y^*.
Is it necessary to compute the matrix product $Y = P_3 Y^*$?

(3) How does one proceed to obtain $Y_0, Y_1, \ldots, Y_7$ in this order?

(4) How must one modify the algorithm to compute the inverse discrete Fourier transform?

Exercise 9.5 Show that it is possible to compute the periodic convolution (10.1) using a form of the FFT that does not involve the mirror permutation.

Lesson 10

Using the FFT for Numerical Computations

We present several examples to indicate the many possible numerical applications of the fast Fourier transform (FFT). It is widely used in signal processing for spectral analysis and for computing convolutions. We will see other important uses in computations involving high-degree polynomials and in interpolation problems.

10.1 Computing a periodic convolution

10.1.1 Complex data

Let $(x_n)_{n\in\mathbb{Z}}$ and $(h_q)_{q\in\mathbb{Z}}$ be two complex periodic sequences having the same period N. The *periodic convolution* of these two sequences is the complex sequence $(y_n)_{n\in\mathbb{Z}}$ defined by

$$y_n = \sum_{q=0}^{N-1} h_q x_{n-q} = \sum_{q=0}^{N-1} h_{n-q} x_q. \tag{10.1}$$

This sequence is clearly periodic with period N. The transformation defined by (10.1) is also a linear transformation $X \mapsto Y = HX$ of C^N into itself with $X = (x_0, x_1, \ldots, x_{N-1})^t$ and $Y = (y_0, y_1, \ldots, y_{N-1})^t$. The matrix of this transformation is called a *circulant* matrix and is given by

$$H = \begin{bmatrix} h_0 & h_{N-1} & h_{N-2} & \ldots & h_1 \\ h_1 & h_0 & h_{N-1} & \ldots & h_2 \\ h_2 & h_1 & h_0 & \ldots & h_3 \\ \vdots & & & & \\ h_{N-1} & h_{N-2} & h_{N-3} & \ldots & h_0 \end{bmatrix}.$$

Computing the convolution directly from the definition (10.1) requires

$$\begin{aligned} &N^2 \text{ complex multiplications,} \\ &N(N-1) \text{ complex additions.} \end{aligned} \tag{10.2}$$

Theorem 8.2.3 points to another way to proceed: Let (X_k), (H_k), and (Y_k) be the discrete Fourier transforms of the sequences (x_n), (h_n), and (y_n). Equation (10.1) becomes

$$Y_k = NH_kX_k, \quad k = 0, 1, \ldots, N-1. \tag{10.3}$$

If we assume that the length N of complex vectors is a power of 2, $N = 2^p$, then this computation proceeds as follows:

	Computation	Cost
Step 1:	Compute the transforms $\mathscr{F}_N$ $(x_n) \overset{\mathscr{F}_N}{\longmapsto} (X_k)$ $(h_n) \overset{\mathscr{F}_N}{\longmapsto} (H_k)$	$[N(p-2);\ 2Np]$
Step 2:	Compute the products (10.3)	$[N;\ 0]$
Step 3:	Compute the transform $\mathscr{F}_N^{-1}$ $(Y_k) \overset{\mathscr{F}_N^{-1}}{\longmapsto} (y_n)$	$[(N/2)(p-2);\ Np]$

The total cost is

$$\begin{aligned} &\frac{N}{2}(3\log_2 N - 4) \text{ complex multiplications,} \\ &3N\log_2 N \text{ complex additions,} \end{aligned} \tag{10.4}$$

which is an appreciable savings compared with (10.2). If $N = 64$, the cost is $[448;\ 1,152]$ complex operations in place of $[4,096;\ 4,032]$. (Note that here and elsewhere we neglect the constant term that appears in (9.2).)

10.1.2 Real data

In this case, the two discrete Fourier transforms (DFT) in Step 1 can be computed with a single DFT of order N on complex data (see Section 8.3). In Step 3 the inverse DFT acts on the periodic sequence (Y_k) that satisfies $Y_{-k} = \overline{Y}_k$, so this can be computed with a single DFT of order $N/2$. Since the products in (10.3) can be obtained with $N/2$ complex multiplications, the total cost is

$$\begin{aligned} &\frac{N}{4}(3\log_2 N - 5) \text{ complex multiplications,} \\ &\frac{N}{2}(3\log_2 N - 1) \text{ complex additions.} \end{aligned}$$

One complex multiplication can be done with 4 real multiplications and 2 real additions. Thus in the real case the cost is

$$\begin{aligned} &N(3\log_2 N - 5) \text{ real multiplications,} \\ &\frac{N}{2}(9\log_2 N - 7) \text{ real additions.} \end{aligned}$$

For $N = 64$, the cost is $[832;\ 1,504]$ versus $[16,384;\ 16,256]$ given by (10.2).

10.2 Nonperiodic convolution

Let $(x_n)_{n\in\mathbb{Z}}$ and $(h_n)_{n\in\mathbb{Z}}$ be two nonperiodic signals that have compact support. In particular we assume that

$$\begin{aligned} x_n &= 0 \quad \text{if} \quad n < 0 \quad \text{or} \quad n \geq M, \\ h_n &= 0 \quad \text{if} \quad n < 0 \quad \text{or} \quad n \geq Q \quad (Q \leq M). \end{aligned}$$

The problem is to compute the nonperiodic convolution

$$y_n = \sum_{q=0}^{Q-1} h_q x_{n-q}. \tag{10.5}$$

The y_n are zero if $n < 0$ or if $n \geq M + Q - 1$. Let N be the smallest power of 2 such that $N \geq M + Q - 1$. By making the original sequences periodic with period N, we come back to the problem of computing a periodic convolution using the FFT, where the cost is given by (10.4):

$$\begin{aligned} &\frac{N}{2}(3\log_2 N - 4) \text{ complex multiplications,} \\ &3N \log_2 N \text{ complex additions.} \end{aligned}$$

EXAMPLE: Take $Q = 200$, $M = 500$, and $N = 1024$. Then the cost of computing (10.5) is

$$\begin{aligned} &MQ = 10^5 \text{ multiplications,} \\ &MQ - (M + Q - 1) \approx 10^4 \text{ additions.} \end{aligned}$$

The cost using the FFT is

$$\begin{aligned} &1024 \times 13 \approx 1.3 \cdot 10^4 \text{ multiplications,} \\ &1024 \times 30 \approx 3 \cdot 10^4 \text{ additions.} \end{aligned}$$

We see that the FFT method is still advantageous in this case. On the other hand, this advantage is lost when the lengths of the two signals are disproportionate. This happens frequently in "real-time" signal processing, where the sequence (x_n) is practically "infinite" and where the support of the filter (h_q) is relatively small. This is the case, for example, when one smoothes data with a "sliding window."

EXAMPLE: Suppose

$$y_n = \sum_{q=0}^{4} h_q x_{n-q} \tag{10.6}$$

with $Q = 5$ and $M = 1000$. The cost of computing (10.6) is

5000 multiplications,
4000 additions.

The cost using the FFT method with N of the order 1024 is

$1.3 \cdot 10^4$ multiplications,
$3 \cdot 10^4$ additions.

The direct application of the FFT method is clearly more costly. There are, however, specific methods for this case. They involve cutting the vector (x_n) into shorter pieces (see, for example, [CLW67] and [Nus81]).

10.3 Computations on high-order polynomials

A polynomial P of degree less than or equal to p can be coded in several different ways. The most common of these is to represent P as the vector of its coordinates in a given basis, for example, the canonical basis $(1, x, \ldots, x^p)$. Thus

$$P(x) = a_0 + a_1 x + \cdots + a_p x^p$$

is represented by

$$(a_0, a_1, \ldots, a_p). \tag{10.7}$$

This representation is convenient for computing the values of P for different values of x (Horner's algorithm). It is less convenient if one wishes to compute the product of P and another polynomial

$$Q(x) = b_0 + b_1 x + \cdots + b_q x^q.$$

The coefficients of the product

$$PQ(x) = c_0 + c_1 x + \cdots + c_{p+q} x^{p+q}$$

are given by the convolution

$$c_k = \sum_{n=0}^{k} a_n b_{k-n}, \quad k = 0, 1, \ldots, p+q. \tag{10.8}$$

This computation requires

$\frac{1}{2}(p+q+1)(p+q+2)$ multiplications,
$\frac{1}{2}(p+q)(p+q+1)$ additions.

There is another representation that is better adapted to the computation of a product. We know that a polynomial of degree less than or equal to $N-1$ is uniquely determined by its values at N distinct points in the complex plane:

$$y_j = P(x_j), \quad j = 0, 1, \ldots, N-1. \tag{10.9}$$

In this case, the product PQ is simply coded by the numbers

$$P(x_j)Q(x_j), \quad j = 0, 1, \ldots, N-1.$$

(We have assumed that $p+q \leq N-1$.) On the other hand, the computation of $P(x)$ for an arbitrary value x is more complicated in this representation.

We are going to examine these two representations and the problem of going from one to the other.

10.3.1 Polynomials represented in the canonical basis

Formula (10.8) expresses the coordinates of the product PQ as a nonperiodic convolution. Thus it can be computed using the method described in Section 10.2: The vectors $(a_0, a_1, \ldots, a_p)$ and $(b_0, b_1, \ldots, b_q)$ are extended with zeros to obtain two vectors of length N, where N is a power of 2 and $N \geq p+q+1$. These vectors are extended periodically to all of $\mathbb{Z}$ to obtain two N-periodic sequences; formula (10.8) then becomes

$$c_k = \sum_{n=0}^{N-1} a_n b_{k-n}, \quad k = 0, 1, \ldots, N-1.$$

The computation using the FFT technique costs

$$\frac{N}{2}(3\log_2 N - 4) \text{ multiplications,}$$
$$3N\log_2 N \text{ additions.}$$

EXAMPLE: Take $p = 13$, $q = 15$, and $N = 32$. The direct method costs

435 multiplications,
406 additions,

while the FFT method costs

176 multiplications,
320 additions.

However, as indicated in Section 10.2, the FFT method loses its advantage when p and q are not of the same order of magnitude.

10.3.2 Polynomials represented by N values

A choice for the points x_k in (10.9) that is particularly interesting is

$$x_k = e^{2i\pi \frac{k}{N}} = \omega_N^k. \tag{10.10}$$

The two representations (10.7) and (10.9) are then related by the equations

$$y_k = \sum_{n=0}^{N-1} a_n \, \omega_N^{nk}, \quad k = 0, 1, \ldots, N-1. \tag{10.11}$$

This is a discrete Fourier transform whose inverse is given by

$$a_n = \frac{1}{N} \sum_{k=0}^{N-1} y_k \, \omega_N^{-nk}, \quad n = 0, 1, \ldots, N-1.$$

We know from Section 9.2 that the cost of going from one representation to the other is

$$\begin{aligned} &\frac{N}{2}(\log_2 N - 2) \text{ multiplications,} \\ &N \log_2 N \text{ additions.} \end{aligned}$$

10.4 Polynomial interpolation and the Chebyshev basis

We will see that the FFT can be used to reduce the cost of computing a polynomial interpolation. This is made possible by representing the polynomial in the Chebyshev basis.

10.4.1 The Chebyshev polynomials

The Chebyshev polynomials are the polynomials $T_0, T_1, T_2, \ldots$ that are defined for all $\theta \in [0, \pi]$ by the relations

$$T_n(\cos\theta) = \cos n\theta. \tag{10.12}$$

That T_n is a polynomial follows from the de Moivre formulas. Furthermore, the degree of T_n is exactly n, and its coefficients are integers.

$$\begin{aligned} T_0(x) &= 1, \\ T_1(x) &= x, \\ T_2(x) &= 2x^2 - 1, \\ T_3(x) &= 4x^3 - 3x, \\ T_4(x) &= 8x^4 - 8x^2 + 1, \quad \text{etc.} \end{aligned}$$

For example, the expression for T_3 comes from the identity

$$\cos 3\theta = 4\cos^3\theta - 3\cos\theta.$$

We are concerned here with the vector space of polynomials with real coefficients. This is a vector space over $\mathbb{R}$, and the T_n form a basis for this space. More precisely, if P has degree less than or equal to N, it can be represented uniquely as

$$P(x) = \sum_{n=0}^{N} a_n T_n(x). \tag{10.13}$$

This representation in terms of the T_n and that of $P(x)$ in terms of its values at $N+1$ points are widely used in pseudo-spectral methods for approximating solutions of certain partial differential equations. The polynomial P is an approximation of the unknown function. It is obtained by computing its values y_k at $N+1$ points. When using this technique, one must constantly pass from one to the other of the two representations (a_n) and (y_k).

10.4.2 Choosing the x_k

Since we wish to remain in the real domain, we cannot use the x_k given by (10.10) directly, but our choices are derived from (10.10). In particular, we take the abscissas x_k (which are called the Chebyshev abscissas) to be

$$x_k = \cos\left(k\frac{\pi}{N}\right), \quad k = 0, 1, \ldots, N. \tag{10.14}$$

From (10.12),

$$T_n(x_k) = \cos\left(nk\frac{\pi}{N}\right), \tag{10.15}$$

and

$$y_k = P(x_k) = \sum_{n=0}^{N} a_n \cos\left(nk\frac{\pi}{N}\right). \tag{10.16}$$

This formula is not exactly a DFT, but it is not far from one. By writing the cosines in terms of exponents, we see that

$$y_k = \frac{1}{2}\sum_{n=0}^{N} a_n \omega_{2N}^{nk} + \frac{1}{2}\sum_{n=0}^{-N} a_{-n}\omega_{2N}^{nk} = \sum_{n=-N}^{N} c_n \omega_{2N}^{nk}, \tag{10.17}$$

where

$$c_n = \begin{cases} \frac{1}{2}\, a_n & \text{if } \ 0 < n \le N, \\ a_0 & \text{if } \ n = 0, \\ \frac{1}{2}\, a_{-n} & \text{if } \ -N \le n < 0. \end{cases} \tag{10.18}$$

The expression (10.16) defines y_k for $0 \leq k \leq N$, but since the functions $\cos(nk\frac{\pi}{N})$ are defined for all $k \in \mathbb{Z}$, we can use the right-hand side of (10.16) to extend the function $k \mapsto y_k$ to all $k \in \mathbb{Z}$. Furthermore, the functions $\cos(nk\frac{\pi}{N})$ are even and $2N$-periodic, so the same is true for the sequence $(y_k)_{k\in\mathbb{Z}}$. Thus $y_{-k} = y_k$ and $y_{2N-k} = y_k$, and the system (10.17) can be written

$$y_k = \sum_{n=-N}^{N} c_n \omega_{2N}^{nk}, \quad k = 0, 1, \ldots, 2N-1. \tag{10.19}$$

Applying the technique used in Section 8.1, we try to invert this system by computing

$$\gamma_p = \sum_{k=0}^{2N-1} y_k \omega_{2N}^{-pk}, \quad p = 0, 1, \ldots, N.$$

By substituting (10.19) for y_k, this becomes

$$\gamma_p = \sum_{k=0}^{2N-1} \sum_{n=-N}^{N} c_n \omega_{2N}^{(n-p)k} = \sum_{n=-N}^{N} c_n \sum_{k=0}^{2N-1} \omega_{2N}^{(n-p)k}.$$

The last sum is equal to $2N$ if $n \equiv p \pmod{2N}$ and 0 otherwise, so that

$$\gamma_p = 2Nc_p, \quad p = 0, 1, 2, \ldots, N-1,$$
$$\gamma_N = 2N(c_{-N} + c_N) = 4Nc_N.$$

Finally, in view of (10.18), we have

$$a_n = \frac{1}{N\varepsilon_n} \sum_{k=0}^{2N-1} y_k \omega_{2N}^{-nk}, \quad n = 0, 1, \ldots, N, \tag{10.20}$$

with

$$\varepsilon_0 = \varepsilon_N = 2; \ \varepsilon_1 = \cdots = \varepsilon_{N-1} = 1.$$

Formulas (10.16) and (10.20) are the reciprocals of each other; they resolve theoretically the problem of going from one representation to the other.

10.4.3 Practical computation

The work proceeds as follows:

(a) COMPUTE (a_n) KNOWING (y_k):

Step 1: Compute $y_{2N-k} = y_k$ for $k = 1, 2, \ldots, N-1$.

Step 2: The FFT algorithm of order $2N$:

$$(y_0, y_1, \ldots, y_{2N-1}) \overset{\mathcal{F}_{2N}}{\longmapsto} (Y_0, Y_1, \ldots, Y_{2N-1}).$$

Step 3: $a_n = Y_n$ if $n = 1, 2, \ldots, N-1$; $a_0 = \frac{1}{2}Y_0$; $a_N = \frac{1}{2}Y_N$.

(We leave it to the reader to explain why the Y_n are real.)

(b) COMPUTE (y_k) KNOWING (a_n):

Step 1: Compute the c_n from (10.18).

Step 2:

$$Y_n = \begin{cases} c_n & \text{if} \quad 0 \leq n \leq N, \\ c_{n-2N} & \text{if} \quad N < n \leq 2N-1. \end{cases}$$

Step 3: The inverse FFT algorithm of order $2N$:

$$(Y_0, Y_1, \ldots, Y_{2N-1}) \overset{\mathscr{F}_{2N}^{-1}}{\longmapsto} (y_0, y_1, \ldots, y_{2N-1}).$$

10.4.4 Cost of the computation

If we compute the cost with reference to Section 9.1, the total cost of the computation is

$$\begin{aligned} &N(\log_2 N - 1) \text{ complex multiplications,} \\ &2N(\log_2 N + 1) \text{ complex additions.} \end{aligned}$$

It is possible to reduce this cost by taking into consideration the fact that the sequences are real and even: In (a) $y_{2N-k} = y_k$, and in (b) $Y_{2N-n} = Y_n$. Thus the computation can be done with a discrete Fourier transform of order $N/2$ rather than order $2N$. The cost is

$$\begin{aligned} &N(\log_2 N - 3) \text{ real multiplications,} \\ &\frac{N}{2}(3\log_2 N - 5) \text{ real additions.} \end{aligned}$$

EXAMPLE: Take $N = 32$.
The cost using formula (10.16) is $32^2 = 1024$ real multiplications.
The cost using the reduced FFT is 64 real multiplications.

10.4.5 A trigonometric interpolation problem

The function

$$f(\theta) = P(\cos\theta)$$

has by (10.12) and (10.13) an expression of the form

$$f(\theta) = \sum_{n=0}^{N} a_n \cos n\theta.$$

The a_n are the Fourier coefficients of the even trigonometric polynomial f, which has degree less than or equal to N. Equation (10.16) becomes

$$f\left(k\frac{\pi}{N}\right) = y_k, \quad k = 0, 1, \ldots, N,$$

and (10.20) shows us how to find the coefficients a_n of f given the values y_k and thus how to solve numerically this particular interpolation problem.

10.4.6 Theorem *Given any $N+1$ real numbers $y_0, y_1, \ldots, y_N$, there exists a unique trigonometric polynomial of the form*

$$f(\theta) = \sum_{n=0}^{N} a_n \cos n\theta$$

that satisfies

$$f\left(k\frac{\pi}{N}\right) = y_k, \quad k = 0, 1, \ldots, N.$$

The coefficients a_n are given by (10.20).

10.5 Exercises

Exercise 10.1 Take two vectors of lengths $M = 4$ and $Q = 3$. Extend them as described in Section 10.2 and verify that the computations with the new vectors do indeed give the original convolution.

Exercise 10.2 Let $f(\theta)$ be the trigonometric polynomial in Theorem 10.4.6. Show that integrating f on $[0, 2\pi]$ using the trapezoidal method yields the correct value of the integral; that is,

$$\int_0^{2\pi} f(\theta)\, d\theta = \frac{\pi}{N} \sum_{k=0}^{2N-1} f\left(k\frac{\pi}{N}\right).$$

Exercise 10.3

(a) Show that the function $f(\theta)$ in Theorem 10.4.6 can be written

$$f(\theta) = \sum_{k=0}^{N} f(\theta_k) g_k(\theta),$$

where

$$g_k(\theta) = \sum_{n=0}^{N} a_n^k \cos n\theta, \quad \theta_k = k\frac{\pi}{N}, \quad \text{and} \quad a_n^k = \frac{1}{N\varepsilon_n} \cos n\theta_k,$$

with $\varepsilon_0 = \varepsilon_N = 2$ and $\varepsilon_1 = \cdots = \varepsilon_{N-1} = 1$.

(b) Find expressions for the g_k in terms of tangents and cotangents.

Chapter IV

The Lebesgue Integral

Lesson 11

From Riemann to Lebesgue

We are going to introduce the Lebesgue integral here and in the next three lessons. Experience has shown that this notion of integration is particularly well suited for operations such as

- taking limits under the integral sign,
- taking derivatives under the integral sign,
- interchanging the order of integration.

We do not intend to give a complete and rigorous development of the theory of Lebesgue integration. Readers wishing a deeper understanding of the fundamentals can consult any of numerous references such as [KF74], [Hal64], and [Roy63]. We wish to go as directly as possible to the applications while at the same time presenting the essential ideas.

11.1 Some history

The idea of integration is based intuitively on the notion of area. Given a positive, continuous function f defined on $[a, b]$, the integral

$$\int_a^b f(x)dx$$

is the area of the region bounded by the curve $y = f(x)$, the x-axis, and the two lines $x = a$ and $x = b$. It was in 1853 that Bernhard Riemann gave a rigorous definition of the integral that bears his name.

For a fixed n, consider a partition Δ_n of the interval $[a, b]$ of the form

$$a = x_0 < x_1 < x_2 < \cdots < x_{n-1} < x_n = b$$

and define the size of the partition to be

$$\mu(\Delta_n) = \sup_i (x_i - x_{i-1}).$$

Form the sums

$$I(\Delta_n) = \sum_i^n f(\xi_i)(x_i - x_{i-1}),$$

where $\xi_i \in (x_{i-1}, x_i)$ but is otherwise arbitrary. The integral of f in the sense of Riemann, or the *Riemann integral* of f, is the limit, if it exists, of the $I(\Delta_n)$ as $n \to \infty$ and $\mu(\Delta_n) \to 0$.

Riemann showed that the partial sums $I(\Delta_n)$ (called *Riemann sums*) have a limit not only for continuous functions but also for a larger class of functions, some of which have an infinite number of discontinuities. He introduced a generalized integral of a function that is unbounded in the neighborhood of an isolated point $c \in [a, b]$ as the limit, if it exists, of

$$\int_a^{c-\alpha} f(x)dx + \int_{c+\beta}^b f(x)dx$$

as $\alpha, \beta \to 0$, $\alpha, \beta > 0$. Riemann also proved that the Fourier coefficients of a periodic, integrable function tend to zero. This result was later generalized by Henri Lebesgue.

Riemann's work stimulated numerous studies aimed at generalizing the definition of the integral to include the widest possible class of functions (T. Stieltjes, E. Borel, H. Lebesgue, F. Riesz). It was around 1900 that Lebesgue proposed his theory of integration. The Riemann sums are well behaved for only a particular class of discontinuous functions; they require that $f(x)$ not vary too much in the intervals (x_{i-1}, x_i). Lebesgue inverted the situation: He considered the range of values $[m, M]$ taken by $f(x)$ and partitioned this interval into segments (y_{i-1}, y_i). He then considered the set of x such that $y_{i-1} \leq f(x) < y_i$. He gave this set a *measure* m_i and formed the sums

$$\sum_i m_i \eta_i \quad \text{where} \quad y_{i-1}\eta_i < y_i.$$

The integral of f on $[a, b]$ is obtained by passing to the limit as the size of the partition of $[m, M]$ tends to zero.

This description, while simplistic, introduces the notion of the measure of a set of real numbers whose structure can be quite complicated. In fact, Lebesgue's work on integration was a natural continuation of the theory of measure initiated by E. Borel and the researches of R. Baire on the structure of sets. Classical presentations of the theory of integration follow this historical development: measure theory followed by integration theory. We will follow this plan.

11.2 Another point of view

It is very convenient when using arguments that involve taking limits to be working in complete spaces. If not, the "limit" of a Cauchy sequence is a foreign object. The situation for arguments about sets of functions is analogous to that for sets of numbers: While practical computations may be made within the rationals $\mathbb{Q}$ (or even a subset of $\mathbb{Q}$), the complete set of reals $\mathbb{R}$ is very useful, indeed almost indispensable, for studying convergent sequences. In the same way, we use the Riemann integral for practical computations. On the other hand, taking limits and differentiating under the integral sign are theoretical operations that are greatly facilitated with the Lebesgue integral.

The space $C^0[a,b]$ of continuous functions on $[a,b]$ endowed with the norm

$$\|f\|_1 = \int_a^b |f(x)|dx$$

is not complete. For continuous functions, the Riemann and Lebesgue integrals are the same. To complete this space it is necessary to extend the class of function for which

$$\int_a^b |f(x)|dx$$

exists and is finite. The set of Riemann integrable functions is included in this completion, but there are Cauchy sequences of continuous functions in the norm $\|\cdot\|_1$ that do not converge to Riemann integrable functions. The theory of Lebesgue integration provides a "constructive" process to complete $C^0[a,b]$ in the $\|\cdot\|_1$ norm.

11.3 By way of transition

At the end of the day, the waiter, B. Riemann, and the owner of a cafe, H. Lebesgue, both verify the day's receipts. M. Riemann has kept a copy of each customer's bill. He computes the day's total income as

$$S_{\mathrm{R}} = t_1 + t_2 + \cdots + t_N.$$

M. Lebesgue, on the other hand, must prepare the bank deposit, and he sorts the money from the cash register by denomination. (It's a French cafe that rounds things off to the nearest franc, so only 1 F, 2 F, 5 F, etc. appear in the register.) Thus, M. Lebesgue counts how many items n_k there are of each denomination D_k. He computes the total for the deposit as

$$S_{\mathrm{L}} = n_0 D_0 + n_1 D_1 + \cdots + n_M D_M.$$

Assuming no pilfering or added tips, M. Riemann and M. Lebesgue will agree that $S_{\mathrm{R}} = S_{\mathrm{L}}$.

Figures 11.1 and 11.2 represent the two situations. We see that the two men measure the x-axis differently. In particular, M. Lebesgue measures $f^{-1}(1)$, $f^{-1}(2)$, $f^{-1}(5)$, $\ldots$. More precisely, he counts these sets; for example, his measure for $f^{-1}(10)$ is 3.

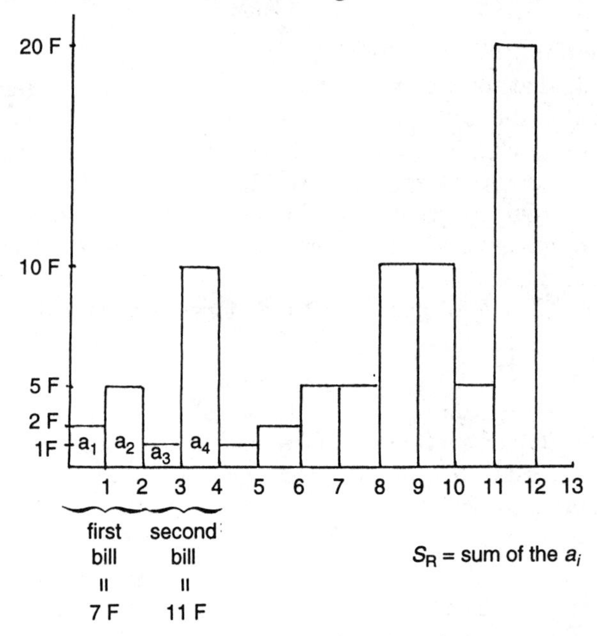

FIGURE 11.1. Riemann's point of view.

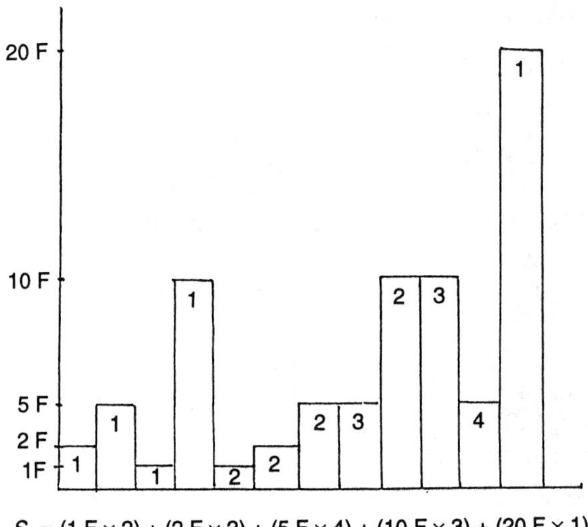

$S_L = (1\text{ F} \times 2) + (2\text{ F} \times 2) + (5\text{ F} \times 4) + (10\text{ F} \times 3) + (20\text{ F} \times 1)$

FIGURE 11.2. Lebesgue's point of view.

Lesson 12

Measuring Sets

The goal of measure theory is to extend the notions of length (of an interval in $\mathbb{R}$), of area (of a rectangle in $\mathbb{R}^2$), of volume, and so on, to more complex sets. In general, given a set X (which in applications will be a part of $\mathbb{R}^n$), one considers a restricted family of subsets of X called a σ-algebra, and it is on the elements of this σ-algebra that one defines a measure. A measure is a function with values in $\mathbb{R}_+ \cup \{+\infty\} = \overline{\mathbb{R}}_+$, where $\mathbb{R}_+ = \{x \in \mathbb{R} \mid x \geq 0\}$, that has certain desirable additivity properties.

12.1 Measurable sets and measure

Intuitively, if the sets D_1 and D_2 have a measure, then the sets $D_1 \cup D_2$, $D_1 \cap D_2$, $D_1 \setminus D_2$, and $D_2 \setminus D_1$ should also have a measure.

12.1.1 Definition Given a set X, let $\mathscr{P}(X)$ be the class of all subsets of X. $\mathscr{T} \subset \mathscr{P}(X)$ is a σ-algebra if the following hold:

(i) $\emptyset$ and X are in $\mathscr{T}$;

(ii) $S \in \mathscr{T}$ implies $X \setminus S \in \mathscr{T}$;

(iii) $S_1, S_2, \ldots \in \mathscr{T} \Rightarrow \bigcup_{n=1}^{\infty} S_n \in \mathscr{T}$.

$\mathscr{T}$ is said to be closed under the formation of complements and countable unions. Note that $\mathscr{P}(X)$ is itself a σ-algebra, but in general it contains too many sets to be of interest.

12.1.2 Definition Let $\mathscr{T}$ be a σ-algebra on X. The elements of $\mathscr{T}$ are called measurable sets and the pair $(X, \mathscr{T})$ is called a measurable space.

A σ-algebra is typically generated from a given collection of subsets of X called the generators of the algebra. Let $\mathscr{E}$ be a subset of $\mathscr{P}(X)$. The intersection of all the σ-algebras containing $\mathscr{E}$ is a σ-algebra. It is the smallest σ-algebra containing $\mathscr{E}$, and this leads to the next definition.

12.1.3 Definition Suppose $\mathscr{E} \subset \mathscr{P}(X)$. The smallest σ-algebra containing $\mathscr{E}$ is called the σ-algebra generated by $\mathscr{E}$.

As an example, take $X = \mathbb{R}^n$ and let $\mathscr{E}$ be the collection of open sets in $\mathbb{R}^n$. $\mathscr{E}$ is not a σ-algebra because the complement of an open set is not open. The elements of the σ-algebra generated by $\mathscr{E}$ are called Borel sets. It is not possible to give an explicit description of this algebra using intersections and unions of open sets.

We now come to the definition of a measure on a σ-algebra.

12.1.4 Definition Let $\mathscr{T}$ be a σ-algebra on X. A measure on $\mathscr{T}$ is a function $\mu : \mathscr{T} \to \overline{\mathbb{R}}_+$ having the following properties:

(i) $\mu(\emptyset) = 0$.

(ii) If S_n is a sequence of disjoint measurable sets, then

$$\mu\Big(\bigcup_{n=1}^{\infty} S_n\Big) = \sum_{n=1}^{\infty} \mu(S_n).$$

The triple $(X, \mathscr{T}, \mu)$ is called a measure space.

12.1.5 Examples Let X be an arbitrary set, $\mathscr{T}$ a σ-algebra on X, and a an element of X. Define μ_a and μ_d as follows:

(a) $$\mu_a(S) = \begin{cases} 1 & \text{if } a \in S, \\ 0 & \text{otherwise.} \end{cases}$$

(b) $$\mu_d(S) = \begin{cases} \text{the number of elements of } S \text{ if } S \text{ is finite,} \\ +\infty \text{ otherwise.} \end{cases}$$

Then μ_a and μ_d are measures on $\mathscr{T}$. However, these measures do not generalize the notion of length when $X = \mathbb{R}$!

In practice, measures are not generally defined directly on a σ-algebra. Usually a measure is defined on a smaller collection of sets and then extended to the σ-algebra generated by these sets. This is the case for Lebesgue measure on $\mathbb{R}^n$.

12.1.6 Definition Given a set X, consider $\mathscr{A} \subset \mathscr{P}(X)$. $\mathscr{A}$ is an algebra if and only if it satisfies the following conditions:

(i) $\emptyset$ and X are in $\mathscr{A}$;

(ii) $S \in \mathscr{A}$ implies $X \setminus S \in \mathscr{A}$;

(iii) $S_1, S_2 \in \mathscr{A}$ implies $S_1 \cup S_2 \in \mathscr{A}$.

Thus an algebra is closed under complements and finite unions. A function ν defined on an algebra $\mathscr{A}$ with values in $\overline{\mathbb{R}}_+$ is said to be a measure on $\mathscr{A}$ if it satisfies the following conditions:

(i) $\nu(\emptyset) = 0$.

(ii) If S_n is a sequence of pairwise disjoint sets in $\mathscr{A}$ such that $\bigcup_{n=1}^{\infty} S_n$ is in $\mathscr{A}$, then

$$\nu\Big(\bigcup_{n=1}^{\infty} S_n\Big) = \sum_{n=1}^{\infty} \nu(S_n).$$

Given a measure on an algebra $\mathscr{A}$, it can be extended to a σ-algebra containing $\mathscr{A}$. This is the content of the following theorem, which we will assume (see, for example, [Hal64]).

12.1.7 Theorem *Let ν be a measure on an algebra $\mathscr{A}$ of X. There exists a measure μ and a σ-algebra $\mathscr{T}$ containing $\mathscr{A}$ with the following properties:*

(i) *For all $S \in \mathscr{A}$, $\nu(S) = \mu(S)$.*

(ii) *If $S_1 \in \mathscr{T}$ and $\mu(S_1) = 0$, then for all $S_2 \subset S_1$, $S_2 \in \mathscr{T}$ and $\mu(S_2) = 0$.*

(iii) *If $X = \bigcup_{n=1}^{\infty} S_n$ with $S_n \in \mathscr{T}$ and $\mu(S_n) < +\infty$, then μ is uniquely determined on the smallest σ-algebra that contains $\mathscr{A}$ and satisfies condition* (ii).

This procedure is used on $\mathbb{R}$ to define Lebesgue measure. One can take the elements of $\mathscr{A}$ to be the sets that are finite unions of intervals. The measure ν is defined for intervals by

$$\nu(a,b) = \begin{cases} |b-a| & \text{if } (a,b) \text{ is a bounded interval,} \\ +\infty & \text{otherwise.} \end{cases} \tag{12.1}$$

For a finite union of disjoint intervals I_n, ν is defined by

$$\nu\Big(\bigcup_{n=1}^{p} I_n\Big) = \sum_{n=1}^{p} \nu(I_n).$$

It is possible to show that ν is a measure on $\mathscr{A}$. (Note that the delicate part of the argument is to prove condition (ii), the countable additivity.)

From Theorem 12.1.7 we know that ν can be extended to a measure μ on a σ-algebra $\mathscr{L}$ containing $\mathscr{A}$ and hence containing the Borel sets. This measure μ is constructed by defining

$$\mu(A) = \inf\Big\{\sum_{n=1}^{\infty} \nu(I_n) \;\Big|\; A \subset \bigcup_{n=1}^{\infty} I_n\Big\},$$

where I_n is an open interval in $\mathbb{R}$.

$\mathscr{L}$ denotes the σ-algebra of Lebesgue-measurable subsets of $\mathbb{R}$, and μ is Lebesgue measure on $\mathbb{R}$. Recall from (ii) of Theorem 12.1.7 that every subset T of a measurable set S for which $\mu(S) = 0$ is a measurable set and $\mu(T) = 0$. Since the sets of measure zero play such an important role in integration theory, we devote a section to discussing them.

12.2 Sets of measure zero

Given a measure space $(X, \mathscr{T}, \mu)$, if $S_1,\ S_2 \in \mathscr{T}$ and if $S_1 \subset S_2$, then it follows from Definition 12.1.4 that $\mu(S_1) \le \mu(S_2)$. Now suppose that $S \in \mathscr{T}$ and that $\mu(S) = 0$. If $G \subset S$ belongs to $\mathscr{T}$, then G is also a set of measure zero. However, if G is not a member of $\mathscr{T}$, we can say nothing about $\mu(G)$ because it is not defined. The measure space $(X, \mathscr{T}, \mu)$ is said to be *complete* if all the subsets of sets of measure zero are measurable, that is, if $G \subset S$ and $\mu(S) = 0$ imply that $G \in \mathscr{T}$. The Lebesgue measure space $(\mathbb{R}, \mathscr{L}, \mu)$ is complete. On the other hand, the measure space $(\mathbb{R}, \mathscr{B}, \mu)$, where $\mathscr{B}$ is the σ-algebra of Borel sets and μ is Lebesgue measure, is not complete.

In practice, we deal mostly with sets of measure zero (null sets) that are finite or countably infinite. There are, however, null sets that have the cardinality of the continuum; the Cantor middle third set is an example.

12.2.1 Examples of sets of measure zero in $\mathbb{R}$

(a) $S = \{a\}$, $a \in \mathbb{R}$. Let $I_n = (a - 1/n, a + 1/n)$. $S = \bigcap_{n=1}^{\infty} I_n$ and $\mu(I_n) = 2/n$ by (12.1). Since $\mu(S) \le 2/n$ for all $n \in \mathbb{N}$, $\mu(S) = 0$.

(b) $S = \{a_1, a_2, \ldots, a_p\}$, $a_n \in \mathbb{R}$. Then $\mu(S) = \sum_{n=1}^{p} \mu(a_n) = 0$.

(c) $S = \bigcup_{n=1}^{\infty} S_n$, where $\mu(S_n) = 0$ and $S_i \cap S_j = \emptyset$ for all $i \ne j$. From Definition 12.1.4(ii) we have

$$\mu(S) = \sum_{n=1}^{\infty} \mu(S_n) = 0.$$

For example, the set of rational numbers, $\mathbb{Q}$, has measure zero.

12.2.2 Proposition *If S_n is a sequence of sets of measure zero, then $S = \bigcup_{n \in \mathbb{N}} S_n$ has measure zero.*

Proof. We reduce this to the case where the sets are pairwise disjoint by defining

$$T_1 = S_1, \quad T_n = S_n \setminus \bigcup_{k=1}^{n-1} S_k, \quad n \ge 2.$$

We have $S = \bigcup_{n=1}^{\infty} T_n$ and $\mu(S) = \sum_{n=1}^{\infty} \mu(T_n)$. By construction, $T_n \subset S_n$, so $\mu(S_n) \ge \mu(T_n) = 0$. Hence $\mu(S) = 0$. □

12.2.3 Definition We say that a property P is true (holds) almost everywhere, which we denote by a.e., if the set where P is not true (does not hold) is a set of measure zero.

For example, a function is said to be zero a.e. if $S = \{x \in X \mid f(x) \neq 0\}$ is measurable and $\mu(S) = 0$. The function defined by

$$f(x) = \begin{cases} 1 & \text{if} \quad x \in \mathbb{Q}, \\ 0 & \text{otherwise} \end{cases}$$

is zero almost everywhere. We write $f = 0$ a.e.

We see in this example that it is necessary to measure sets of the form $f^{-1}(S) = \{x \in X \mid f(x) \in S\}$. In fact, this requirement appeared in the brief description of the Lebesgue integral given in Lesson 11.

12.3 Measurable functions

It is useful to consider functions taking values in $\overline{\mathbb{R}} = \mathbb{R} \cup \{+\infty\} \cup \{-\infty\}$. We add the following conventions to the usual operations:

(i) For all $a \in \mathbb{R}$, $\quad a \pm \infty = \pm\infty$.

(ii) For all $a > 0$, $\quad a \times (\pm\infty) = \pm\infty$.

(iii) $0 \times (\pm\infty) = 0$.

The operations $+\infty + (-\infty)$ and $+\infty - (+\infty)$ remain undefined.

We deal with complex-valued functions f by decomposing them into their real and imaginary parts: $f = g + ih$.

12.3.1 Definition Let $(X, \mathscr{T})$ be a measurable space.

(i) A function $f : X \to \overline{\mathbb{R}}$ is said to be measurable if for all real α the set

$$f^{-1}((\alpha, +\infty]) = \big\{x \in X \mid f(x) > \alpha\big\} \in \mathscr{T} .$$

(ii) A function $f : X \to \mathbb{C}$ of the form $g + ih$ is measurable if g and h are measurable.

12.3.2 Proposition *Let $\mathscr{B}$ denote the σ-algebra of Borel sets on $\mathbb{R}^n$, and let $f : \mathbb{R}^n \to \mathbb{R}$ be a continuous function. Then f is measurable.*

Proof. Since f is continuous, $f^{-1}((\alpha, +\infty]) = f^{-1}((\alpha, +\infty))$ is an open set in $\mathbb{R}^n$; hence it is a Borel set. □

This proves that all continuous functions are measurable on $(\mathbb{R}, \mathscr{L})$, where $\mathscr{L}$ is the Lebesgue σ-algebra. We note, however, that there exist many measurable functions that are not continuous.

12.3.3 Example The *characteristic function* of a measurable set A is denoted by χ_A and is defined by

$$\chi_A(x) = \begin{cases} 1 & \text{if} \quad x \in A, \\ 0 & \text{if} \quad x \notin A. \end{cases}$$

In practice, it is easy to verify that the functions being used are measurable. The following properties are particularly useful in this regard.

12.3.4 Proposition *Let $(X, \mathscr{T})$ be a measurable space and suppose that f and g are measurable functions on X with values in $\overline{\mathbb{R}}$. With the conventions for computing in $\overline{\mathbb{R}}$, the following functions are also measurable:*

(i) αf *for all* α *in* $\mathbb{R}$;

(ii) $f + g$ *(when this sum is defined);*

(iii) fg;

(iv) $\max(f, g)$ *and* $\min(f, g)$;

(v) $f^+ = \max(f, 0)$ *and* $f^- = \max(-f, 0)$;

(vi) $|f|$.

Proof. All of these functions can be shown to be measurable by direct reference to Definition 12.3.1. □

The next result addresses limit processes and measurability.

12.3.5 Proposition *Let $(X, \mathscr{T})$ be a measurable space and assume that f_n, $n \in \mathbb{N}$, is a sequence of measurable functions from X to $\overline{\mathbb{R}}$. Then we have the following results:*

(i) $f(x) = \inf_{n\in\mathbb{N}} f_n(x)$ *and* $g(x) = \sup_{n\in\mathbb{N}} f_n(x)$ *are measurable functions.*

(ii) *If the pointwise limit of f_n exists, it is measurable.*

Proof. We show as an example that g is measurable. Note first that g is defined for all $x \in X$ (possibly with $g(x) = +\infty$). The set

$$S = \{x \in X \mid g(x) > \alpha\} = \bigcup_{n=1}^{\infty} \{x \in X \mid f_n(x) > \alpha\}$$

is the union of a countable number of measurable sets. Hence $S \in \mathscr{T}$, and g is measurable. □

The class of measurable step functions plays a fundamental role in the theory of integration.

12.3.6 Definition Let $(X, \mathscr{T})$ be a measurable space. A function e defined on X is called a step function, or simple function, if there exists a finite number of measurable sets $S_1, S_2, \ldots, S_n$ and n finite real values $\alpha_1, \alpha_2, \ldots, \alpha_n$ such that

$$e = \sum_{i=1}^{n} \alpha_i \chi_{S_i}, \tag{12.2}$$

where χ_{S_i} is the characteristic function of S_i.

Note that the representation (12.2) is by no means unique. A step function is a measurable function that takes a finite number of values. The sum, product, absolute value, etc. of step functions are again step functions. The next result highlights the importance of step functions.

12.3.7 Proposition *Let $(X, \mathscr{T})$ be a measurable space and assume that the function $f : X \to \overline{\mathbb{R}}$ is measurable. Then the following hold:*

(i) *f is the pointwise limit of a sequence e_n of step functions. If f is bounded, the e_n can be chosen such that the limit is uniform.*

(ii) *If f is positive, the e_n can be chosen to be positive and increasing.*

Proof. Assume that f is positive. Define the sequence e_n by

$$e_n(x) = \begin{cases} 2^{-n}k & \text{if} \quad 2^{-n}k \le f(x) < 2^{-n}(k+1), \\ & \qquad \text{where } k = 0, 1, \ldots, 2^{2n} - 1, \\ 2^n & \text{if} \quad 2^n \le f(x). \end{cases}$$

Then e_n is a step function that takes at most $2^{2n} + 1$ positive values. It is easy to see that the sequence e_n is increasing. If $f(x) < +\infty$, then there is an N such that for all $n \ge N$, $0 \le f(x) - e_n(x) \le 1/2^n$, which proves convergence. If $f(x) = +\infty$, we have $e_n(x) = 2^n$ for all $n \in \mathbb{N}$, and again $\lim_{n\to\infty} e_n(x) = f(x)$.

This proves (ii). To establish (i), it is sufficient to decompose f into its positive and negative parts, $f = f^+ - f^-$. We leave it to the reader to show that the convergence is uniform when f is bounded. □

The integration of measurable functions will be studied in the next lesson. The set of measurable functions is extensive, and the functions that occur in practice are always measurable. The question of measurability is not a point of difficulty for the practical application of the theorems of integration.

12.4 Exercises

Exercise 12.1 Let X be a set and $\mathscr{E} \subset \mathscr{P}(X)$. Show that the intersection of all the σ-algebras containing $\mathscr{E}$ is a σ-algebra.

Exercise 12.2

(a) Show that every closed set F of $\mathbb{R}^n$ is a Borel set of $\mathbb{R}^n$. For example, $a \in \mathbb{R}^n$ is a Borel set.

(b) Show that $\mathbb{Q}$ and $\mathbb{R}\backslash\mathbb{Q}$ are Borel sets in $\mathbb{R}$.

Exercise 12.3 Show that the Borel σ-algebra on $\mathbb{R}$ is generated by each of the following families of sets in $\mathbb{R}$, $a, b \in \mathbb{R}$:

(a) $(-\infty, a)$; (b) $(-\infty, a]$; (c) $[a, b)$; (d) the closed sets.

***Exercise 12.4** Let $(X, \mathscr{T}, \mu)$ be a measure space.

(a) Let A_n be a decreasing sequence of measurable sets. Write

$$A = \bigcap_{n=1}^{\infty} A_n,$$

and assume that $\mu(A_1) < \infty$. Show that $\mu(A) = \lim_{n\to\infty} \mu(A_n)$.

(b) Let B_n an increasing sequence of measurable sets. Show that

$$\mu\Big(\bigcup_{n=1}^{\infty} B_n\Big) = \lim_{n\to\infty} \mu(B_n).$$

Hint for (a): Write A_n as a countable union of pairwise disjoint measurable sets:

$$A_n = (A_n \backslash A_{n+1}) \cup (A_{n+1} \backslash A_{n+2}) \cup \cdots \cup A.$$

Exercise 12.5 Show that in $\mathbb{R}$ the family $\mathscr{A}$ of unions of intervals is an algebra.

Exercise 12.6 Verify that μ_a and μ_d defined in Section 12.1.5 are measures.

Exercise 12.7 Let $(X, \mathscr{T}, \mu)$ be a measure space. Consider the space E of measurable functions with values in $\overline{\mathbb{R}}$. Show that the relation $f \sim g$ if $f - g = 0$ almost everywhere (a.e.) is an equivalence relation on E.

Exercise 12.8 What is the Lebesgue measure of $\mathbb{Q} \cap [0, 1]$? Of the irrationals in $[0, 1]$?

Exercise 12.9 Let $(X, \mathscr{T})$ be a measurable space and let f be a measurable function from X to $\overline{\mathbb{R}}$. Show that the following sets are measurable for all $\alpha \in \mathbb{R}$:

$$\{x \in X \mid f(x) \leq \alpha\}; \quad \{x \in X \mid f(x) \geq \alpha\};$$

$$\{x \in X \mid f(x) < \alpha\}; \quad \{x \in X \mid f(x) = \alpha\}.$$

Exercise 12.10 Let $(X, \mathscr{T})$ be a measurable space and assume that f, g are measurable functions from X to $\overline{\mathbb{R}}$. Show that the following sets are measurable:

$$S_1 = \{x \in X \mid f(x) < g(x)\};$$
$$S_2 = \{x \in X \mid f(x) = g(x)\};$$
$$S_3 = \{x \in X \mid f(x) \leq g(x)\}.$$

Hint: Write

$$S_1 = \bigcup_{q \in \mathbb{Q}} \left(\{x \in X \mid f(x) < q\} \cap \{x \in X \mid g(x) > q\} \right).$$

Exercise 12.11 Use Exercise 12.9 to prove Proposition 12.3.4.

Hint for (iii): Write

$$fg = \frac{1}{4}\left((f+g)^2 - (f-g)^2\right).$$

Exercise 12.12 Show that χ_A is measurable if and only if A is measurable.

Lesson 13

Integrating Measurable Functions

Measure theory is the difficult part of developing the Lebesgue integral. Now that we have a measure space $(X, \mathscr{T}, \mu)$ at our disposal, we are going to define the integral for measurable functions on X with real or complex values. Here and in the rest of the book, when we speak of a function we will mean a measurable function; most of the time we will not mention specifically that the function is measurable. This lesson contains the elementary properties of the Lebesgue integral, including a statement of the monotone convergence theorem.

13.1 Constructing the integral

We are going to develop the integral in the following three steps:

(i) Define the integral of a nonnegative simple function.

(ii) Define The integral of a function with values in $\overline{\mathbb{R}}_+$.

(iii) Define the integral of real- or complex-valued functions.

13.1.1 Definition Assume that $(X, \mathscr{T}, \mu)$ be a measure space and that $e = \sum_{i=1}^n \alpha_i \chi_{S_i}$ is a nonnegative simple function on X $(e(x) \geq 0)$. The integral of e on X with respect to the measure μ is the nonnegative number (possibly $+\infty$) denoted by $\int_X e\, d\mu$ and defined by

$$\int_X e\, d\mu = \sum_{i=1}^n \alpha_i \mu(S_i).$$

We say that e is integrable if $\int_X e\, d\mu$ is finite.

We will assume that the value of $\sum_{i=1}^n \alpha_i \mu(S_i)$ does not depend on the particular representation of e (see Definition 12.3.6), hence that the integral of e is well-defined.

If $E \subset X$ is measurable, the integral of e on E is defined by

$$\int_E e\, d\mu = \int_X e\chi_E\, d\mu = \sum_{i=1}^{n} \alpha_i \mu(E \cap S_i).$$

For the characteristic function of E we have

$$\int_E d\mu = \int_X \chi_E\, d\mu = \mu(E).$$

Proposition 12.3.7 is used to extend this definition to a function defined on X with values in $\overline{\mathbb{R}}_+ = \mathbb{R}_+ \cup \{+\infty\}$.

13.1.2 Definition (integral of a nonnegative function)

Let $(X, \mathscr{T}, \mu)$ be a measure space and assume that $f : X \to \overline{\mathbb{R}}_+$ is a nonnegative measurable function. The integral of f with respect to the measure μ is the nonnegative number (possibly $+\infty$) denoted by $\int_X f\, d\mu$ and defined by

$$\int_X f\, d\mu = \sup\left\{ \int_X e\, d\mu \;\middle|\; 0 \le e \le f,\ e \text{ simple} \right\}.$$

We say that f is integrable if $\int_X f\, d\mu$ is finite.

If E is measurable, we define the integral of f on E as before:

$$\int_E f\, d\mu = \int_X f\chi_E\, d\mu.$$

13.1.3 Examples The measure μ is Lebesgue measure in the following examples.

(a) $X = [0, 1]$ and

$$f(x) = \begin{cases} 0 & \text{if } x \in \mathbb{Q}, \\ 1 & \text{otherwise.} \end{cases}$$

$\int_X f\, d\mu = 1$, and one can show that f is not integrable in the sense of Riemann.

(b) $X = [0, 1]$ and

$$f(x) = \begin{cases} 1/x & \text{if } x > 0, \\ 0 & \text{if } x = 0. \end{cases}$$

One can show that $\int_X f\, d\mu = +\infty$ by applying Definition 13.1.2.

(c) $X = \mathbb{R}$, $f : X \to \overline{\mathbb{R}}_+$, and E is a set of measure zero. Then it follows from the definition that $\int_E f\, d\mu = 0$: Given a simple function e such that $0 \le e \le f$, one has $\int_E e\, d\mu = 0$.

The integral of a function defined on X with values in $\overline{\mathbb{R}}$ is obtained by decomposing f into its positive and negative parts: $f = f^+ - f^-$. The functions f^+ and f^- are measurable and nonnegative (Proposition 12.3.4). It is then natural to define the integral of f on E by

$$\int_E f\,d\mu = \int_E f^+\,d\mu - \int_E f^-\,d\mu$$

for all measurable sets E. The two integrals on the right make sense; however, in case they are both $+\infty$, $\int_E f\,d\mu$ is not defined. This leads to the following definition.

13.1.4 Definition Let $(X, \mathscr{T}, \mu)$ be a measure space and $f : X \to \overline{\mathbb{R}}$ a measurable function. f is said to be integrable on the set E if $\int_E f^+\,d\mu$ and $\int_E f^-\,d\mu$ are finite. In this case, the integral on E is defined by

$$\int_E f\,d\mu = \int_E f^+\,d\mu - \int_E f^-\,d\mu.$$

When f is complex-valued, the integral of $f = g + ih$ on E is defined by

$$\int_E f\,d\mu = \int_E g\,d\mu + i\int_E h\,d\mu.$$

13.2 Elementary properties of the integral

The usual properties of the integral (linearity, monotonicity, order) follow directly from the definitions for simple functions. These properties are extended to nonnegative measurable functions by using Definition 13.1.2 and by taking limits using Proposition 12.3.7 and the monotone convergence theorem (Theorem 13.2.2). This theorem is one of the central results in Lebesgue integration theory.

13.2.1 Proposition *Assume that $(X, \mathscr{T}, \mu)$ is a measure space, f and g are defined on X with values in $\overline{\mathbb{R}}_+$, and E and F are in $\mathscr{T}$.*

(i) *If $0 \le f \le g$, then $0 \le \int_E f\,d\mu \le \int_E g\,d\mu$.*

(ii) *If $E \cap F = \emptyset$, then $\int_{E\cup F} f\,d\mu = \int_E f\,d\mu + \int_F f\,d\mu$.*

(iii) *If $E \subset F$, then $\int_E f\,d\mu \le \int_F f\,d\mu$.*

The proof is a direct application of Definition 13.1.2 and is left as an exercise.

Proving the linearity of the integral for nonnegative functions is more delicate. One first establishes the result for simple functions; then one uses Proposition 12.3.7 and the next theorem to prove the result for arbitrary nonnegative measurable functions.

13.2.2 Theorem (monotone convergence) *Let $(X, \mathscr{T}, \mu)$ be a measure space. Suppose that there is a sequence of measurable functions on X with values in $\overline{\mathbb{R}}_+$ such that $0 \leq f_n(x) \leq f_{n+1}(x)$ for all $x \in X$ and all $n \in \mathbb{N}$. Then the function f defined by $f(x) = \lim_{n\to\infty} f_n(x)$ is nonnegative and measurable, and for all $E \in \mathscr{T}$ we have*

$$\lim_{n\to\infty} \int_E f_n \, d\mu = \int_E f \, d\mu.$$

The technique for proving this theorem can be found, for example, in [Ber70]. This result is used to prove linearity for nonnegative functions, which in turn is use to prove the next result.

13.2.3 Proposition (linearity) *Let $(X, \mathscr{T}, \mu)$ be a measure space. If f and g are integrable on the set $E \subset \mathscr{T}$, then we have the following results that express the linearity of the integral:*

(i) $\displaystyle\int_E \alpha f \, d\mu = \alpha \int_E f \, d\mu$ for all $\alpha \in \mathbb{R}$;

(ii) $\displaystyle\int_E (f+g) \, d\mu = \int_E f \, d\mu + \int_E g \, d\mu$.

Proof. To prove (i), write $f = f^+ - f^-$ and use the linearity of the integral for nonnegative functions. For (ii), write $h = f+g$. Then $0 \leq h^+ \leq f^+ + g^+$ and $0 \leq h^- \leq f^- + g^-$. It follows from Proposition 13.2.1(i) that h is integrable. Since

$$h = f^+ - f^- + g^+ - g^- = h^+ - h^-,$$

we have

$$f^+ + g^+ + h^- = h^+ + f^- + g^-.$$

Using the linearity of the integral for nonnegative functions and rearranging terms shows that

$$\int_E h \, d\mu = \int_E f \, d\mu + \int_E g \, d\mu. \qquad \square$$

13.2.4 Proposition *Let $(X, \mathscr{T}, \mu)$ be a measure space and $E \in \mathscr{T}$.*

(i) *If f and g are integrable and $f \leq g$, then*

$$\int_E f \, d\mu \leq \int_E g \, d\mu.$$

(ii) *If f is integrable, then* $\left|\int_E f\,d\mu\right| \leq \int_E |f|\,d\mu.$

(iii) *If there exists an integrable function g such that $|f| \leq g$, then f is integrable and*

$$\int_E |f|\,d\mu \leq \int_E g\,d\mu.$$

Proof. These results follow from Proposition 13.2.1 and linearity. $\square$

These results lead to the next proposition, which is another key result in the Lebesgue theory.

13.2.5 Proposition *Let $(X, \mathscr{T}, \mu)$ be a measure space. If $f : X \to \overline{\mathbb{R}}$ is a measurable function, then f is integrable if and only if $|f|$ is integrable.*

Proof. By definition, $\int_X f^+\,d\mu$ and $\int_X f^-\,d\mu$ are finite if f is integrable. Since $|f| = f^+ + f^-$, $\int_X |f|\,d\mu < +\infty$ by linearity.

Conversely, since $0 \leq f^+ \leq |f|$ and $0 \leq f^- \leq |f|$, the integrability of $|f|$ implies that $\int_X f^+\,d\mu < +\infty$ and $\int_X f^-\,d\mu < +\infty$ (Proposition 13.2.1), and hence f is integrable. $\square$

13.2.6 Remark Proposition 13.2.5 is false for the Riemann integral. A simple counterexample is given by the function f defined on $[0, 1]$ by

$$f(x) = \begin{cases} +1 & \text{if} \quad x \notin \mathbb{Q}, \\ -1 & \text{if} \quad x \in \mathbb{Q}. \end{cases}$$

It is clear that f is not Riemann integrable. On the other hand, $|f| = 1$ on $[0, 1]$ and is Riemann integrable.

13.3 The integral and sets of measure zero

Given two measurable functions f and g, we say that $f = g$ almost everywhere (a.e.) if the set on which the functions differ has measure zero. Thus if $M = \{x \mid f(x) \neq g(x)\}$, $f = g$ a.e if and only if $\mu(M) = 0$. We will see that when f is integrable, g is also integrable and their integrals are equal.

13.3.1 Proposition *Let $(X, \mathscr{T}, \mu)$ be a measure space. If $f : X \to \overline{\mathbb{R}}_+$ is a nonnegative measurable function, then $\int_X f\,d\mu = 0$ if and only if $f = 0$ almost everywhere.*

Proof. Let $N = \{x \in X \mid f(x) \neq 0\}$ and write $X = (X \setminus N) \cup N$. If $f = 0$ a.e., then $\mu(N) = 0$; this and the fact that f is zero on $X \setminus N$ imply that $\int_X f\,d\mu = 0$ (Proposition 13.2.1(ii) and Section 13.1.3(c)).

Now suppose that $\int_X f\,d\mu = 0$ and write $N = \bigcup_{n=1}^{\infty} S_n$, where S_n is $\{x \in X \mid f(x) > 1/n\}$. Then $0 = \int_X f\,d\mu \geq \int_{S_n} f\,d\mu \geq \mu(S_n)/n$, and hence $\mu(S_n) = 0$ for all $n \in \mathbb{N}$. We conclude from Proposition 12.2.2 that $\mu(N) = 0$ and hence that $f = 0$ a.e. □

One consequence of this proposition is that an integrable function can be modified on a set of measure zero without changing the value of its integral. From the point of view of the integral, we cannot distinguish two functions that are equal almost everywhere. Proposition 13.3.1 shows that "equal a.e." is an equivalence relation on the vector space of integrable functions on a measure space $(X, \mathscr{T}, \mu)$.

13.3.2 Definition Assume that $(X, \mathscr{T}, \mu)$ is a measure space. Define $L^1(X, \mathscr{T}, \mu)$ to be the vector space of (classes) of measurable functions defined on X and integrable with respect to μ. We also write $L^1(X)$, or even L^1, when there is no chance of misunderstanding.

The quantity $\int_X |f|\,d\mu$ is a norm on $L^1(X, \mathscr{T}, \mu)$. Technically, this norm is defined on the equivalence classes. We will not distinguish between the class of functions for which f is a representative and the function f.

13.4 Comparing the Riemann and Lebesgue integrals

We consider $\mathbb{R}$ to be endowed with Lebesgue measure. We have seen that a function can be Lebesgue integrable without being Riemann integrable. On the other hand, one can prove the following result (see, for example, [KF74]).

13.4.1 Theorem *If the Riemann integral $\int_a^b f(x)\,dx$ exists, then the Lebesgue integral $\int_{[a,b]} f\,d\mu$ exists and the two integrals are equal.*

The proof of this theorem is based on the definitions of the two integrals and on the monotone convergence theorem. The following sufficient condition for a function to be Lebesgue integrable is much easier to establish.

13.4.2 Proposition *Let $(X, \mathscr{T}, \mu)$ be a finite measure space, which means that $\mu(X) < +\infty$. If $f : X \to \overline{\mathbb{R}}$ is bounded almost everywhere on a measurable set E, then f is Lebesgue integrable on E.*

The proof is left as an exercise.

Beware that the converse is not true: f integrable does not imply the existence of a number M such that $|f| \leq M$ a.e. Consider, for example, the function $f(x) = 1/\sqrt{x}$ on $(0, 1)$. However, we do have the following result.

13.4.3 Proposition *Let $(X, \mathscr{T}, \mu)$ be a measure space. If $f : X \to \overline{\mathbb{R}}$ is Lebesgue integrable on X, then f is finite a.e.*

Proof. Let $N = \{x \in X \mid |f(x)| = +\infty\}$. If $\mu(N) \neq 0$, we would have $\int_X |f| \, d\mu \geq \int_N |f| \, d\mu = +\infty$. □

It follows directly from the definition of the Riemann integral in terms of Riemann sums that an unbounded function cannot be Riemann integrable. As we have seen, certain unbounded functions are Lebesgue integrable; furthermore, their Lebesgue integrals can often be computed using Riemann integrals.

13.4.4 Proposition *Let $[a, b]$ be a bounded interval of $\mathbb{R}$. If the function $f : [a, b] \to \mathbb{R}_+$ is such that for all $\varepsilon > 0$ the Riemann integral $I_\varepsilon = \int_{a+\varepsilon}^b f(x) \, dx$ exists and if $\lim_{\varepsilon \to 0} I_\varepsilon = I < +\infty$, then f is Lebesgue integrable on $[a, b]$ and $\int_{[a,b]} f \, d\mu = I$.*

Proof. Take $\varepsilon_n > 0$, $\varepsilon_n \to 0$, and define $f_n = f \cdot \chi_{[a+\varepsilon_n, b]}$. Clearly, $f_n(x)$ converges to $f(x)$ as $n \to \infty$ for all $x \in [a, b]$. Furthermore, the sequence f_n is increasing. By the monotone convergence theorem,

$$\int_{[a,b]} f \, d\mu = \lim_{n \to \infty} \int_{[a,b]} f_n \, d\mu = \lim_{n \to \infty} \int_{a+\varepsilon_n}^b f(x) \, dx = I. \qquad \square$$

When f is not positive, we use this result for $|f|$ to conclude integrability but not, for the moment, to compute the value of $\int_{[a,b]} f \, d\mu$. We will see further results relating the Riemann and Lebesgue integrals in Lesson 14.

13.4.5 Examples

(a) $f(x) = \dfrac{1}{\sqrt{x}}$ on $(0, 1]$ is not Riemann integrable on $[0, 1]$. It is integrable on every interval $[\varepsilon, 1]$ with $0 < \varepsilon < 1$, and

$$\int_\varepsilon^1 \frac{dx}{\sqrt{x}} = [2\sqrt{x}]_\varepsilon^1 = 2 - 2\sqrt{\varepsilon}.$$

Thus f is Lebesgue integrable on $[0, 1]$ and

$$\int_{[0,1]} f \, d\mu = 2.$$

(b) $f(x) = \dfrac{1}{\sqrt{x}} \sin \dfrac{1}{x}$ on $(0, 1]$ is Lebesgue integrable on $[0, 1]$ because

$$|f(x)| \leq \frac{1}{\sqrt{x}}.$$

13.4.6 Remark If f is such that

$$\lim_{\varepsilon_n \to 0+} \int_{a+\varepsilon_n}^{b} |f(x)|\,dx = +\infty,$$

then f is not Lebesgue integrable on $[a,b]$. In the case where f is nonnegative, it is clear that the (generalized) Riemann integral

$$\int_a^b f(x)\,dx = \lim_{\varepsilon_n \to 0+} \int_{a+\varepsilon_n}^{b} f(x)\,dx$$

does not exist either. If, however, f takes both positive and negative values, the generalized Riemann integral can exist without f being Lebesgue-integrable. Take as an example $f(x) = \frac{1}{x}\sin\frac{1}{x}$. Set

$$I_n = \int_{\varepsilon_n}^{1} \frac{1}{x}\sin\frac{1}{x}\,dx = \int_1^{1/\varepsilon_n} \frac{\sin u}{u}\,du = \left[\frac{-\cos u}{u}\right]_1^{1/\varepsilon_n} - \int_1^{1/\varepsilon_n} \frac{\cos u}{u^2}\,du.$$

The term

$$\left[-\frac{\cos u}{u}\right]_1^{1/\varepsilon_n} = -\varepsilon_n \cos\frac{1}{\varepsilon_n} + \cos 1$$

converges as $\varepsilon_n \to 0$. The integral $\displaystyle\int_1^{1/\varepsilon_n} \frac{\cos u}{u^2}\,du$ converges absolutely because

$$\int_1^{1/\varepsilon_n} \frac{|\cos u|}{u^2}\,du \le \left[-\frac{1}{u}\right]_1^{1/\varepsilon_n} = 1 - \varepsilon_n.$$

Thus I_n converges as $\varepsilon_n \to 0$. On the other hand, $J_\varepsilon = \displaystyle\int_1^{1/\varepsilon} \left|\frac{\sin u}{u}\right| du$ tends to $+\infty$ as $\varepsilon \to 0$ because

$$J_\varepsilon \ge \int_\pi^{n\pi} \left|\frac{\sin u}{u}\right| du, \quad n\pi \le \frac{1}{\varepsilon} \le (n+1)\pi,$$

and

$$\int_\pi^{n\pi} \left|\frac{\sin u}{u}\right| du \ge \sum_{k=1}^{n-1} \frac{1}{k\pi} \int_{k\pi}^{(k+1)\pi} |\sin u|\,du = \frac{2}{\pi}\sum_{k=1}^{n-1}\frac{1}{k}.$$

Proposition 13.4.4 can be proved for a generalized Riemann integral on an interval $[a,+\infty]$, and Remark 13.4.6 is also true in this case.

13.4.7 A convention

For the Riemann integral, the symbol $\int_a^b f(x)\,dx$ makes sense when $b < a$ by the relation $\int_a^b f(x)\,dx = -\int_b^a f(x)\,dx$, which comes directly from the Riemann sums. On the other hand, the Lebesgue integral is taken over a nonoriented set (a,b). When the integrals in a given context are all Lebesgue integrals, we will adopt this sign convention. For example, $\int_1^0 f(x)\,dx$ will denote $-\int_{[0,1]} f\,d\mu$.

13.5 Exercises

***Exercise 13.1** Prove Proposition 13.2.1 for simple functions. Extend this result to nonnegative measurable functions.

Exercise 13.2 Use Theorem 13.2.2 to prove the linearity of the integral for nonnegative functions.

Exercise 13.3 Let A_n be a sequence of disjoint measurable sets and let A denote the union $\bigcup_{n=1}^{\infty} A_n$. If f is a nonnegative measurable function on A, show that

$$\int_A f\,d\mu = \sum_{n=1}^{\infty} \int_{A_n} f\,d\mu.$$

Exercise 13.4 (absolute continuity) Let $(X, \mathscr{T}, \mu)$ be a measure space and let f be integrable on $A \in \mathscr{T}$. Show that for all $\varepsilon > 0$, there exists a $\delta > 0$ such that for all measurable sets E in A with $\mu(E) < \delta$, one has

$$\left| \int_E f\,d\mu \right| < \varepsilon.$$

Hints:

(a) Establish the result for f bounded.

(b) Write $A = B \cup \left(\bigcup_{n=0}^{\infty} A_n \right)$ with

$$B = \{x \in A \mid |f(x)| = +\infty\}$$

and

$$A_n = \{x \in A \mid n \leq |f(x)| < n+1\},$$

and decompose A as $A = B \cup B_N \cup (A \backslash (B \cup B_N))$ with $B_N = \bigcup_{n=0}^{N} A_n$.

Exercise 13.5 (Chebyshev's inequality) Let f be a nonnegative function defined on a measurable set E. For $\alpha > 0$ show that

$$\mu\{x \in E \mid f(x) \geq \alpha\} \leq \frac{1}{\alpha} \int_E f\,d\mu.$$

Exercise 13.6 Show that $f(x) = x^{\alpha}$ is Lebesgue integrable on

(a) $[0, 1]$ for $\alpha > -1$;

(b) $[1, +\infty)$ for $\alpha < -1$.

Exercise 13.7 (Beppo–Levi's theorem) Let f_n be an increasing sequence of integrable functions on a measurable set E such that for some $M > 0$,

$$\int_E f_n \, d\mu \leq M$$

for all $n \in \mathbb{N}$. Show that the sequence f_n converges almost everywhere to an integrable function f on E and that

$$\int_E f \, d\mu = \lim_{n \to \infty} \int_E f_n \, d\mu.$$

Hint: Write $g_n = f_n - f_1$ and use Theorem 13.2.2.

Exercise 13.8 Let f_n be a sequence of nonnegative measurable functions on a measurable set E. Assume that $\sum_{n=1}^{\infty} \int_E f_n \, d\mu < +\infty$. Show that the series $\sum_{n=1}^{\infty} f_n(x)$ converges almost everywhere and that

$$\int_E \left(\sum_{n=1}^{\infty} f_n \right) d\mu = \sum_{n=1}^{\infty} \int_E f_n \, d\mu.$$

***Exercise 13.9 (Fatou's lemma)** Let f_n be sequence of nonnegative measurable functions defined on a measurable set E.

(a) Show that

$$\int_E \liminf_{n \to \infty} f_n \, d\mu \leq \liminf_{n \to \infty} \int_E f_n \, d\mu.$$

Hint: Use Theorem 13.2.2 on the functions $g_n = \inf_{k \geq n} f_k$. Recall that for a sequence of real numbers a_n, $n \in \mathbb{N}$, the limit inferior, denoted by $\liminf_{n \to \infty} a_n$, is the quantity $\sup_{k \in \mathbb{N}} \{\inf_{\ell \in \mathbb{N}} a_{k+\ell}\}$.

(b) Investigate the sequence $f_n = -\frac{1}{n} \chi_{[0,n]}$, which does not satisfy the nonnegativity hypothesis, and verify that Fatou's inequality does not hold.

Exercise 13.10 Consider the σ-algebra $\mathscr{P}(\mathbb{N})$ of all subsets of $\mathbb{N}$ and endow the elements E of $\mathscr{P}(\mathbb{N})$ with the counting measure defined by

$$\mu(E) = \text{the number of integers in } E.$$

Show that $f : \mathbb{N} \to \mathbb{R}$ is integrable with respect to the measure μ if and only if

$$\sum_{n=0}^{\infty} |f(n)| < +\infty.$$

Exercise 13.11 Let $(X, \mathscr{T}, \mu)$ be a measure space. For all $E \in \mathscr{T}$ define

$$\sigma(E) = \int_E f \, d\mu,$$

where f is a given nonnegative integrable function. Show that σ is a measure on $\mathscr{T}$. (Use Exercise 13.3.)

Lesson 14

Integral Calculus

This lesson contains the essential tools for putting into practice integral computations: It is the Lebesgue version of integral calculus. We present rules for manipulating integrals that depend on a parameter. In particular, we discuss continuity and derivation with a view toward applications to the Fourier transform. The lesson also contains the formulas for changing variables and the rules for interchanging the order of integration in double integrals, the celebrated Fubini's theorem.

14.1 Lebesgue's dominated convergence theorem

We saw one way to pass to a limit under the integral sign in Lesson 13 (Theorem 13.2.2). Note that this result applies only to an increasing sequence of nonnegative functions. One should take care not to confuse the theorem on monotone convergence with the following more powerful result.

14.1.1 Theorem (Lebesgue) *Let $(X, \mathscr{T}, \mu)$ be a measure space. Let f_n, $n \in \mathbb{N}$, be a sequence of measurable functions defined on X that converges almost everywhere to a function f. Suppose that there exists an integrable function g such that for each $n \in \mathbb{N}$, $|f_n(x)| \leq g(x)$ a.e. on X. Then*

(i) *f is integrable;*

(ii) $\lim_{n\to\infty} \int_E f_n \, d\mu = \int_E f \, d\mu$ *for all $E \in \mathscr{T}$.*

The proof of (i) is immediate: f is measurable because it is the limit a.e. of measurable functions; f is integrable because $|f|$ is bounded (dominated) a.e. by an integrable function g. We will assume (ii). The proof, which is more technical, can be found, for example, in [Ber70].

14.1.2 Remark This theorem has two consequences: an integrability criterion and a method to compute the integral.

We use an example to illustrate the advantage of the Lebesgue integral over the Riemann integral as regards passing to the limit under the integral sign. Take $X = [0,1]$, let μ be Lebesgue measure, and suppose that the rationals in $[0,1]$ are ordered in a sequence $q_1, q_2, \ldots, q_n, \ldots$. Define

$$f_n(x) = \begin{cases} +1 & \text{if } x \in \{q_1, q_2, \ldots, q_n\}, \\ 0 & \text{otherwise.} \end{cases}$$

The sequence converges pointwise to the function

$$f(x) = \begin{cases} +1 & \text{if } x \in \mathbb{Q} \cap [0,1], \\ 0 & \text{otherwise.} \end{cases}$$

We have $|f_n| \leq 1$ for all $n \in \mathbb{N}$. Theorem 14.1.1 implies that f is integrable and that

$$\int_X f \, d\mu = \lim_{n \to \infty} \int_X f_n \, d\mu = 0.$$

(Of course, we already knew this because $f = 0$ a.e.)

We note that in this example f_n is Riemann integrable for all $n \in \mathbb{N}$, while the limit f is not Riemann integrable. This shows that the pointwise limit of a sequence of Riemann-integrable functions is not always Riemann integrable. In a sense, Lebesgue's dominated convergence theorem resolves this issue. We will see many other applications.

14.2 Integrals that depend on a parameter

We are given a measure space $(X, \mathscr{T}, \mu)$ and an arbitrary interval (a,b), bounded or not, of $\mathbb{R}$. Let f be defined on $(a,b) \times X$ with values in $\mathbb{R}$ or $\mathbb{C}$. Assume that for all $t \in (a,b)$ the function $x \mapsto f(t,x)$ is integrable. We define

$$I(t) = \int_X f(t,x) \, d\mu, \quad t \in (a,b).$$

Example: The Fourier transform of a function of $L^1(\mathbb{R})$,

$$\widehat{f}(t) = \int_{\mathbb{R}} e^{-2i\pi tx} f(x) \, dx.$$

We intend to examine the continuity and differentiability of the function

$$I : t \mapsto \int_X f(t,x) \, d\mu.$$

14.2.1 Proposition (continuity) *If for almost all $x \in X$ the function $t \mapsto f(t,x)$ is continuous at $t^* \in (a,b)$ and if there exists an integrable function g such that for all t in a neighborhood V of t^**

$$|f(t,x)| \leq g(x) \quad \text{a.e.},$$

then I is continuous at t^*.

Proof. Let t_n be an arbitrary sequence in V that converges to t^*. Define $f_n(x) = f(t_n, x)$. From the hypotheses, $\lim_{n\to\infty} f_n(x) = f(t^*, x)$ for almost all x in X and $|f_n(x)| = |f(t_n, x)| \leq |g(x)|$ a.e. Applying Theorem 14.1.1 shows that

$$\lim_{n\to\infty} \int_X f_n(x)\, d\mu = \int_X \lim_{n\to\infty} f_n(x)\, d\mu,$$

or

$$\lim_{n\to\infty} I(t_n) = \int_X f(t^*, x)\, d\mu = I(t^*).$$

□

14.2.2 Proposition (derivation) *Suppose that V is a neighborhood of t^*, $V \subset (a,b)$, such that the following two conditions hold:*

(i) *For almost all x, $t \mapsto f(t,x)$ is continuously differentiable on V.*

(ii) *There exists an integrable function g such that for all $t \in V$,*

$$\left|\frac{\partial f}{\partial t}(t,x)\right| \leq g(x) \quad \text{a.e.}$$

Then I is differentiable at t^*, and $I'(t^*) = \int_X \frac{\partial f}{\partial t}(t^*, x)\, d\mu$.

Proof. The proof is essentially the same as the one above. Here we write

$$f_n(x) = \frac{f(t_n, x) - f(t^*, x)}{t_n - t^*}$$

and use the mean value theorem. □

14.2.3 Remark The last two results apply to complex-valued functions by taking real and imaginary parts.

14.2.4 Example Suppose that $f : \mathbb{R} \to \mathbb{R}$ is integrable. The Fourier transform

$$\widehat{f}(t) = \int_{\mathbb{R}} e^{-2i\pi tx} f(x)\, dx$$

is well-defined for all $t \in \mathbb{R}$ because $|e^{-2i\pi tx} f(x)| \leq |f(x)|$. Formally, the derivative of $\widehat{f}$ is

$$\widehat{f}\,'(t) = \int_{\mathbb{R}} e^{-2i\pi tx}(-2i\pi x) f(x)\, dx.$$

To apply Proposition 14.2.2 we must show that the right-hand side of

$$|e^{-2i\pi tx}(-2i\pi x)f(x)| \leq 2\pi|xf(x)|$$

is dominated by an integrable function. A simple sufficient condition is that $x \mapsto xf(x)$ be integrable. In this case we have

$$\widehat{f}\,'(t) = -\widehat{2i\pi xf(x)}(t).$$

14.3 Fubini's theorem

This section deals with rules for interchanging the order of integration in double integrals. Fubini's theorem, which addresses this problem, will be essential for our work on the Fourier transform and convolutions. We will be concerned with functions of two variables and Lebesgue measure and integration on $\mathbb{R}^2$. The development of these theories for $\mathbb{R}^2$ is similar in most respects to that for $\mathbb{R}$. In the case of $\mathbb{R}^2$ one begins with a measure ν defined on rectangles $[a_1, b_1] \times [a_2, b_2]$ (recall (12.1)).

14.3.1 Theorem (Fubini) *Assume that $f : \mathbb{R} \times \mathbb{R} \to \overline{\mathbb{R}}$ is measurable and that $E \times F$ is a measurable set in $\mathbb{R} \times \mathbb{R}$.*

(i) *If f is nonnegative on $E \times F$, then*

$$\int_{E \times F} f(x, y)\, dx\, dy = \int_E dx \int_F f(x, y)\, dy = \int_F dy \int_E f(x, y)\, dx. \tag{14.1}$$

The three integrals can possibly be equal to $+\infty$.

(ii) *If f is integrable on $E \times F$, the function $x \mapsto f(x, y)$ is integrable for almost every y, the function $y \mapsto f(x, y)$ is integrable for almost every x, and the three integrals in (14.1) are finite and equal.*

(iii) *f is integrable if and only if*

$$\int_E dx \int_F |f(x, y)|\, dx \quad \text{or} \quad \int_F dy \int_E |f(x, y)|\, dx$$

is finite.

For the proof see [Hal64] or [Roy63].

The practical aspect is that one can compute the double integral by choosing a convenient order of integration if at least one of the iterated integrals of $|f(x, y)|$ exists. Note, however, that the existence of the two integrals

$$\int_E dx \int_F f(x, y)\, dy \quad \text{and} \quad \int_F dy \int_E f(x, y)\, dx$$

does not imply the integrability of f on $E \times F$.

14.4 Changing variables in an integral

We give the formula for $\mathbb{R}^n$ knowing that in practice it is used mostly for $1 \leq n \leq 4$.

Let Δ and Ω be two domains in $\mathbb{R}^n$ related by a 1-to-1 mapping Φ. Suppose that Φ and Φ^{-1} are continuously differentiable on Ω and Δ, respectively. The Jacobian of Φ at $x = (x_1, x_2, \ldots, x_n)$ is the matrix of partial derivatives of $\Phi = (\varphi_1, \varphi_2, \ldots, \varphi_n)$ with respect to x:

$$\operatorname{Jac}\Phi(x) = \begin{bmatrix} \frac{\partial \varphi_1}{\partial x_1}(x) & \frac{\partial \varphi_1}{\partial x_2}(x) & \cdots & \frac{\partial \varphi_1}{\partial x_n}(x) \\ \frac{\partial \varphi_2}{\partial x_1}(x) & \frac{\partial \varphi_2}{\partial x_2}(x) & \cdots & \frac{\partial \varphi_2}{\partial x_n}(x) \\ \vdots & & & \vdots \\ \frac{\partial \varphi_n}{\partial x_1}(x) & \cdots & \cdots & \frac{\partial \varphi_n}{\partial x_n}(x) \end{bmatrix}.$$

$J_\Phi(x)$ denotes the determinant of the Jacobian, and $|J_\Phi(x)|$ denotes its absolute value.

14.4.1 Theorem *Suppose that f is defined on Δ.*

(i) *f is measurable if and only if $f \circ \Phi$ is measurable.*

(ii) *If f is measurable and positive, then*

$$\int_\Delta f(y)\, d\mu(y) = \int_\Omega f(\Phi(x))|J_\Phi(x)|\, d\mu(x).$$

(iii) *f is integrable on Δ if and only if $(f \circ \Phi)|J_\Phi|$ is integrable on Ω, in which case*

$$\int_\Delta f(y)\, d\mu(y) = \int_\Omega f(\Phi(x))|J_\Phi(x)|\, d\mu(x).$$

See [Her86] for a proof.

If E is a measurable subset of Ω, its image under Φ is also measurable, and

$$\mu(\Phi(E)) = \int_{\Phi(E)} d\mu(y) = \int_E |J_\Phi(x)|\, d\mu(x).$$

In particular, this shows that Lebesgue measure is invariant under translation and symmetry.

14.5 The indefinite Lebesgue integral and primitives

In this section we consider $\mathbb{R}$ with Lebesgue measure. Given a function $f \in L^1(a,b)$, we are going to study the function

$$I(x) = \int_{[a,x]} f \, d\mu,$$

which we denote by $I(x) = \displaystyle\int_a^x f(t)\,dt$. I is called the *indefinite Lebesgue integral* of f.

The following results are true for the Riemann integral:

(i) If f is continuous on $[a,b]$, then

$$J(x) = \int_a^x f(t)\,dt$$

is differentiable for all $x \in [a,b]$ and $J'(x) = f(x)$ for all $x \in [a,b]$.

(ii) If f is continuously differentiable on $[a,b]$, then for all $x \in (a,b)$

$$f(x) = f(a) + \int_a^x f'(t)\,dt. \tag{14.2}$$

What is the situation for the Lebesgue integral?

The first result generalizes to Lebesgue-integrable functions thanks to the following result [KF74].

14.5.1 Theorem *A function that is monotone on an interval $[a,b]$ is differentiable almost everywhere on $[a,b]$.*

When f is nonnegative, I is monotone. For an arbitrary f we make the usual decomposition $f = f^+ - f^-$ and see that I is the difference of two monotone functions. Hence I is differentiable a.e.

14.5.2 Proposition *Suppose that $f : [a,b] \to \mathbb{R}$ is integrable. The function I defined on $[a,b]$ by $I(x) = \int_a^x f(t)\,dt$ is differentiable a.e., and $I'(x) = f(x)$ a.e.*

A proof can be found in [KF74] and other books on integration.

We now turn to formula (14.2). For this to make sense, f must be differentiable almost everywhere with f' integrable on $[a,b]$. The conclusion is that f is continuous on $[a,b]$. Beware! These conditions are not sufficient. It is possible to construct a function $f : [0,1] \to [0,1]$ that is continuous and strictly increasing, with $f(0) = 0$ and $f(1) = 1$, and such that f' exists and is zero almost everywhere. (The classic example, due to Cantor, is called The Devil's Staircase.) In this case, (14.2) is clearly not true. The problem posed by this situation leads to the introduction of a new class of functions.

14.5.3 Definition A function $f : [a,b] \to \mathbb{R}$ is said to be absolutely continuous (AC) on $[a,b]$ if it satisfies the following conditions:

(i) f is differentiable a.e.

(ii) f' is Lebesgue integrable on $[a,b]$.

(iii) For all $x \in [a,b]$, $f(x) = f(a) + \displaystyle\int_a^x f'(t)\,dt$.

An AC function is continuous; the converse is false. A continuously differentiable function is absolutely continuous; again the converse is false. (Take $f(x) = xu(x)$, where u is the Heaviside function.)

14.5.4 Proposition *Suppose that $f : [a,b] \to \mathbb{R}$ is integrable. The function I defined on $[a,b]$ by $I(x) = \int_a^x f(t)\,dt$ is absolutely continuous.*

Proof. This follows directly from Proposition 14.5.2 and the definition. □

14.5.5 Definition A function F is said to be a primitive of f on (a,b) if F is absolutely continuous and $F' = f$ a.e.

We apply these results to integration by parts. The formula that is true for continuously differentiable functions extends to AC functions.

14.5.6 Theorem (integration by parts) *Let u and v be two AC functions on (a,b). Then*

$$\int_a^b u(x)v'(x)\,dx = u(b)v(b) - u(a)v(a) - \int_a^b u'(x)v(x)\,dx.$$

Proof. $u(x) = u(a) + \int_a^x u'(t)\,dt$. Multiplying both sides by v' and integrating with respect to x shows that

$$\int_a^b u(x)v'(x)\,dx = u(a)[v(b) - v(a)] + \int_a^b v'(x)\Big(\int_a^x u'(t)\,dt\Big)\,dx.$$

Write

$$\int_a^b v'(x)\Big(\int_a^x u'(t)\,dt\Big)dx = \int_a^b v'(x)\Big(\int_a^b \chi_{[a,x]}(t)u'(t)\,dt\Big)dx.$$

Apply Fubini's theorem and interchange the order of integration; the last integral becomes

$$\begin{aligned}\int_a^b u'(t)\Big(\int_t^b v'(x)\,dx\Big)\,dt &= \int_a^b u'(t)[v(b) - v(t)]\,dt \\ &= u(b)v(b) - u(a)v(b) - \int_a^b u'(t)v(t)\,dt,\end{aligned}$$

and this proves the result. (We leave it to the reader to show that the hypotheses of Fubini's theorem are fulfilled.) □

14.6 Exercises

Exercise 14.1 Consider the sequence of functions $f_n : \mathbb{R} \to \overline{\mathbb{R}}_+$ defined by

$$f_n(x) = \begin{cases} \frac{1}{|x|} & \text{if } |x| \leq \frac{1}{n}, \\ 0 & \text{otherwise.} \end{cases}$$

Discuss this sequence in the context of Lebesgue's Theorem 14.1.1.

Exercise 14.2 Assume that f is Lebesgue integrable on $\mathbb{R}$.

(a) Let (a_n) and (b_n) be any two sequences of real numbers such that a_n and b_n tend to $+\infty$ as $n \to +\infty$. Show that

$$\lim_{n\to\infty} \int_{-a_n}^{b_n} f(x)\,dx = \int_{\mathbb{R}} f(x)\,dx.$$

Give an example to show that the converse is false.

(b) Show that

$$\int_{\mathbb{R}} f(x)\,dx = \sum_{n=-\infty}^{\infty} \int_{na}^{(n+1)a} f(x)\,dx$$

for all $a > 0$.

(c) Define $I(t) = \displaystyle\int_{-\infty}^{t} f(x)\,dx$. Show that I is continuous on $\mathbb{R}$ and that

$$\lim_{t\to+\infty} I(t) = \int_{\mathbb{R}} f(x)\,dx.$$

Exercise 14.3 Let $f_n : \mathbb{R} \to \mathbb{R}$ be a sequence of Lebesgue-integrable functions on $\mathbb{R}$. Show that if

$$\sum_{n=1}^{\infty} \int_{\mathbb{R}} |f_n(x)|\,dx < +\infty,$$

then the series $\sum\limits_{n=1}^{\infty} f_n(x)$ converges for almost all x and

$$\sum_{n=1}^{\infty} \int_{\mathbb{R}} f_n(x)\,dx = \int_{\mathbb{R}} \Big(\sum_{n=1}^{\infty} f_n(x)\Big)\,dx.$$

Exercise 14.4 Show that

$$\lim_{n\to\infty} \int_0^{\infty} e^{-nx} x^{-\frac{1}{2}}\,dx = 0.$$

Exercise 14.5 Let Δ be a bounded domain in $\mathbb{R}^2$. Discuss

$$\lim_{n\to\infty} \iint_{\Delta} \left(1 + \frac{x+y}{n}\right)^n dx\,dy.$$

Exercise 14.6 Let $f : \mathbb{R} \to \mathbb{R}$ be such that the function $x \to x^n f(x)$ is integrable on $\mathbb{R}$ for all $n \geq 0$. Show that the Fourier transform

$$\widehat{f}(\xi) = \int_{\mathbb{R}} e^{-2i\pi\xi x} f(x)\, dx$$

is infinitely differentiable.

***Exercise 14.7** Show that $f(x,t) = e^{-tx}$ is integrable on $\mathbb{R}_+$ for all $t > 0$. Use the following two methods to verify that $I(t) = \int_0^\infty e^{-tx}\, dx$ is infinitely differentiable:

(a) Compute $I(t)$ explicitely.

(b) Apply Proposition 14.2.2.

Use this result to deduce that

$$\int_0^\infty x^n e^{-x}\, dx = n!.$$

Exercise 14.8 Consider the function

$$f(x,y) = \frac{xy}{(x^2+y^2)^2}$$

on the square $\Delta = [-1,1] \times [-1,1]$. Show that

$$\int_{-1}^{1} dx \int_{-1}^{1} f(x,y)\, dy = \int_{-1}^{1} dy \int_{-1}^{1} f(x,y)\, dx.$$

Show that f is not Lebesgue integrable on Δ (use polar coordinates). Conclusion?

Exercise 14.9 Consider the function

$$f(x,y) = \frac{x^2 - y^2}{(x^2+y^2)^2}$$

defined on $\Delta = [0,1] \times [0,1]$ (except at $(0,0)$). Show that the iterated integrals exist and are different. Conclusion?

Exercise 14.10 Show that if f and g are in $L^1(\mathbb{R})$, then $h(x,y) = f(x)g(y)$ is in $L^1(\mathbb{R}^2)$.

Exercise 14.11 Compute the integral of

$$f(x,y) = e^{-x-y}\,(x+y)^n$$

on $\Delta = \mathbb{R}_+ \times \mathbb{R}_+$ by making the change of variables $u = x$ and $v = x + y$.

Exercise 14.12 Let $f : \mathbb{R}^n \to \mathbb{R}$ be defined by

$$f(x_1, x_2, \ldots, x_n) = \frac{1}{(x_1^2 + x_2^2 + \cdots + x_n^2)^{\frac{\alpha}{2}}} \quad \text{with} \quad \alpha \in \mathbb{R}.$$

Change variables x_i to polar coordinates:

$$\begin{aligned}
x_1 &= \rho \cos\theta_1 \cos\theta_2 \cdots \cos\theta_{n-2} \cos\theta_{n-1}, \\
x_2 &= \rho \cos\theta_1 \cos\theta_2 \cdots \cos\theta_{n-2} \sin\theta_{n-1}, \\
x_3 &= \rho \cos\theta_1 \cos\theta_2 \cdots \cos\theta_{n-3} \sin\theta_{n-2}, \\
&\vdots \\
x_{n-1} &= \rho \cos\theta_1 \sin\theta_2, \\
x_n &= \rho \sin\theta_1,
\end{aligned}$$

with $\rho > 0$, $-\frac{\pi}{2} < \theta_1, \ldots, \theta_{n-2} < \frac{\pi}{2}$ and $0 < \theta_{n-1} < 2\pi$. Let

$$B = \left\{ (x_1, x_2, \ldots, x_n) \in \mathbb{R}^n \;\middle|\; \sum_{i=1}^{n} x_i^2 < 1 \right\}.$$

Show that f is integrable on B if and only if $\alpha < n$. Show that f is integrable on $\mathbb{R}^n \backslash B$ if and only if $\alpha > n$.

Hint: The determinant of the Jacobian of the transformation is

$$(-1)^{n-1} \rho^{n-1} \cos^{n-2}\theta_1 \cos^{n-3}\theta_2 \cdots \cos\theta_{n-2}.$$

Chapter V

Spaces

Lesson 15

Function Spaces

We have collected in this lesson the definitions and essential results for the commonly encountered spaces of functions (function spaces or functional spaces). The lesson is somewhat like a catalog, and it can be used as a reference to find one's way around function spaces that are perhaps unfamiliar. Several proofs are technical and can safely be skipped on first reading.

15.1 Spaces of differentiable functions

15.1.1 Definition Let I be an arbitrary interval (bounded or not) of $\mathbb{R}$. For $p \in \mathbb{N}$ we define the space of functions $C^p(I)$ by

$$C^p(I) = \big\{f : I \to \mathbb{R}\,(\text{or } \mathbb{C}) \mid f \text{ is } p\text{-times continuously differentiable}\big\}.$$

When $f \in C^p(I)$, we say that f is of class C^p.

When we say that f is continuously differentiable, we mean that f is differentiable and that $x \mapsto f'(x)$ is continuous. If f has values in $\mathbb{C}$, we are speaking of the differentiability of $\mathrm{Re}(f)$ and $\mathrm{Im}(f)$. $f \in C^0(I)$ simply means that f is continuous. For functions of several variables, the continuity of the derivative of order p is replaced by the continuity of all the partial derivatives whose total order is p. Finally, the adjective *regular* is often used to indicate some (unspecified) degree of differentiability of a function.

15.1.2 Proposition *If I is a closed and bounded interval of $\mathbb{R}$, then $C^p(I)$ is a complete normed vector space in each of the following norms:*

(i) $N_1(f) = \sum_{k=0}^{p} \|f^{(k)}\|_\infty$;

(ii) $N_2(f) = \Big(\sum_{k=0}^{p} \|f^{(k)}\|_\infty^2\Big)^{1/2}$;

(iii) $N_\infty(f) = \max_{k=0,1,\dots,p} \|f^{(k)}\|_\infty,$ *where* $\|f^{(k)}\|_\infty = \max_{x \in I} |f^{(k)}(x)|$.

Proof. This is a classical result. The completeness of these spaces is based on the fact that $C^0(I)$ is complete in the norm for uniform convergence, $\|\cdot\|_\infty$ (see, for example, [KF74]). □

15.1.3 Remark When I is not a bounded interval, for example $I = \mathbb{R}$, none of the norms above make sense if $f, f', \ldots, f^{(p)}$ are not bounded. Take, for example, $f(x) = e^x$. The same is true if I is bounded but not closed ($f(x) = 1/x$ on $(0,1]$).

15.1.4 Definition A function f is said to be infinitely differentiable, or of class C^∞, on I if f is in $C^p(I)$ for all $p \in \mathbb{N}$. This space is denoted by $C^\infty(I)$, which is read "C-infinity."

We are often going to need the notion of the support of a function. Here we define the support of a continuous functions. We will extend the definition to measurable functions in Lesson 20, where we will need it for studying convolution.

15.1.5 Definition Assume that $f : I \mapsto \mathbb{C}$ is continuous. The support of f, which is denoted by supp(f), is defined to be the complement of the largest open set on which f is zero.

Since f is continuous, $\text{supp}(f) = \overline{\{x \mid x \in I, f(x) \neq 0\}}$, the closure of the set where $f(x) \neq 0$.

15.1.6 Definition $C_c^p(I)$ denotes the space of functions in $C^p(I)$ that have bounded support in I.

To introduce distributions, we will need to use functions with bounded support that have as much regularity as possible.

15.1.7 Definition $\mathscr{D}(\mathbb{R})$ (or $\mathscr{D}(I)$) will denote the space of functions in $C^\infty(\mathbb{R})$ (or $C^\infty(I)$) that have bounded support.

15.1.8 Example The function

$$f(x) = \begin{cases} e^{-\frac{1}{1-x^2}} & \text{if } \ |x| \leq 1, \\ 0 & \text{otherwise} \end{cases}$$

is in $\mathscr{D}(\mathbb{R})$.

In the last definition it is important to be precise about the interval I. For example, if $I = [a,b]$, one can have $f \in \mathscr{D}(I)$ without f being zero at a and b. If $I = (a,b)$, the support of f must be a closed set in (a,b). In this case, f can be extended by continuity to all of $\mathbb{R}$ ($f(x) = 0,\ x \in \mathbb{R}\backslash\text{supp}\,(f)$), and we have $\mathscr{D}((a,b)) \subset \mathscr{D}(\mathbb{R})$.

15.2 Spaces of integrable functions

In Lesson 13 we introduced the vector space of Lebesgue-integrable functions, or more precisely, classes of functions. Recall that two functions that are equal almost everywhere are equivalent from the point of view of integration. In what follows we will not make a distinction between the class of functions and a representative; we will speak simply of integrable functions. However, if the class is "continuous," that is, if there is a continuous function f in the class, we will generally choose f as the representative.

15.2.1 Definition Let $p > 0$ be an arbitrary positive number and let I be an interval (bounded or not) of $\mathbb{R}$. Then $L^p(I)$ denotes the space of measurable functions $f : I \to \mathbb{R}$ (or $\mathbb{C}$) for which $|f(x)|^p$ is integrable on I.

15.2.2 Definition Let I be an interval (bounded or not) of $\mathbb{R}$. Then $L^\infty(I)$ denotes the space of measurable functions $f : I \to \mathbb{R}$ (or $\mathbb{C}$) that are bounded almost everywhere.

15.2.3 Proposition *The spaces $L^p(I)$, $1 \leq p \leq +\infty$, are complete normed linear vector spaces when endowed with the following norms:*

(i) $\|f\|_p = \Big(\int_I |f(t)|^p \, dt\Big)^{1/p}$ *if* $1 \leq p < +\infty$;

(ii) $\|f\|_\infty = \inf\big\{c \mid \text{measure } \{x \mid |f(x)| \geq c\} = 0\big\}$.

When $p = \infty$ it is easy to show that $|f(x)| \leq \|f\|_\infty$ except on a set of measure zero.

We assume from now on that $1 \leq p, q \leq +\infty$. The proof of Proposition 15.2.3 is not immediate; we first establish the following result.

15.2.4 Lemma (Hölder's inequality) *Assume that $f \in L^p(I)$ and $g \in L^q(I)$, where $\frac{1}{p} + \frac{1}{q} = 1$. Then $fg \in L^1(I)$, and*

$$\int_I |f(t)g(t)| \, dt \leq \|f\|_p \|g\|_q.$$

Proof. If $p = 1$ or $p = +\infty$, the inequality is clearly true. Assume that $1 < p < +\infty$. By Young's inequality,

$$ab \leq \frac{1}{p}\, a^p + \frac{1}{q}\, b^q$$

for all $a \geq 0$ and $b \geq 0$. We apply this to $|f(t)||g(t)|$ and integrate both sides over I to obtain the inequality

$$\int_I |f(t)g(t)| \, dt \leq \frac{1}{p}\|f\|_p^p + \frac{1}{q}\|g\|_q^q.$$

By replacing f by αf, $\alpha > 0$, we see that

$$\int_I |f(t)g(t)|\,dt \leq \frac{\alpha^{p-1}}{p}\|f\|_p^p + \frac{1}{q\alpha}\|g\|_q^q;$$

taking $\alpha = \|g\|_q^{q/p}/\|f\|_p$ yields Hölder's inequality. □

Proof of Proposition 15.2.3. It follows immediately that $L^p(I)$ is a normed vector space when $p = 1$ or $p = +\infty$, so we assume that $1 < p < \infty$. Suppose $f\, g \in L^p(I)$. Then $f + g$ is in $\mathrm{L}^p(I)$ since by the convexity of x^p on $\mathbb{R}_+$ we have

$$\big(|f(t)| + |g(t)|\big)^p \leq 2^{p-1}\big(|f(t)|^p + |g(t)|^p\big) \quad \text{a.e.}$$

To show that $\|\cdot\|_p$ is a norm on $\mathrm{L}^p(I)$, we need only to prove the triangle inequality; the other properties of $\|\cdot\|_p$ are obvious. Write

$$\begin{aligned}\|f+g\|_p^p &= \int_I |f(t)+g(t)|^{p-1}|f(t)+g(t)|\,dt \\ &\leq \int_I |f(t)+g(t)|^{p-1}|f(t)|\,dt + \int_I |f(t)+g(t)|^{p-1}|g(t)|\,dt.\end{aligned}$$

Note that $|f+g|^{p-1} \in L^q$ when $\dfrac{1}{p} + \dfrac{1}{q} = 1$; thus by Hölder's inequality

$$\begin{aligned}\|f+g\|_p^p &\leq \Big(\int_I |f(t)+g(t)|^{q(p-1)}\,dt\Big)^{\frac{1}{q}}\Big(\|f\|_p + \|g\|_p\Big) \\ &= \|f+g\|_p^{p-1}\Big(\|f\|_p + \|g\|_p\Big),\end{aligned}$$

from which we have $\|f+g\|_p \leq \|f\|_p + \|g\|_p$.

Showing that the spaces $L^p(I)$ are complete is more technical; we refer to [Bre83], for example, for a proof. □

15.2.5 Remark When I is not a bounded interval, for example $I = \mathbb{R}$, the bounded functions are not in $L^p(I)$. Take, for example, $f \equiv 1$. The bounded functions are, however, integrable on all bounded intervals of $\mathbb{R}$.

15.2.6 Definition $L^p_{\mathrm{loc}}(\mathbb{R})$, $p \geq 1$, denotes the space of measurable functions from $\mathbb{R}$ to $\mathbb{R}$ or $\mathbb{C}$ such that $|f|^p$ is integrable on every bounded interval of $\mathbb{R}$.

Clearly, $L^p(\mathbb{R}) \subset L^p_{\mathrm{loc}}(\mathbb{R})$. The periodic functions studied in the earlier lessons on Fourier series with $p = 1$ or 2 are examples of functions in $L^p_{\mathrm{loc}}(\mathbb{R})$.

15.2.7 Definition $L^1_{\mathrm{p}}(0, a)$ denotes the space of functions with period a that are integrable on $(0, a)$. Similarly, $L^2_{\mathrm{p}}(0, a)$ denotes the space of a-periodic functions that are square integrable.

The vector space $L^1_{\mathrm{p}}(0,a)$ was introduced in Lesson 5. We now know that

$$L^1_{\mathrm{p}}(0,a) = \left\{ f : \mathbb{R} \to \mathbb{C} \;\middle|\; f \text{ has period } a \text{ and } \int_0^a |f(t)|\,dt < +\infty \right\}$$

is complete in the norm $\|f\|_1 = \int_0^a |f(t)|\,dt$. Since f is periodic, this integral can be computed over any interval of the form $(\alpha, \alpha + a)$.

15.3 Inclusion and density

We are going to indicate the inclusion relations among the spaces we have just defined. These relations for the spaces of differentiable functions need no proof.

15.3.1 Proposition *The following inclusions hold for all $p \in \mathbb{N}$ and for all intervals I of $\mathbb{R}$:*

(i) $C^{p+1}(I) \subset C^p(I)$.

(ii) $C^\infty(I) \subset C^p(I)$.

(iii) $\mathscr{D}(I) \subset C^p_c(I)$.

It is necessary to consider the measure of I when discussing the inclusions of the L^p spaces.

15.3.2 Proposition *Assume that $\mu(I) < +\infty$. Then the following relations hold:*

(i) *$L^\infty(I) \subset L^p(I)$ for all $p \geq 1$.*

(ii) *$L^q(I) \subset L^p(I)$ for all $q \geq p \geq 1$.*

(iii) *There exists a constant $c = c(p,q)$ such that $\|h\|_p \leq c\|h\|_q$ for all $h \in L^q(I)$, $1 \leq p \leq q \leq +\infty$.*

Proof.

(i) For $f \in L^\infty(I)$, $\int_I |f(t)|^p\,dt \leq \mu(I)\|f\|^p_\infty < +\infty$.

(ii) Take $p < q$ and $f \in L^q(I)$. Write $S = \{t \in I \mid |f(t)| \geq 1\}$. For $t \in S$, $|f(t)|^p \leq |f(t)|^q$ on S; thus

$$\begin{aligned}\int_I |f(t)|^p\,dt &\leq \int_S |f(t)|^q\,dt + \int_{I\setminus S} |f(t)|^p\,dt \leq \int_I |f(t)|^q\,dt + \int_{I\setminus S} dt \\ &\leq \|f\|^q_q + \mu(I) < +\infty,\end{aligned}$$

which proves that $f \in L^p(I)$.

(iii) The inequality $\|h\|_p \le c\|h\|_q$ is another way of saying that the injection of $L^q(I)$ in $L^p(I)$ is continuous. Take h in $L^q(I)$ and apply Hölder's inequality with

$$f(t) = |h(t)|^p, \quad g(t) = 1, \quad r = \frac{q}{p}, \quad \text{and} \quad s = \frac{r}{r-1}.$$

Since $\dfrac{1}{r} + \dfrac{1}{s} = 1$, we see that

$$\int_I |h(t)|^p \, dt \le \Big(\int_I |f(t)|^r \, dt \Big)^{\frac{1}{r}} \Big(\int_I dt \Big)^{\frac{1}{s}}.$$

It follows that $\|h\|_p \le c\|h\|_q$ with $c = (\mu(I))^{\frac{q-p}{pq}}$. □

When $\mu(I) = +\infty$ these results are false. The spaces $L^1(\mathbb{R})$ and $L^2(\mathbb{R})$ are not comparable.

Turning to the integrability properties of the regular functions, we see immediately that if I is not bounded, then there is no inclusion of $C^m(I)$ in $L^p(I)$ for any $m \ge 0$ and $p \ge 1$. It is sufficient to take $f \equiv 1$. If I is closed and bounded, then $C^m(I) \subset L^\infty(I) \subset L^p(I)$ for all $m \ge 0$ and all $p \ge 1$.

The following theorem about approximation is more precise and will be used often in the rest of the book.

15.3.3 Theorem *Let I be an open (bounded or unbounded) interval of $\mathbb{R}$. The space $C_c^0(I)$ is dense in $L^1(I)$. In other words, given $f \in L^1(I)$ and $\varepsilon > 0$ there exists $\varphi_\varepsilon \in C_c^0(I)$ such that $\|f - \varphi_\varepsilon\|_1 < \varepsilon$.*

Proof. Assume that I is bounded. Given $f \in L^1(I)$ and $\varepsilon > 0$, it is possible to find a closed interval $K \subset I$ such that $\int_I |f - \chi_K f| \, d\mu < \varepsilon/4$. By definition of the Lebesgue integral, we know that the simple functions are dense in $L^1(K)$. Let $s = \sum_{n=1}^p \alpha_n \chi_{S_n}$ be a simple function such that $\int_K |f - s| \, d\mu < \varepsilon/4$. Then $\int_I |f - s| \, d\mu < \varepsilon/2$.

The next step is to show that we can approximate the functions χ_{S_n} with continuous functions $\varphi_n \in C_c^0(I)$ so that $\int_I |s - \sum \alpha_n \varphi_n| \, d\mu < \eta$, where $\eta > 0$ is arbitrary. If we do this with $\eta = (\varepsilon/2) \sum |\alpha_n|$, we will have shown that $\|f - \varphi_\varepsilon\|_1 < \varepsilon$ for $\varphi_\varepsilon = \sum \alpha_n \varphi_n$ and proved the theorem.

To simplify the notation, let E denote any one of the S_n and take $\eta > 0$ as indicated above. E is a measurable set in K. From the construction of Lebesgue measure, we know that there exists an open set Ω and a closed set F such that $F \subset E \subset \Omega \subset I$, where the last inclusion is proper and such that $\mu(\Omega/F) < \eta/2$. Consider the function g defined by

$$g(x) = \frac{d(x, I \setminus \Omega)}{d(x, I \setminus \Omega) + d(x, F)},$$

where $d(x, A) = \inf\{d(x, a) \mid a \in A\}$ is the distance from x to A. The denominator of this fraction cannot vanish, because $\overline{I} \setminus \Omega$ and F are disjoint, closed, and bounded sets. The function $x \mapsto d(x, A)$ is continuous; thus g is continuous. For all $x \in I$, we have $0 \leq g(x) \leq 1$, and $\chi_E(x) - g(x) = 0$ if $x \in F$ or if $x \in I \setminus \Omega$. It follows that

$$\int_I |\chi_E - g| \, d\mu = \int_{\Omega \setminus F} |\chi_E - g| \, d\mu \leq 2\mu(\Omega \setminus F) < \eta$$

This proves the result for the case when I is bounded. If I is unbounded, we first approximate f in the L^1 norm with $\chi_J f$, where J is a bounded open interval. □

15.3.4 Remark Theorem 15.3.3 is also true for $1 < p < +\infty$. It is false for $p = +\infty$, as can be seen by taking $I = (0, 1)$ and $f \equiv 1$ (see [Bre83]). We will see later (Lesson 21) that $\mathscr{D}(\mathbb{R})$ is dense in $L^1(\mathbb{R})$. For this we will use the convolution.

The display below summarizes the inclusion relations among the function spaces that we have introduced so far.

$$\begin{array}{ccccccccc} C^\infty(I) & \subset \cdots \subset & C^{p+1}(I) & \subset & C^p(I) & \subset \cdots \subset & C^0(I) \\ \cup & & \cup & & \cup & & \cup \\ \mathscr{D}(I) & \subset \cdots \subset & C_c^{p+1}(I) & \subset & C_c^p(I) & \subset \cdots \subset & C_c^0(I) \end{array}$$

If $\mu(I) < +\infty$,

$$L^\infty(I) \subset \cdots \subset L^p(I) \subset L^2(I) \subset L^1(I) \subset L^1_{\text{loc}}(I)$$

The space $L^2(\mathbb{R})$ plays a central role in Fourier analysis. Its norm is derived from a scalar product, and it is a Hilbert space. This will be the subject of the next lesson.

15.4 Exercises

Exercise 15.1 Let Ω be an open set in $\mathbb{R}^n$ and let $f : \Omega \to \mathbb{R}$ be measurable. Assume that f satisfies the following property $(\mathcal{B})$:

$(\mathcal{B})$ There exists $C > 0$ such that $|f(x)| \leq C$ for almost all $x \in \Omega$.

Define

$$\|f\|_\infty = \inf \left\{ \alpha \mid |f(x)| \leq \alpha \text{ a.e. on } \Omega \right\}.$$

(a) Show that $|f(x)| \leq \|f\|_\infty$ a.e. on Ω.

(b) Let $L^\infty(\Omega)$ be the set of functions defined on Ω with values in $\mathbb{R}$ that satisfy $(\mathcal{B})$. Show that $\|\cdot\|_\infty$ is a norm on $L^\infty(\Omega)$.

Exercise 15.2 Let $f : [-1,1] \to \mathbb{R}$ be defined by $f(x) = x^2$ and let g be equal to f except at $x = 0$, where $g(0) = 2$, and at $x = \pm\frac{1}{2}$, where $g\big(\pm\frac{1}{2}\big) = 4$. Compare $\sup_{-1\le x\le 1} |f(x)|$ with $\sup_{-1\le x\le 1} |g(x)|$ and $\|f\|_\infty$ with $\|g\|_\infty$. Conclusion?

Exercise 15.3 Let f and g be in $L^p(I)$, $1 \le p \le \infty$. Show that $\|f-g\|_p = 0$ if and only if $f = g$ a.e.

Exercise 15.4 Show that if f and g are in $L^2(I)$, the product fg is integrable. Give an example where $f, g \in L^1(I)$ but fg is not integrable on I.

Exercise 15.5 Let $f : \mathbb{R} \to \mathbb{R}^+$ be an integrable function on $\mathbb{R}$. Show that if $g : \mathbb{R} \to \mathbb{R}$ is equivalent to f at $+\infty$ ($\lim_{x\to+\infty} g(x)/f(x) = \lambda \neq 0$), then g is integrable in a neighborhood of $+\infty$.

Exercise 15.6 Let $f(x) = P(x)/Q(x)$ be a rational function with coefficients in $\mathbb{C}$. Assume that Q has no real roots. Show that:

(a) $\deg(P) \le \deg(Q) \implies f \in L^\infty(\mathbb{R})$.

(b) $\deg(P) \le \deg(Q-1) \implies f \in L^2(\mathbb{R})$.

(c) $\deg(P) \le \deg(Q-2) \implies f \in L^1(\mathbb{R})$.

Study the implications in the other direction. (deg = degree.)

****Exercise 15.7 ($L^1(\mathbb{R})$ is complete)** Let f_n be a Cauchy sequence in $L^1(\mathbb{R})$.

(a) Show that one can extract a subsequence $f_{\sigma(n)}$ such that for all $n \in \mathbb{N}$,

$$\|f_{\sigma(n+1)} - f_{\sigma(n)}\|_1 < \frac{1}{2^n}\,.$$

(b) Write

$$g_n(x) = |f_{\sigma(1)}(x)| + \sum_{k=1}^{n} |f_{\sigma(k+1)}(x) - f_{\sigma(k)}(x)|.$$

Show that g_n converges almost everywhere to a function g of $L^1(\mathbb{R})$.

(c) Show that $f_{\sigma(n)}$ converges almost everywhere to a function f in $L^1(\mathbb{R})$. Verify that $\lim_{n\to\infty} \|f_{\sigma(n)} - f\|_1 = 0$ and deduce that $\lim_{n\to\infty} \|f_n - f\|_1 = 0$. Conclusions?

****Exercise 15.8 (L^∞ is complete)** Let f_n be a Cauchy sequence in $L^\infty(\mathbb{R})$.

(a) Use the Cauchy criterion to show that the sequence f_n converges uniformly on $\mathbb{R}\backslash M$, where M is a set of measure zero. Show that the limit f is in $L^\infty(\mathbb{R})$.

(b) Prove that $\lim_{n\to\infty} \|f - f_n\|_\infty = 0$.

Lesson 16

Hilbert Spaces

In this lesson we present the basic elements of the theory of Hilbert spaces. These spaces generalize several aspects of $\mathbb{R}^n$. Hilbert spaces are endowed with a "Euclidean" geometry in the sense that there is a distance function and the notion of angle between two vectors. Hilbert spaces are complete, and this allows one to develop the notion of an infinite-dimensional basis. The prototypic Hilbert space is $L^2(I)$. Some of the points introduced in Lesson 4, including orthogonal projections, will be formalized.

16.1 Definitions and geometric properties

16.1.1 Definition Let E be a vector space over K ($K = \mathbb{R}$ or $\mathbb{C}$). A scalar product on E is a mapping from $E \times E$ to K, denoted by $(\cdot, \cdot)$, that satisfies the following properties for all $x, y, z \in E$ and $\alpha \in K$:

(S1) $(x, x) \geq 0 \quad \text{and} \quad (x, x) = 0 \;\Rightarrow\; x = 0$;

(S2) $(x, y) = \overline{(y, x)}$;

(S3) $(x + y, z) = (x, z) + (y, z) \quad \text{and} \quad (\alpha x, y) = \alpha(x, y)$.

In (S2), $\overline{(y, x)}$ is the complex conjugate of (y, x). (S2) and (S3) imply that $(x, \alpha y) = \overline{\alpha}(x, y)$. When $K = \mathbb{R}$, the conjugation bars are clearly superfluous. Since we will be working most of the time in $\mathbb{C}$, we present the results for this field.

16.1.2 Definition A vector space endowed with a scalar product is called a pre-Hilbert space.

Define $\|x\| = \sqrt{(x, x)}$. We will see that this is a norm once the following lemma is established .

16.1.3 Lemma *Let H be a pre-Hilbert space over $\mathbb{C}$. The following geometric relations hold for all $x, y \in H$:*

(i) *Schwarz inequality:*

$$|(x, y)| \leq \|x\| \|y\|.$$

(ii) *Parallelogram identity:*

$$\|x + y\|^2 + \|x - y\|^2 = 2\,(\|x\|^2 + \|y\|^2).$$

Proof. If $y = 0$, then (i) is trivial, so assume $y \neq 0$. We must show that

$$\left|\left(x, \frac{y}{\|y\|}\right)\right| \leq \|x\|.$$

This reduces to the case $\|y\| = 1$; but if $\|y\| = 1$, then

$$\begin{aligned} 0 \leq \|x - (x, y)y\|^2 &= \|x\|^2 + |(x, y)|^2 - (x, (x, y)y) - ((x, y)y, x) \\ &= \|x\|^2 - |(x, y)|^2, \end{aligned}$$

which proves (i).

To prove (ii), just expand the left-hand side:

$$\begin{aligned} \|x + y\|^2 + \|x - y\|^2 =& \|x\|^2 + (x, y) + (y, x) + \|y\|^2 \\ &+ \|x\|^2 - (x, y) - (y, x) + \|y\|^2 \\ =& 2(\|x\|^2 + \|y\|^2). \end{aligned}$$

□

16.1.4 Proposition *A pre-Hilbert space is a normed vector space with the norm* $\|f\| = \sqrt{(f, f)}$.

Proof. The only axiom that is not obvious is the triangle inequality. We have $\|f + g\|^2 = \|f\|^2 + \|g\|^2 + 2\mathrm{Re}(f, g)$. From the Schwarz inequality, we deduce that $2\mathrm{Re}(f, g) \leq 2\|f\|\|g\|$; hence, $\|f + g\|^2 \leq (\|f\| + \|g\|)^2$. □

16.1.5 Definition A pre-Hilbert space H that is complete with respect to its norm $\|f\| = \sqrt{(f, f)}$ is called a Hilbert space.

16.1.6 Examples

Hilbert spaces:

- $\mathbb{R}^n$ with the scalar product $(x, y) = \sum_{i=1}^n x_i y_i$, where $x = (x_1, \ldots, x_n)$ and $y = (y_1, \ldots, y_n)$.
- $\mathbb{C}^n$ with the scalar product $(x, y) = \sum_{i=1}^n x_i \bar{y}_i$.
- $L^2_{\mathrm{p}}(0, a)$ with the scalar product $(f, g) = \int_0^a f(t)\bar{g}(t)\, dt$ (see Section 4.1). $L^2_{\mathrm{p}}(0, a)$ is complete by Proposition 15.2.3.

Pre-Hilbert spaces:

- $C^0([a, b]; \mathbb{R})$ with the scalar product $(f, g) = \int_a^b f(t)g(t)\, dt$.
- $C^0([a, b]; \mathbb{C})$ with the scalar product $(f, g) = \int_a^b f(t)\bar{g}(t)\, dt$.

16.1.7 Definition Let H be a pre-Hilbert space. We say that x and $y \in H$ are orthogonal if $(x, y) = 0$. Let S be a subset of H. The orthogonal complement of S is the set $S^\perp$ defined by

$$S^\perp = \{y \in H \mid (x, y) = 0 \text{ for all } x \in S\}.$$

16.1.8 Proposition (Pythagorean identity) *Let H be a pre-Hilbert space. If x and y are orthogonal, then*

$$\|x + y\|^2 = \|x\|^2 + \|y\|^2.$$

More generally, if $\phi_1, \phi_2, \ldots, \phi_n$ are pairwise orthogonal, then

$$\|\sum_{k=1}^{n} c_k \phi_k\|^2 = \sum_{k=1}^{n} |c_k|^2 \|\phi_k\|^2.$$

Proof. It is sufficient to expand the left-hand sides of these equations. □

16.1.9 Remark When $K = \mathbb{R}$, the Pythagorean relation implies that x and y are orthogonal. Thus $(x, y) = 0 \Leftrightarrow \|x + y\|^2 = \|x\|^2 + \|y\|^2$. The converse is false if $K = \mathbb{C}$, since $\mathrm{Re}(x, y)$ can be zero without (x, y) being zero. (Take $x = (1, i)$ and $y = (-i, 1)$).

16.2 Best approximation in a vector subspace

Given a pre-Hilbert space H, a linear subspace $V \subset H$, and an element $f \in H$, we can ask the following questions:

(i) Does there exist an $f^* \in V$ such that

$$\|f - f^*\| = \min_{v \in V} \|f - v\|?$$

(ii) Can we characterize f^*?

If f^* exists, it is called the best approximation of f in V. For example, take $H = L^2_{\mathrm{p}}(0, 2\pi)$ and V the subspace of trigonometric polynomials generated by 1, $\sin x$, $\cos x$, ..., $\sin nx$, $\cos nx$ (see Section 4.2).

The next theorem answers the first question.

16.2.1 Theorem (orthogonal projection) *Suppose H is a pre-Hilbert space and V is a complete linear subspace of H. Given $f \in H$, there exists a unique $f^* \in V$ such that*

$$\|f - f^*\| = \min_{v \in V} \|f - v\|.$$

Proof. Let $d = \inf\{\|f - v\| \mid v \in V\}$. V is complete and hence closed. If $d = 0$, we have $f = f^*$. Thus assume $d > 0$ and consider the sets

$$C_n = \left\{v \in V \mid \|v - f\| \leq d + \frac{1}{n}\right\}.$$

We wish to estimate the size of C_n, so let v_1 and v_2 be arbitrary elements of C_n. By the parallelogram identity,

$$\|v_1 + v_2 - 2f\|^2 + \|v_1 - v_2\|^2 = 2\left(\|v_1 - f\|^2 + \|v_2 - f\|^2\right).$$

The right hand-side is bounded above by $4d^2 + 8d/n + 4/n^2$. On the other side, $\|v_1 + v_2 - 2f\|^2 \geq 4d^2$ because $(v_1 + v_2)/2 \in V$ and hence is a contender in the definition of d. These inequalities imply that $\|v_1 - v_2\|^2 \leq 8d/n + 4/n^2$, and thus the diameter of C_n tends to 0 as $n \to \infty$.

Since C_n is not empty, we can choose $v_n \in C_n$. The sequence v_n is a Cauchy sequence because $\|v_n - v_{n+p}\|^2 \leq 8d/n + 4/n^2$ for all $p \geq 0$. Thus v_n tends to a limit f^* in V, and $\|f - f^*\| \leq d$, from which it follows that

$$\|f - f^*\| = d = \min_{v \in V} \|f - v\|.$$

The uniqueness comes from the scalar product via the parallelogram identity. Suppose f_1 and f_2 are two solutions; then $d = \|f - f_1\| = \|f - f_2\|$. Since $(f_1 + f_2)/2$ is in V,

$$d \leq \|f - \frac{1}{2}(f_1 + f_2)\| \leq \frac{1}{2}\|f - f_1\| + \frac{1}{2}\|f - f_2\| = d.$$

By writing $u = \frac{1}{d}(f - f_1)$ and $v = \frac{1}{d}(f - f_2)$ we see that

$$\|u\| = \|v\| = 1 \quad \text{and} \quad \|u + v\| = 2,$$

and $\|u - v\| = 0$ by the parallelogram identity; hence $f_1 = f_2$. □

16.2.2 Remark Note that the proof of Theorem 15.2.1 remains valid when V is any set of vectors that is convex and closed in the norm.

Computing the best approximation is easy when working with a norm derived from a scalar product. The success of least squares techniques rests on the next result.

16.2.3 Proposition *Let H be a pre-Hilbert space and let V be a linear subspace of H. If $f \in H$, then f^* is the best approximation of f in V if and only if*

$$(f - f^*, v) = 0 \quad \text{for all} \quad v \in V. \tag{16.1}$$

Proof. We first prove the sufficient condition. Since $f^* - v \in V$ for all $v \in V$, $(f - f^*, f^* - v) = 0$, and by the Pythagorean identity,

$$\|f - v\|^2 = \|f - f^*\|^2 + \|f^* - v\|^2.$$

Hence $\|f - f^*\| \leq \|f - v\|$ for all $v \in V$, and f^* is the best approximation.

The necessary condition is obtained by examining vectors in V in a neighborhood of f^*. The idea is that if $(f - f^*, v) \neq 0$, there are perhaps vectors nearby that lower the value of $\|f - f^*\|$. Thus take $v_\alpha = f^* + \alpha(v - f^*)$, where $\alpha \in \mathbb{C}$ and $v \in V$ are arbitrary. By the definition of f^* we have

$$\begin{aligned}\|f - f^*\|^2 \leq \|f - v_\alpha\|^2 =& \|f - f^*\|^2 + |\alpha|^2 \|v - f^*\|^2 \\ &- \alpha(v - f^*, f - f^*) - \overline{\alpha}(f - f^*, v - f^*).\end{aligned}$$

Consequently,

$$\alpha(v - f^*, f - f^*) + \overline{\alpha}\overline{(v - f^*, f - f^*)} \leq |\alpha|^2 \|v - f^*\|^2$$

for all $v \in V$ and all $\alpha \in \mathbb{C}$. Write $\alpha = |\alpha| e^{i\theta}$ and $w = v - f^*$. Dividing both sides by $|\alpha|$ and letting $|\alpha| \to 0$, we have

$$e^{i\theta}(w, f - f^*) + e^{-i\theta}\overline{(w, f - f^*)} \leq 0$$

for all θ and all $w \in V$. This implies that $(w, f - f^*) = 0$, which can be seen by taking $\theta = 0$, $\pi/2$, π, and $3\pi/2$. □

This necessary and sufficient condition allows one to compute f^* when V is a subspace of dimension n. If $\phi_1, \phi_2, \ldots, \phi_n$ is a basis for V and we write

$$f^* = \sum_{k=1}^{n} \lambda_k \phi_k,$$

then condition (16.1) translates into a system of linear equations in the λ_k:

$$\sum_{k=1}^{n} \lambda_k (\phi_k, \phi_j) = (f, \phi_j), \quad j = 1, \ldots, n.$$

The matrix G of this system, with $G_{ij} = (\phi_i, \phi_j)$, $i, j = 1, \ldots, n$, is called the *Gram matrix* associated with the basis $\phi_1, \ldots, \phi_n$. This clearly shows one reason why we want to have an orthogonal basis: In this case the matrix is diagonal. (Recall the orthogonal polynomials in Section 6.2.)

16.2.4 Proposition *Assume that the ϕ_i form a basis for V. The Gram matrix with general term (ϕ_i, ϕ_j), $i, j = 1, \ldots, n$, is Hermitian and positive definite.*

Proof. G is clearly Hermitian: $G_{ij} = (\phi_i, \phi_j) = \overline{(\phi_j, \phi_i)} = \overline{G_{ji}}$. Now let $X = (x_1, \dots, x_n)$ be an element of $\mathbb{C}^n$ and compute (X, GX). Since

$$\overline{(GX)}_i = \sum_{j=1}^{n} \overline{(\phi_i, \phi_j)}\overline{x}_j = \Big(\sum_{j=1}^{n} \overline{x}_j \phi_j, \phi_i\Big),$$

it follows that

$$(X, GX) = \sum_{i=1}^{n} x_i (\overline{GX})_i = \sum_{i=1}^{n} x_i \Big(\sum_{j=1}^{n} \overline{x}_j \phi_j, \phi_i\Big) = \|\sum_{i=1}^{n} \overline{x}_i \phi_i\|^2 \geq 0.$$

If $(X, GX) = 0$, then $\sum_{i=1}^{n} \overline{x}_i \phi_i = 0$, and hence $X = 0$ because the ϕ_i are linearly independent. □

This proof provides another way to show that the best approximation exists and is unique when V has finite dimension. When the basis is orthogonal, f^* is given by

$$f^* = \sum_{i=1}^{n} \frac{(f, \phi_i)}{(\phi_i, \phi_i)} \phi_i. \tag{16.2}$$

16.2.5 Definition The quantity $c_i(f) = (f, \phi_i)/(\phi_i, \phi_i)$ is called the Fourier coefficient of f relative to ϕ_i. The coefficient $c_i(f) = (f, \phi_i)$ if the ϕ_i are orthonormal.

This definition is a generalization of what was developed in Section 4.2 for the basis $\phi_k(x) = e^{2i\pi k \frac{x}{a}}$, which is indeed orthogonal. Thus (16.2) is the trigonometric polynomial that best approximates f in the quadratic norm. It is also the truncated Fourier series of f. We will see how these ideas generalize to an arbitrary Hilbert space.

16.3 Orthogonal systems and Hilbert bases

In this section we generalize to Hilbert spaces the notions of orthogonal and orthonormal bases found in the Euclidean spaces $\mathbb{R}^n$.

16.3.1 Definition Assume that H is a pre-Hilbert space. A countable subset $\mathscr{B} = \{\phi_n \mid n \in \mathbb{N}\}$ is

orthogonal if $(\phi_n, \phi_m) = 0$ for all $n \neq m$;

orthonormal if $(\phi_n, \phi_m) = \begin{cases} 1 & \text{if } n = m, \\ 0 & \text{otherwise;} \end{cases}$

total if $\mathscr{B}^{\perp} = \{0\}$.

The last condition means that $(f, \phi_n) = 0$ for all $n \in \mathbb{N}$ implies $f = 0$. When $\mathscr{B}$ is an orthogonal system, the numbers $c_n(f) = (f, \phi_n)/(\phi_n, \phi_n)$ are called the Fourier coefficients of f relative to $\mathscr{B}$.

16.3.2 Proposition (Bessel's inequality) *Suppose that H is a pre-Hilbert space and that $\mathscr{B} = \{\phi_n \mid n \in \mathbb{N}\}$ is an orthogonal system. For all $f \in H$,*

$$\sum_{n=1}^{\infty} |c_n(f)|^2 \|\phi_n\|^2 \leq \|f\|^2.$$

Proof. Let $f_p = \sum_{n=1}^{p} c_n(f)\phi_n$. Then $(f - f_p, f_p) = 0$ for all $p \in \mathbb{N}$ by Proposition 16.2.3. The Pythagorean relation implies that

$$\|f - \sum_{n=1}^{p} c_n(f)\phi_n\|^2 = \|f\|^2 - \sum_{n=1}^{p} |c_n(f)|^2 \|\phi_n\|^2 \geq 0$$

for all $p \in \mathbb{N}$, and this proves the result. □

Based on the generalization of the idea of a basis, we would like to write

$$f = \sum_{n=1}^{\infty} c_n(f)\phi_n.$$

Writing this expression immediately raises two questions:

Q1: Does the sequence $S(f) = \sum_{n=1}^{\infty} c_n(f)\phi_n$ converge? More precisely, do the partial sums $S_p = \sum_{n=1}^{p} c_n(f)\phi_n$ converge in H?

Q2: If the answer to Q1 is yes, is $f = S(f)$?

16.3.3 Definition (Fourier series) Let H be a pre-Hilbert space and let $\mathscr{B} = \{\phi_n \mid n \in \mathbb{N}\}$ be an orthogonal system. $S(f) = \sum_{n=1}^{\infty} c_n(f)\phi_n$ is called the Fourier series of f.

The answer to Q1 is given in the general context of Hilbert spaces by the next result.

16.3.4 Proposition *Let H be a Hilbert space, $\mathscr{B} = \{\phi_n \mid n \in \mathbb{N}\}$ an orthogonal system, and α_n, $n \in \mathbb{N}$, a sequence of scalars. The series $\sum_{n=1}^{\infty} \alpha_n\phi_n$ converges if and only if $\sum_{n=1}^{\infty} |\alpha_n|^2 \|\phi_n\|^2 < +\infty$.*

Proof. Write $S_p = \sum_{n=1}^{p} \alpha_n\phi_n$ and assume that S_p converges to some element g in H. We have

$$\sum_{n=1}^{p} |\alpha_n|^2 \|\phi_n\|^2 = \|\sum_{n=1}^{p} \alpha_n\phi_n\|^2 = \|S_p\|^2 \leq \left(\|S_p - g\| + \|g\|\right)^2$$

for all $p \in \mathbb{N}$. Letting p tend to $+\infty$ shows that

$$\sum_{n=1}^{+\infty} |\alpha_n|^2 \|\phi_n\|^2 \leq \|g\|^2 < +\infty.$$

To prove the converse, assume that $\sum_{n=1}^{\infty} |\alpha_n|^2 \|\phi_n\|^2 < +\infty$. We wish to show that S_p is a Cauchy sequence in H. For $p < q$,

$$\|S_q - S_p\|^2 = \| \sum_{n=p+1}^{q} \alpha_n \phi_n \|^2 = \sum_{n=p+1}^{q} |\alpha_n|^2 \|\phi_n\|^2.$$

The right-hand side is the Cauchy remainder of a convergent series in $\mathbb{R}$. Consequently, S_p is a Cauchy sequence in H, and since H is complete, it converges to some $g \in H$. □

16.3.5 Proposition *Let $\mathscr{B} = \{\phi_n \mid n \in \mathbb{N}\}$ be an orthogonal system in a Hilbert space H. The Fourier series $S(f) = \sum_{n=1}^{\infty} c_n(f)\phi_n$ converges for all $f \in H$.*

Proof. By Bessel's inequality, $\sum_{n=1}^{\infty} |c_n(f)|^2 \|\phi_n\|^2 \leq \|f\|^2 < +\infty$; the Fourier series converges by Proposition 16.3.4. □

We now move to question Q2.

16.3.6 Definition An orthogonal system $\mathscr{B} = \{\phi_n \mid n \in \mathbb{N}\}$ in a Hilbert space H is said to be an orthogonal basis (or a Hilbert basis) if for all $f \in H$ we have $f = \sum_{n=1}^{\infty} c_n(f)\phi_n$.

The next theorem characterizes these bases.

16.3.7 Theorem *Let H be a Hilbert space and $\mathscr{B} = \{\phi_n \mid n \in \mathbb{N}\}$ an orthogonal system. The following assertions are equivalent:*

(i) *The system $\mathscr{B}$ is total.*

(ii) *The finite linear combinations of elements of $\mathscr{B}$ are dense in H.*

(iii) *$\mathscr{B}$ is an orthogonal basis.*

(iv) *For all $f \in H$, $\|f\|^2 = \sum_{n=1}^{\infty} |c_n(f)|^2 \|\phi_n\|^2$ (Parseval's equality).*

Proof.

(i) ⇒ (ii) Let $[\mathscr{B}]$ be the set of finite linear combinations of elements of $\mathscr{B}$. Suppose that $[\mathscr{B}]$ is not dense in H. Then there exists $f \in H$ such that $f \notin \overline{[\mathscr{B}]}$, the closure of $[\mathscr{B}]$. $\overline{[\mathscr{B}]}$ is a closed subspace of H, so by Theorem 16.2.1 there exists a unique f^* in $\overline{[\mathscr{B}]}$ such that $f - f^* \in (\overline{[\mathscr{B}]})^{\perp}$. Since $\mathscr{B} \subset \overline{[\mathscr{B}]}$, we have $(\overline{[\mathscr{B}]})^{\perp} \subset \mathscr{B}^{\perp}$, and since $\mathscr{B}$ is total, $\mathscr{B}^{\perp} = 0$. Thus $f = f^*$, contradicting the assumption $f \notin \overline{[\mathscr{B}]}$.

(ii) $\Rightarrow$ (iii) Take $f \in H = \overline{[\mathscr{B}\,]}$ and $\varepsilon > 0$. There exists a finite number of scalars $\alpha_1, \alpha_2, \ldots, \alpha_m$ such that

$$\|f - \sum_{k=1}^{m} \alpha_k \phi_k\| < \varepsilon.$$

By Theorem 16.2.1,

$$\|f - \sum_{k=1}^{m} c_k(f)\phi_k\| \leq \|f - \sum_{k=1}^{m} \alpha_k \phi_k\| < \varepsilon.$$

If we let $\alpha_k = 0$ for $k = m+1, \ldots, p$, this relation remains true for all $p \geq m$, again by Theorem 16.2.1. Thus

$$\|f - \sum_{k=1}^{p} c_k(f)\phi_k\| < \varepsilon$$

for all $p \geq m$, which proves that the Fourier series $S(f)$ converges to f. Since f was arbitrary, this means that $\mathscr{B}$ is an orthogonal basis.
(iii) $\Rightarrow$ (iv) We saw in the proof of Proposition 16.3.2 that

$$\|f - \sum_{n=1}^{p} c_n(f)\phi_n\|^2 = \|f\|^2 - \sum_{n=1}^{p} |c_n(f)|^2 \|\phi_n\|^2.$$

The left-hand term tends to 0 as p tends to $+\infty$. In the limit we have Parseval's relation:

$$\|f\|^2 = \sum_{n=1}^{\infty} |c_n(f)|^2 \|\phi_n\|^2.$$

(iv) $\Rightarrow$ (i) Take $f \in \mathscr{B}^{\perp}$. Then for all $n \in \mathbb{N}$, $(f, \phi_n) = 0$, and Parseval's relation implies $\|f\| = 0$. Hence $f = 0$, which proves that $\mathscr{B}$ is a total system. □

16.3.8 Corollary *Two elements in H having the same Fourier coefficients are equal.*

Proof. If suffices to show that an element f of H for which $c_n(f) = 0$ for all $n \in \mathbb{N}$ is the zero element. This follows from Parseval's identity. □

In summary,

$$f = \sum_{n=1}^{\infty} \frac{(f, \phi_n)}{(\phi_n, \phi_n)} \phi_n$$

if and only if ϕ_n is an orthogonal basis. In general, the difficult step in this theory is to show that a given family of functions is a basis, for example, to show that the $\phi_n(x) = e^{2i\pi nx}$ form a basis in $H = L^2_{\mathrm{p}}(0, 1)$.

16.3.9 Theorem *The trigonometric system $\{e^{2i\pi n\frac{x}{a}}\}_{n\in\mathbb{Z}}$ is a basis for the Hilbert space $L^2_{\mathrm{p}}(0,a)$.*

Proof. Take $f \in L^2_{\mathrm{p}}(0,a)$. Let f_N denote the best approximation of f in the finite-dimensional subspace generated by $e^{2i\pi n\frac{x}{a}}$, $n = -N, \dots, +N$. We know from Bessel's inequality that the series whose general term is $|c_n(f)|^2$ is summable and from the Pythagorean relation that

$$a \sum_{n=-N}^{+N} |c_n(f)|^2 + \int_0^a |f(t) - f_N(t)|^2 \, dt = \int_0^a |f(t)|^2 \, dt. \tag{16.3}$$

Assume for the moment that f is continuous on $\mathbb{R}$. The function

$$\varphi(x) = \int_0^a f(x+t)\overline{f}(t) \, dt$$

is a-periodic and continuous on $\mathbb{R}$ by Proposition 14.2.1. We compute the Fourier coefficients of φ:

$$\begin{aligned} c_n(\varphi) &= \frac{1}{a}\int_0^a \Big(\int_0^a f(x+t)\overline{f}(t)\,dt\Big) e^{-2i\pi n\frac{x}{a}}\,dx \\ &= \frac{1}{a}\int_0^a \overline{f}(t)\Big(\int_0^a f(x+t)e^{-2i\pi n\frac{x}{a}}\,dx\Big)\,dt \\ &= \frac{1}{a}\int_0^a \overline{f}(t)\Big(\int_t^{t+a} f(s)e^{-2i\pi n\frac{s}{a}}\,ds\Big)e^{2i\pi n\frac{t}{a}}\,dt \\ &= a|c_n(f)|^2. \end{aligned}$$

This implies by Corollary 5.3.2 that the Fourier series of φ converges uniformly to some continuous, periodic function ψ. Then φ and ψ have the same Fourier coefficients. Since they are both continuous, we know from Exercise 4.7 (which is completely independent of other results) that $\varphi(x) = \psi(x)$ for all $x \in \mathbb{R}$. In particular,

$$\sum_{n=-\infty}^{+\infty} c_n(\varphi)e^{2i\pi n\frac{x}{a}} = \varphi(x) = \int_0^a f(x+t)\overline{f}(t)\,dt.$$

By taking $x = 0$, we have Parseval's relation

$$\int_0^a |f(t)|^2\,dt = \sum_{n=-\infty}^{+\infty} c_n(\varphi) = a\sum_{n=-\infty}^{+\infty} |c_n(f)|^2;$$

from this and (16.3) we deduce that $\lim_{N\to\infty} \int_0^a |f(t) - f_N(t)|^2\,dt = 0$.

The result for $f \in L^2_{\mathrm{p}}(0,a)$ follows from the fact that $C^0_{\mathrm{c}}(a,0)$ is dense in $L^2_{\mathrm{p}}(0,a)$ (Section 15.3.4). □

16.3.10 Corollary *Each function $f \in L^2_{\mathrm{p}}(0,a)$ can be written as a unique Fourier series*

$$f(t) = \sum_{n=-\infty}^{\infty} c_n(f) e^{2i\pi n \frac{t}{a}}$$

that converges in the norm of $L^2_{\mathrm{p}}(0,a)$.

As promised in Lesson 4, this provides a proof of Theorem 4.3.1.

16.4 Exercises

Exercise 16.1 Verify that $C^0[-1,+1]$ endowed with the scalar product

$$(f,g) = \int_{-1}^{1} f(t)g(t)\,dt$$

is not a Hilbert space over $\mathbb{R}$. (Construct a counterexample.)
Remark: Consider the sequence f_n defined by

$$f_n(x) = \begin{cases} 0 & \text{if} \quad \dfrac{1}{n} \le x \le 1, \\ 1 - nx & \text{if} \quad 0 \le x \le \dfrac{1}{n}. \end{cases}$$

It is instructive to see that this is not a counterexample.

****Exercise 16.2** Define

$$l^2(\mathbb{N}) = \Big\{ f : \mathbb{N} \to \mathbb{C} \;\Big|\; \sum_{n=1}^{\infty} |f(n)|^2 < +\infty \Big\}.$$

Define a scalar product on $l^2(\mathbb{N})$ by

$$(f,g) = \sum_{n=1}^{\infty} f(n)\overline{g}(n).$$

We are going to show that $l^2(\mathbb{N})$ is a Hilbert space:

(a) Let f_k a Cauchy sequence in $l^2(\mathbb{N})$. Show that for all $n \in \mathbb{N}$, the sequence $f_k(n)$ converges in $\mathbb{C}$. Denote this limit by $f(n)$.

(b) Verify that $n \mapsto f(n)$ is in $l^2(\mathbb{N})$ and show that f_k converges to f in $l^2(\mathbb{N})$.

***Exercise 16.3** Let H be a Hilbert space and let S be a subset of H.

(a) Show that

$$S^\perp = \big\{ y \in H \mid (s,y) = 0 \text{ for all } s \in S \big\}$$

is a closed linear subspace of H.

(b) Show that $(S^\perp)^\perp = \bar{S}$ if S is a linear subspace.

Exercise 16.4 Show that if $\phi_1, \phi_2, \dots, \phi_n$ are n nonzero elements of a pre-Hilbert space that are pairwise orthogonal, then they are linearly independent.

***Exercise 16.5 (the parallelogram identity)** Let E be a normed vector space over $\mathbb{C}$. Assume that the norm satisfies the parallelogram identity (Section 16.1.3(ii)).

(a) Show that the quantity

$$(x, y) = \frac{1}{4}\left(\|x+y\|^2 - \|x-y\|^2 + i\|x+iy\|^2 - i\|x-iy\|^2\right)$$

defines a scalar product on E.

(b) Show that among the spaces $L^p(0,1)$, $1 \leq p < \infty$, the only pre-Hilbert space is for $p = 2$. (Consider the functions $x(t) = t$ and $y(t) = 1 - t$.)

Exercise 16.6 Show that a Hilbert space H that has a countable orthonormal basis is isomorphic to $l^2(\mathbb{N})$. (Consider the mapping $\Phi : H \to l^2(\mathbb{N})$ defined by $\Phi(f) = c_n(f)$.)

Exercise 16.7 Let ϕ_n be an orthonormal basis for the Hilbert space H. Show that for all f and g in H,

$$(f, g) = \sum_{n=1}^{\infty} c_n(f)\,\overline{c}_n(g).$$

Chapter VI

Convolution and the Fourier Transform of Functions

Lesson 17

The Fourier Transform of Integrable Functions

In this lesson we begin to develop properties of the Fourier transform of functions defined on $\mathbb{R}$. Our main concern is with the basic rules for manipulating these integrals. The inverse Fourier transform and properties involving the convolution will be studied later.

17.1 The Fourier transform on $L^1(\mathbb{R})$

17.1.1 Definition Given $f \in L^1(\mathbb{R})$ we write

$$\mathscr{F} f(\xi) = \widehat{f}(\xi) = \int_{\mathbb{R}} e^{-2i\pi\xi x} f(x)\, dx, \tag{17.1}$$

$$\overline{\mathscr{F}} f(\xi) = \int_{\mathbb{R}} e^{2i\pi\xi x} f(x)\, dx. \tag{17.2}$$

By definition, the function $\mathscr{F} f$ is the Fourier transform of f, and $\overline{\mathscr{F}} f$ is the conjugate Fourier transform of f.

These integrals make sense if and only if $f \in L^1(\mathbb{R})$, since $|e^{\pm 2i\pi x\xi}| = 1$. We will see later that $\overline{\mathscr{F}}$ is the inverse of the Fourier transform $\mathscr{F}$ whenever $\mathscr{F} f \in L^1(\mathbb{R})$.

17.1.2 Example Let $f = \chi_{[a,b]}$ be the characteristic function of the interval $[a, b]$. A simple computation shows that

$$\widehat{f}(\xi) = \begin{cases} b - a & \text{if } \xi = 0, \\ \dfrac{\sin \pi(b-a)\xi}{\pi\xi} e^{-i\pi(a+b)\xi} & \text{if } \xi \neq 0. \end{cases}$$

In this case, $\widehat{f}$ is not is $L^1(\mathbb{R})$ because $|\frac{\sin \pi(b-a)\xi}{\pi\xi}|$ is not integrable (Section 13.4.6). We will refer to this example several times.

The following celebrated result describes the general behavior of $\widehat{f}$.

17.1.3 Theorem (Riemann–Lebesgue) *If $f \in L^1(\mathbb{R})$, then $\widehat{f}$ satisfies the following conditions:*

(i) $\mathscr{F} f$ *is continuous and bounded on* $\mathbb{R}$.

(ii) $\mathscr{F}$ *is a continuous linear operator from* $L^1(\mathbb{R})$ *to* $L^\infty(\mathbb{R})$, *and*

$$\|\widehat{f}\|_\infty \leq \|f\|_1. \tag{17.3}$$

(iii) $\lim_{|\xi|\to+\infty} |\widehat{f}(\xi)| = 0.$

Proof.

(i) The continuity of $\widehat{f}$ follows directly from the continuity of the integral (17.1) with respect to the parameter ξ. The function $\xi \mapsto e^{-2i\pi\xi x}f(x)$ is continuous on $\mathbb{R}$ and is dominated by $|f(x)|$, which is in $L^1(\mathbb{R})$. Proposition 14.2.1 applies.

(ii) For all $\xi \in \mathbb{R}$ we have $|\widehat{f}(\xi)| \leq \int |f(x)|\,dx = \|f\|_1$. Thus $\widehat{f}$ is bounded, and $\mathscr{F}$ is continuous from $L^1(\mathbb{R})$ to $L^\infty(\mathbb{R})$.

(iii) For $f = \chi_{[a,b]}$ we have $|\widehat{f}(\xi)| \leq 1/\pi|\xi|$ for $\xi \neq 0$ (Section 17.1.2). Thus $\lim_{|\xi|\to\infty} \widehat{f}(\xi) = 0$; clearly this is true for all simple functions. Now take f in $L^1(\mathbb{R})$. Since the simple functions are dense in $L^1(\mathbb{R})$, there exists a sequence g_n of simple functions such that $\lim_{n\to\infty} \|f - g_n\|_1 = 0$ and, for each fixed n, $\lim_{|\xi|\to\infty} |\widehat{g}_n(\xi)| = 0$. From (17.3), $|\widehat{f}(\xi) - \widehat{g}_n(\xi)| \leq \|f - g_n\|_1$ uniformly in $\xi \in \mathbb{R}$ for each fixed n. It follows that $\lim_{|\xi|\to\infty} \widehat{f}(\xi) = 0$. □

The following formula is essential for introducing the inverse Fourier transform.

17.1.4 Proposition *Let f and g be two functions in $L^1(\mathbb{R})$. Then $f\widehat{g}$ and $\widehat{f}g$ are in $L^1(\mathbb{R})$ and*

$$\int f(t)\widehat{g}(t)\,dt = \int \widehat{f}(x)g(x)\,dx. \tag{17.4}$$

Proof. We saw in the last theorem that $\widehat{g}$ is bounded; thus $f\widehat{g}$ is in $L^1(\mathbb{R})$. Similarly, $\widehat{f}g \in L^1(\mathbb{R})$. Equality (17.4) comes from a direct application of Fubini's theorem (Theorem 14.3.1). Since $e^{-2i\pi tx}f(t)g(x) \in L^1(\mathbb{R}^2)$, we have

$$\begin{aligned}\int f(t)\widehat{g}(t)\,dt &= \int f(t)\Big(\int e^{-2i\pi tx}g(x)\,dx\Big)\,dt \\ &= \int g(x)\Big(\int e^{-2i\pi tx}f(t)\,dt\Big)\,dx = \int g(x)\widehat{f}(x)\,dx.\end{aligned}$$ □

17.1.5 Remark The mapping $\overline{\mathscr{F}}$ has the same properties as those announced for $\mathscr{F}$ in Theorem 17.1.3 and Proposition 17.1.4; to see this, just change i to $-i$.

17.2 Rules for computing with the Fourier transform

One of the remarkable properties of the Fourier transform is the relation between derivation and multiplication by a monomial.

17.2.1 Proposition (derivation)

(i) *If $x^k f(x)$ is in $L^1(\mathbb{R})$, $k = 0, 1, 2, \dots, n$, then $\widehat{f}$ is n times differentiable, and*

$$\widehat{f}^{(k)}(\xi) = [(-2i\pi x)^k f(x)]\widehat{\ }(\xi) \quad \text{for} \quad k = 1, 2, \dots, n, \tag{17.5}$$

where $[(-2i\pi x)^k f(x)]\widehat{\ }(\xi)$ denotes $\mathscr{F}\ [(-2i\pi x)^k f(x)](\xi)$.

(ii) *If $f \in C^n(\mathbb{R}) \cap L^1(\mathbb{R})$ and if all the derivatives $f^{(k)}$, $k = 1, 2, \dots, n$, are in $L^1(\mathbb{R})$, then*

$$\widehat{f^{(k)}}(\xi) = (2i\pi\xi)^k \widehat{f}(\xi) \quad \text{for} \quad k = 1, 2, \dots, n. \tag{17.6}$$

(iii) *If $f \in L^1(\mathbb{R})$ has bounded support, then $\widehat{f} \in C^\infty(\mathbb{R})$.*

Proof.

(i) The function $h : \xi \mapsto e^{-2i\pi\xi x} f(x)$ is infinitely differentiable; furthermore, $h^{(k)}(\xi) = (-2i\pi x)^k e^{-2i\pi\xi x} f(x)$ and $|h^{(k)}(\xi)| \leq 2\pi |x^k f(x)|$. Proposition 14.2.2 applies for $k = 1, 2, \dots, n$, and

$$\widehat{f}^{(k)}(\xi) = \int e^{-2i\pi\xi x} (-2i\pi x)^k f(x)\, dx.$$

(ii) We prove this for $n = 1$; the result for $n \geq 2$ is obtained by induction. Since $f' \in L^1(\mathbb{R})$, we can compute $\widehat{f'}$ by the formula

$$\widehat{f'}(\xi) = \lim_{a \to +\infty} \int_{-a}^{+a} e^{-2i\pi\xi x} f'(x)\, dx.$$

Integrating by parts shows that

$$\int_{-a}^{+a} e^{-2i\pi\xi x} f'(x)\, dx = \left[e^{-2i\pi\xi x} f(x)\right]_{-a}^{+a} + \int_{-a}^{+a} (2i\pi\xi) e^{-2i\pi\xi x} f(x)\, dx. \tag{17.7}$$

Assume for the moment that $f(\pm a)$ has a limit as $a \to +\infty$. Since f is integrable, this limit must be zero. As $a \to +\infty$, (17.7) becomes

$$\int e^{-2i\pi\xi x} f'(x)\, dx = \int (2i\pi\xi) e^{-2i\pi\xi x} f(x)\, dx,$$

which is formula (17.6) for $k = 1$.

It remains to show that $\lim_{a\to+\infty} f(a)$ exists. Since f' is continuously differentiable, we have

$$f(a) = f(0) + \int_0^a f'(t)\,dt.$$

Since $f' \in L^1(\mathbb{R})$, $\lim_{a\to+\infty} \int_0^a f'(t)\,dt$ exists, and hence $\lim_{a\to+\infty} f(a)$ exists. A similar argument shows that $\lim_{a\to+\infty} f(-a)$ exists.
(iii) If $f \in L^1(\mathbb{R})$ has bounded support ($f(x) = 0$ a.e. for $|x|$ greater than some $K > 0$), it is clear that $x^k f(x)$ is integrable for all $k \in \mathbb{N}$; thus by (i), $f \in C^\infty(\mathbb{R})$. □

We are now going to examine how the Fourier transform behaves with respect to translation and parity. We will use the following notation.

17.2.2 Notation

(i) If f has values in $\mathbb{C}$, then $\overline{f}(x) = \overline{f(x)}$, the complex conjugate of $f(x)$.
(ii) f_σ denotes the reflection of f defined by $f_\sigma(x) = f(-x)$.
(iii) The translate $\tau_a f$ of f is defined by $\tau_a f(x) = f(x-a)$.

17.2.3 Proposition (conjugation and parity)

For $f \in L^1(\mathbb{R})$ we have the following relations:
(i) $\overline{\mathscr{F}(f)} = \overline{\mathscr{F}}(\overline{f})$.
(ii) $(\mathscr{F}(f))_\sigma = \overline{\mathscr{F}}(f) = \mathscr{F}(f_\sigma)$.
(iii) *f even (odd) $\Rightarrow \widehat{f}$ even (odd).*
(iv) *f real and even (real and odd) $\Rightarrow \widehat{f}$ real and even (imaginary and odd).*

Proof.
(i) and (ii) follow directly from the definitions.
(iii) f is even $\Leftrightarrow f = f_\sigma$. Thus $\mathscr{F}(f) = \mathscr{F}(f_\sigma) = (\mathscr{F}(f))_\sigma$ from (ii), and $\widehat{f}$ is even. Similarly, f odd implies $\widehat{f}$ is odd.
(iv) Suppose f is real and even. It suffices to show that $\widehat{f}$ is real. We have $\overline{\mathscr{F}(f)} = \overline{\mathscr{F}}(\overline{f}) = \overline{\mathscr{F}}(f) = \mathscr{F}(f_\sigma) = \mathscr{F}(f)$. If f is real and odd, we have $\overline{\mathscr{F}(f)} = \mathscr{F}(f_\sigma) = -\mathscr{F}(f)$, which shows that $\widehat{f}$ is purely imaginary. □

17.2.4 Proposition (translation)

For $f \in L^1(\mathbb{R})$,
(i) $\widehat{\tau_a f}(\xi) = e^{-2i\pi a\xi}\widehat{f}(\xi)$;
(ii) $\tau_a \widehat{f}(\xi) = \mathscr{F}[e^{2i\pi ax} f(x)](\xi) = [e^{2i\pi ax} f(x)]^\wedge(\xi)$.

Proof. To prove (i) we have

$$\widehat{\tau_a f}(\xi) = \int e^{-2i\pi\xi x} f(x-a)\,dx = \int e^{-2i\pi\xi(a+t)} f(t)\,dt = e^{-2i\pi a\xi}\widehat{f}(\xi).$$

The proof of (ii) is similar. □

17.3 Some standard examples

Recall that u denotes the Heaviside function, which is defined by $u(x) = 1$ for $x > 0$ and $u(x) = 0$ for $x \leq 0$. Several useful Fourier transforms are presented in the next section.

17.3.1 Direct computation of the Fourier transform

(a) $f_1(x) = e^{-ax}u(x), \quad \mathrm{Re}(a) > 0.$

$$\widehat{f_1}(\xi) = \int_0^{\infty} e^{-2i\pi x\xi} e^{-ax}\,dx = \lim_{b\to+\infty} \left[\frac{-e^{-x(a+2i\pi\xi)}}{a+2i\pi\xi}\right]_0^b = \frac{1}{a+2i\pi\xi}.$$

(b) $f_2(x) = e^{ax}u(-x), \quad \mathrm{Re}(a) > 0.$

$$\widehat{f_2}(\xi) = \int_{-\infty}^{0} e^{-2i\pi x\xi} e^{ax}\,dx = \frac{-1}{-a+2i\pi\xi}.$$

(c) $f_3(x) = \dfrac{x^k}{k!}\, e^{-ax}u(x), \quad \mathrm{Re}(a) > 0.$

$$f_3(x) = \frac{1}{(-2i\pi)^k}\,\frac{1}{k!}\,(-2i\pi x)^k f_1(x), \text{ and } \widehat{f_3}(\xi) = \frac{1}{k!}\,\frac{1}{(-2i\pi)^k}\widehat{f_1}^{(k)}(\xi)$$

by Proposition 17.2.1(i). Since $\widehat{f_1}^{(k)}(\xi) = k!(a+2i\pi\xi)^{-(k+1)}(-2i\pi)^k$,

$$\widehat{f_3}(\xi) = \frac{1}{(a+2i\pi\xi)^{k+1}}.$$

(d) $f_4(x) = \dfrac{x^k}{k!}\, e^{ax}u(-x), \quad \mathrm{Re}(a) > 0.$

The computation is the same as in (c), and we have

$$\widehat{f_4}(\xi) = \frac{-1}{(-a+2i\pi\xi)^{k+1}}.$$

(e) $f_5(x) = e^{-a|x|}, \quad \mathrm{Re}(a) > 0.$ From (a) and (b) we deduce

$$\widehat{f_5}(\xi) = \frac{2a}{a^2+4\pi^2\xi^2}.$$

(f) $f_6(x) = \mathrm{sign}(x)e^{-a|x|}, \quad \mathrm{Re}(a) > 0.$ We have

$$\widehat{f_6}(\xi) = \frac{-4i\pi\xi}{a^2+4\pi^2\xi^2}.$$

17.3.2 A computation using a differential equation

We are going to compute the Fourier transform of $f(x) = e^{-ax^2}$, $a > 0$. A direct method is to evaluate a contour integral in the complex plane, and this is often the best way to proceed when other attempts fail. We present another way to do this evaluation. Observe that $f'(x) = -2axf(x)$ and take the Fourier transform of both sides of this equality. By using equations (17.5) and (17.6) with $k = 1$, we see that

$$2i\pi\xi\widehat{f}(\xi) = \frac{a}{i\pi}[-2i\pi x f(x)]\widehat{\ } = \frac{a}{i\pi}\widehat{f}\,'(\xi).$$

Thus

$$\widehat{f}\,'(\xi) + \frac{2\pi^2}{a}\xi\widehat{f}(\xi) = 0.$$

A particular solution of (17.8) is $e^{-\frac{\pi^2}{a}\xi^2}$. When we look for a general solution of the form $\widehat{f}(\xi) = K(\xi)e^{-\frac{\pi^2}{a}\xi^2}$, we see that $K'(\xi) = 0$, so $K(\xi) = K$, a constant, and $\widehat{f}(0) = K$. But $\widehat{f}(0) = \int_{\mathbb{R}} e^{-ax^2}\,dx = (\pi/a)^{\frac{1}{2}}$. Hence

$$\widehat{f}(\xi) = \sqrt{\frac{\pi}{a}}e^{-\frac{\pi^2}{a}\xi^2}.$$

17.3.3 Summary of useful formulas

(i)
$$\widehat{f}^{\,(k)}(\xi) = [(-2i\pi x)^k f(x)]\widehat{\ }(\xi)$$
$$\widehat{f^{(k)}}(\xi) = (2i\pi\xi)^k\widehat{f}(\xi)$$

(ii)
$$f(x-a) \overset{\mathcal{F}}{\longmapsto} e^{-2i\pi a\xi}\widehat{f}(\xi)$$
$$e^{2i\pi ax}f(x) \overset{\mathcal{F}}{\longmapsto} \widehat{f}(\xi - a)$$

(iii) $a \neq 0$.
$$f(ax) \overset{\mathcal{F}}{\longmapsto} \frac{1}{|\xi|}\widehat{f}\left(\frac{\xi}{a}\right)$$

(iv) $a \in \mathbb{C},\ \mathrm{Re}(a) > 0,\ k = 0, 1, 2 \ldots.$
$$\frac{x^k}{k!}e^{-ax}u(x) \overset{\mathcal{F}}{\longmapsto} \frac{1}{(a + 2i\pi\xi)^{k+1}}$$
$$\frac{x^k}{k!}e^{ax}u(-x) \overset{\mathcal{F}}{\longmapsto} \frac{-1}{(-a + 2i\pi\xi)^{k+1}}$$
$$e^{-a|x|} \overset{\mathcal{F}}{\longmapsto} \frac{2a}{a^2 + 4\pi^2\xi^2}$$
$$\mathrm{sign}(x)e^{-a|x|} \overset{\mathcal{F}}{\longmapsto} \frac{-4i\pi\xi}{a^2 + 4\pi^2\xi^2}$$

(v) $a \in \mathbb{R},\ a > 0.$
$$e^{-ax^2} \overset{\mathcal{F}}{\longmapsto} \sqrt{\frac{\pi}{a}}e^{-\frac{\pi^2}{a}\xi^2}$$
$$\chi_{[-a,+a]}(x) \overset{\mathcal{F}}{\longmapsto} \frac{\sin 2a\pi\xi}{\pi\xi}$$

17.3.4 Remark When attempts to compute a Fourier transform directly seem to lead nowhere, it is often useful to try to evaluate the integral using the residue theorem and a contour integral in the complex plane. Standard texts on functions of a complex variable discuss this technique, and two of the exercises are done this way.

17.4 Exercises

Exercise 17.1 Assume that $f \in L^1(\mathbb{R})$.

(a) For $\lambda \in \mathbb{R}\backslash 0$, define $g(x) = f(\lambda x)$. Show that

$$\widehat{g}(\xi) = \frac{1}{|\lambda|}\widehat{f}\left(\frac{\xi}{\lambda}\right).$$

(b) For $\lambda \in \mathbb{R}\backslash 0$ and $\mu \in \mathbb{R}$, define $g(x) = f(\lambda x - \mu)$. Show that

$$\widehat{g}(\xi) = \frac{1}{|\lambda|}e^{-2\pi\frac{\mu}{\lambda}\xi}\widehat{f}\left(\frac{\xi}{\lambda}\right).$$

Exercise 17.2 Let $f(t) = (1-t^2)\chi_{[-1,1]}(t)$. Show that

$$\widehat{f}(\xi) = \frac{1}{\pi^2\xi^2}\left(\frac{\sin 2\pi\xi}{2\pi\xi} - \cos 2\pi\xi\right).$$

Exercise 17.3 If $x^k f(x)$ is in $L^1(\mathbb{R})$ for $k = 0, \ldots, p$, the numbers

$$M_k = \int_{\mathbb{R}} x^k f(x)\,dx, \quad k = 0, \ldots, p,$$

exist. M_k is called the moment of order k of f, or the kth moment of f. Show that

$$M_k = \frac{1}{(-2i\pi)^k}\widehat{f}^{(k)}(0).$$

Exercise 17.4 Compute the Fourier transform of $f(x) = \dfrac{1}{a^2+x^2}$, $a > 0$, using the calculus of residues.

***Exercise 17.5** Use contour integration to compute the Fourier transform of $f(x) = e^{-ax^2}$, $a > 0$.

Hint: It is sufficient to compute $\widehat{f}(\xi)$ for $\xi > 0$, since $\widehat{f}$ is even (Proposition 17.2.3). Consider the contour Γ_R formed by the lines joining $-R$ to R, R to $R + i\frac{\pi}{a}\xi$, $R + i\frac{\pi}{a}\xi$ to $-R + i\frac{\pi}{a}\xi$, and $-R + i\frac{\pi}{a}\xi$ to $-R$. By Cauchy's formula, $\int_{\Gamma_R} e^{-az^2}\,dz = 0$. Let R tend to $+\infty$ to show that $\widehat{f}(\xi) = \sqrt{\frac{\pi}{a}}e^{-\frac{\pi^2\xi^2}{a}}$.

Exercise 17.6 Show by a simple example that the hypothesis $f \in C^n(\mathbb{R})$ is essential in Proposition 17.2.1(ii).

Lesson 18

The Inverse Fourier Transform

The Fourier transform allows us to pass from the time domain to the frequency domain. It is remarkable that the inverse operation is obtained very simply from $\mathscr{F}$ itself. In fact, it is just $\overline{\mathscr{F}}$. However, one must be cautious, for as we have seen in the last lesson, f being integrable does not imply that $\widehat{f}$ is integrable (Section 17.1.2). We will need additional hypotheses on f to invert $f \mapsto \widehat{f}$.

18.1 An inversion theorem for $L^1(\mathbb{R})$

18.1.1 Theorem *If f and $\widehat{f}$ are both in $L^1(\mathbb{R})$, then $\overline{\mathscr{F}}\,\widehat{f}(t) = f(t)$ at all points where f is continuous.*

Proof. For each $n > 0$, we introduce the function $g_n(x) = e^{-\frac{2\pi}{n}|x|}$, whose Fourier transform is

$$\widehat{g}_n(\xi) = \frac{1}{\pi}\frac{n}{1 + n^2\xi^2}.$$

The functions g_n and $\widehat{g}_n$ are in $\mathrm{L}^1(\mathbb{R})$. We can apply formula (17.4) to the two functions f and $e^{2i\pi tx}g_n(x)$, which in view of Proposition 17.2.4(ii) is

$$\int_{\mathbb{R}} \widehat{f}(x)g_n(x)e^{2i\pi tx}\,dx = \int_{\mathbb{R}} f(u)\widehat{g}_n(u-t)\,du. \tag{18.1}$$

For all $x \in \mathbb{R}$, $|\widehat{f}(x)g_n(x)e^{2i\pi tx}| \le |\widehat{f}(x)|$, and $\lim_{n\to\infty} g_n(x) = 1$. Since $\widehat{f}$ is in $L^1(\mathbb{R})$, we can apply Lebesgue's theorem and pass to the limit under the integral sign. Thus

$$\lim_{n\to\infty}\int_{\mathbb{R}} \widehat{f}(x)g_n(x)e^{2i\pi tx}\,dx = \int_{\mathbb{R}} \widehat{f}(x)e^{2i\pi tx}\,dx = \overline{\mathscr{F}}\,\widehat{f}(t).$$

Assume that f is continuous at t; we need to show that the integral on the right-hand side of (18.1) tends to $f(t)$. Since $\widehat{g}_n \in L^1(\mathbb{R})$,

$$\int_{\mathbb{R}} \widehat{g}_n(\xi)\,d\xi = \lim_{a\to+\infty} \int_{-a}^{+a} \frac{1}{\pi} \frac{n}{1+n^2\xi^2}\,d\xi = 1.$$

Thus we can write

$$\int_{\mathbb{R}} f(u)\widehat{g}_n(u-t)\,du - f(t) = \int_{\mathbb{R}} (f(\xi+t)-f(t))\widehat{g}_n(\xi)\,d\xi. \qquad (18.2)$$

Given $\varepsilon > 0$, there exists $\eta > 0$ such that $|y-t| \leq \eta$ implies $|f(y)-f(t)| \leq \varepsilon$. We decompose (18.2) as follows:

$$\begin{aligned}\int_{\mathbb{R}} (f(\xi+t)-f(t))\widehat{g}_n(\xi)\,d\xi &= \int_{|\xi|\leq\eta} (f(\xi+t)-f(t))\widehat{g}_n(\xi)\,d\xi \\ &\quad + \int_{|\xi|>\eta} (f(\xi+t)-f(t))\widehat{g}_n(\xi)\,d\xi.\end{aligned}$$

For all $n > 0$,

$$\int_{|\xi|\leq\eta} |f(\xi+t)-f(t)||\widehat{g}_n(\xi)|\,d\xi \leq \varepsilon \int_{|\xi|\leq\eta} |\widehat{g}_n(\xi)|\,d\xi \leq \varepsilon.$$

The last step is to show that $\lim\limits_{n\to\infty} \int_{|\xi|>\eta} (f(t+\xi)-f(t))\widehat{g}_n(\xi)\,d\xi = 0$. For this we have

$$\left| f(t) \int_{|\xi|>\eta} \widehat{g}_n(\xi)\,d\xi \right| = |f(t)|\left(1 - \frac{2}{\pi}\arctan n\eta\right), \qquad (18.3)$$

and since $\widehat{g}_n$ is even and decreasing on $\mathbb{R}_+$,

$$\left| \int_{|\xi|>\eta} f(t+\xi)\widehat{g}_n(\xi)\ d\xi \right| \leq \widehat{g}_n(\eta)\|f\|_1. \qquad (18.4)$$

As n tends to $+\infty$, the right-hand sides of (18.3) and (8.4) tend to 0, and this proves the theorem. □

As a consequence of this result, the Fourier transform $\widehat{f}$ of a function f in $L^1(\mathbb{R})$ that is continuous except for a jump discontinuity at $x = a$ cannot be integrable: If $\widehat{f}$ were in $L^1(\mathbb{R})$, Theorem 18.1.1 implies that $\overline{\mathscr{F}}\widehat{f}(x) = f(x)$ except for $x = a$. But $\overline{\mathscr{F}}\widehat{f}(x)$ is continuous everywhere, so this is impossible (see, for example, Section 17.3.1(a)).

We will see in Lesson 23 that f and $\widehat{f}$ in $L^1(\mathbb{R})$ implies that f is equal almost everywhere to the continuous function $\overline{\mathscr{F}}\widehat{f}$.

The next proposition gives sufficient conditions on f so that $\widehat{f} \in L^1(\mathbb{R})$.

18.1.2 Proposition *If f belongs to $C^2(\mathbb{R})$ and if f, f', and f'' are in $L^1(\mathbb{R})$, then $\widehat{f}$ is integrable.*

Proof. According to (17.6), $\widehat{f''}(\xi) = -4\pi^2\xi^2\widehat{f}(\xi)$. On the other hand, $\lim_{|\xi|\to\infty} |\widehat{f''}(\xi)| = 0$ (Theorem 17.1.3). Thus there is an $M > 0$ such that for all $|\xi| > M$, $4\pi^2|\xi|^2|\widehat{f}(\xi)| \leq 1$. Since $|\widehat{f}(\xi)|$ is continuous on $\mathbb{R}$ and dominated by $1/(4\pi^2|\xi|^2)$ at infinity, it follows that $\widehat{f}$ is in $L^1(\mathbb{R})$. □

18.2 Some Fourier transforms obtained by the inversion formula

When the hypotheses of Theorem 18.1.1 are satisfied, one can often easily compute $\mathscr{F}\ \widehat{f}$.

18.2.1 Proposition *If f continuous and integrable and if $\widehat{f}$ is in $L^1(\mathbb{R})$, then for all $x \in \mathbb{R}$,*

$$\mathscr{F}\ \widehat{f}(x) = f_\sigma(x) = f(-x).$$

Proof. Let $g = \widehat{f}$. We have

$$\widehat{g}(-x) = \int e^{2i\pi x\xi}\widehat{f}(\xi)\,d\xi = \overline{\mathscr{F}}\ \widehat{f}(x).$$

By Theorem 18.1.1, $\widehat{g}(-x) = f(x)$ for all $x \in \mathbb{R}$; thus

$$\mathscr{F}\ \widehat{f}(x) = f_\sigma(x).$$

□

We will now use the results of Section 17.3.4 to compute several inverse transforms. We take $a \in \mathbb{C}$ with $\mathrm{Re}(a) > 0$.

(a) $f_1(x) = \dfrac{x^k}{k!}e^{-ax}u(x)$ is integrable for all $k \in \mathbb{N}$, but it is continuous only for $k \geq 1$. We have $g_1(\xi) = \widehat{f_1}(\xi) = \dfrac{1}{(a+2i\pi\xi)^{k+1}} \in L^1(\mathbb{R})$ for $k \geq 1$. Thus

$$\widehat{g_1}(x) = \frac{(-x)^k}{k!}\ e^{ax}u(-x)$$

for $k \geq 1$.

(b) In the same way, $f_2(x) = \dfrac{x^k}{k!}e^{ax}u(-x)$ for $k \geq 1$, and we have

$$g_2(\xi) = \widehat{f_2}(\xi) = \frac{-1}{(-a+2i\pi\xi)^{k+1}} \quad \text{and} \quad \widehat{g_2}(x) = \frac{(-x)^k}{k!}e^{-ax}u(x).$$

(c) $f_3(x) = e^{-a|x|}$ is in $L^1(\mathbb{R})$ and continuous on $\mathbb{R}$.

$$g_3(\xi) = \widehat{f_3}(\xi) = \frac{2a}{a^2 + 4\pi^2\xi^2}$$

is in $L^1(\mathbb{R})$. Thus

$$\widehat{g_3}(x) = e^{-a|x|}.$$

(d) $f_4(x) = \text{sign}(x)e^{-a|x|}$ is not continuous at 0. We note that

$$\widehat{f_4}(\xi) = \frac{-4i\pi\xi}{a^2 + 4\pi^2\xi^2}$$

is not integrable at infinity.

(e) $f_5(x) = e^{-ax^2}$ is in $L^1(\mathbb{R})$ and continuous on $\mathbb{R}$.

$$g_5(\xi) = \widehat{f_5}(\xi) = \sqrt{\frac{\pi}{a}} e^{-\frac{\pi^2}{a}\xi^2}$$

is in $L^1(\mathbb{R})$. Hence

$$\widehat{g_5}(\xi) = e^{-a\xi^2}.$$

18.2.2 Summary

(i) $a \in \mathbb{C},\ \text{Re}(a) > 0,\ k = 1, 2, \ldots.$

$$\frac{1}{(a + 2i\pi x)^{k+1}} \overset{\mathcal{F}}{\longmapsto} \frac{(-\xi)^k}{k!} e^{a\xi} u(-\xi)$$

$$\frac{-1}{(-a + 2i\pi x)^{k+1}} \overset{\mathcal{F}}{\longmapsto} \tfrac{(-\xi)^k}{k!} e^{-a\xi} u(\xi)$$

$$\frac{1}{a^2 + x^2} \overset{\mathcal{F}}{\longmapsto} \tfrac{\pi}{a} e^{-2\pi a|\xi|}$$

(ii) $a \in \mathbb{R},\ a > 0.$

$$\sqrt{\frac{\pi}{a}} e^{-\frac{\pi^2}{a}x^2} \overset{\mathcal{F}}{\longmapsto} e^{-a\xi^2}$$

18.3 The principal value Fourier inversion formula

We have remarked several times that the Fourier transform of an integrable function is not necessarily integrable. In this is the case, the integral

$$\int_{\mathbb{R}} e^{2i\pi x\xi} \widehat{f}(\xi)\, d\xi$$

is not defined. This does not exclude the possible existence of the limit

$$\lim_{a\to+\infty}\int_{-a}^{+a} e^{2i\pi x\xi}\widehat{f}(\xi)\,d\xi.$$

18.3.1 Theorem *Assume that $f \in L^1(\mathbb{R})$ satisfies the following two conditions:*

(i) *There is a finite number of real numbers $a_1, a_2, \ldots, a_p$ such that f is continuously differentiable on $(-\infty, a_1)$, (a_1, a_2), $\ldots$, $(a_p, +\infty)$.*

(ii) $f' \in L^1(\mathbb{R})$.

Then

$$\lim_{a\to\infty}\int_{-a}^{a} e^{2i\pi t\xi}\widehat{f}(\xi)\,d\xi = \frac{1}{2}\,(f(t+) + f(t-)).$$

Proof. Note first that (i) and (ii) imply that the limits $f(t+)$ and $f(t-)$ exist for all t. Let $g(\xi) = e^{2i\pi t\xi}\chi_{[-a,a]}(\xi)$. Since f and g are in $L^1(\mathbb{R})$, it follows from Theorem 17.1.4 that

$$v(a) = \int_{-a}^{a} e^{2i\pi t\xi}\widehat{f}(\xi)\,d\xi = \int_{\mathbb{R}} g(\xi)\widehat{f}(\xi)\,d\xi = \int_{\mathbb{R}} \widehat{g}(x)f(x)\,dx.$$

We compute $\widehat{g}(x)$ using Proposition 17.2.4(ii) and Section 17.1.2:

$$\widehat{g}(x) = \tau_t\widehat{\chi}_{[-a,a]}(x) = \frac{\sin 2\pi a(x-t)}{\pi(x-t)}.$$

Thus,

$$\begin{aligned} v(a) &= \int_{\mathbb{R}} f(t+u)\frac{\sin 2\pi au}{\pi u}\,du \\ &= \int_0^{\infty} (f(t+u) + f(t-u))\frac{\sin 2\pi au}{\pi u}\,du. \end{aligned} \tag{18.5}$$

The function $\dfrac{\sin x}{x}$ has the following properties:

(i) $$\lim_{R\to\infty}\int_0^R \frac{\sin x}{x}\,dx = \frac{\pi}{2}. \tag{18.6}$$

(ii) $s(y) = \displaystyle\int_y^{\infty}\frac{\sin x}{x}\,dx$ is well-defined and differentiable on $\mathbb{R}$; its derivative is

$$s'(y) = -\frac{\sin y}{y}; \tag{18.7}$$

and

$$\lim_{y\to+\infty} s(y) = 0. \tag{18.8}$$

Consequently, s is bounded on $[0, +\infty)$, and we can write

$$M = \sup_{y \geq 0} |s(y)|. \tag{18.9}$$

We return to (18.5). By hypothesis, the function

$$h_t(u) = f(t+u) + f(t-u)$$

is integrable on $[0, +\infty)$, it has at most a finite number of discontinuities $b_1, b_2, \ldots, b_q$, it is continuously differentiable on $(0, b_1)$, (b_1, b_2), ..., (b_q, ∞) (possibly $b_1 = 0$), and h'_t is integrable. Thus we can integrate (18.5) by parts. By letting $b_0 = 0$ and $b_{q+1} = +\infty$, we see that

$$\begin{aligned} v(a) = \int_{-a}^{+a} e^{2i\pi t\xi} \widehat{f}(\xi)\, d\xi &= -2a \int_0^\infty [f(t+u) + f(t-u)] s'(2\pi au)\, du \\ &= -2a \sum_{j=0}^{q} \int_{b_j}^{b_{j+1}} h_t(u) s'(2\pi au)\, du \\ &= -\frac{1}{\pi} \sum_{j=0}^{q} \big[s(2\pi ab_{j+1}) h_t(b_{j+1}^-) - s(2\pi ab_j) h_t(b_j^+) \\ &\qquad - \int_{b_j}^{b_{j+1}} h'_t(u) s(2\pi au)\, du \big]. \end{aligned}$$

The term $s(2\pi ab_{j+1})h_t(b_{j+1}^-)$ is actually the limit of $s(2\pi au)h_t(u)$ as u tends to $+\infty$. This limit is 0, since both $s(2\pi au)$ and $h_t(u)$ tend to 0 by (18.8) and the proof of Proposition 17.2.1, respectively. For $j = 1, 2, \ldots, q$, $\lim_{a\to\infty} s(2\pi ab_j) = 0$, again by (18.8). Now consider the limits

$$\lim_{a\to\infty} \int_{b_j}^{b_{j+1}} h'_t(u) s(2\pi au)\, du$$

for $j = 1, 2, \ldots, q$. We have $\lim_{a\to\infty} h'_t(u)s(2\pi au) = 0$ for almost every u, and

$$|h'_t(u)s(2\pi au)| \leq M |h'_t(u)|$$

by (18.9). Since f' is integrable, $h'_t(u)$ is in $L^1(\mathbb{R})$; we can apply the dominated convergence theorem and conclude that

$$\lim_{a\to\infty} \int_{b_j}^{b_{j+1}} h'_t(u) s(2\pi au)\, du = 0.$$

The remaining term is $\dfrac{1}{\pi} s(0) h_t(0^+)$; thus $\lim_{a\to\infty} v(a) = \dfrac{1}{\pi} s(0) h_t(0^+)$. Returning to the original notation, we have

$$\lim_{a\to\infty} \int_{-a}^{a} e^{2i\pi\xi t} \widehat{f}(\xi)\, d\xi = \frac{1}{2}\, (f(t+) + f(t-)),$$

which completes the proof. $\square$

18.3.2 Example We saw in Section 17.1.2 that $\dfrac{\sin\xi}{\xi}$ is the Fourier transform of $f(x) = \pi\chi_{[-\alpha,\alpha]}(x)$ with $\alpha = 1/(2\pi)$. Thus,

$$\lim_{a\to+\infty}\int_{-a}^{a} e^{2i\pi\xi t}\frac{\sin\xi}{\xi}\,d\xi = \begin{cases} \pi & \text{if } |t| < 1/(2\pi), \\ \pi/2 & \text{if } |t| = 1/(2\pi), \\ 0 & \text{if } |t| > 1/(2\pi). \end{cases}$$

18.4 Exercises

Exercise 18.1 Consider the following two statements:

(a) f is equal almost everywhere to a continuous function.

(b) f is continuous almost everywhere.

Show that (b) implies (a) but that the converse is false.

Exercise 18.2 Compute the Fourier transform of $f(x) = \dfrac{1}{1+x^2}$. Deduce from this the transforms of $g(x) = \dfrac{x}{(1+x^2)^2}$ and $h(x) = \dfrac{1}{1+(x-a)^2}$.

Exercise 18.3 Compute $\displaystyle\lim_{R\to+\infty}\int_{-R}^{R} e^{2i\pi\xi t}\frac{1}{a+2i\pi\xi}\,d\xi$, $a \in \mathbb{C}$, $\mathrm{Re}(a) > 0$.

Hint: Use Section 17.3.1(a) and Theorem 18.3.1. The result is 1/2.

Exercise 18.4 The computation in Section 17.3.2 of the Fourier transform of $f(x) = e^{-ax^2}$, $a > 0$, showed that $\widehat{f}(\xi) = Ke^{-\frac{\pi^2}{a}\xi^2}$. Determine the constant K using Proposition 18.2.1, and use this result to evaluate $\displaystyle\int_{\mathbb{R}} e^{-ax^2}\,dx$ directly.

***Exercise 18.5 (Shannon's formula for $f \in L^1 \cap C^0(\mathbb{R})$)** Assume that $f \in L^1(\mathbb{R}) \cap C^0(\mathbb{R})$ and that $\mathrm{supp}(\widehat{f}) \subset [-\lambda_c, \lambda_c]$, $\lambda_c > 0$. Let a be real with $0 < a \le 1/(2\lambda_c)$. Define g to be the function with period $1/a$ that coincides with $\widehat{f}$ on $(-1/(2a), 1/(2a))$.

(a) Show that the Fourier coefficients of g are

$$c_n(g) = af(-na), \quad n \in \mathbb{Z}.$$

(b) For t real and fixed, let h be the function with period $1/a$ defined by

$$h(\lambda) = e^{2i\pi t\lambda}, \quad \lambda \in \left(-\frac{1}{2a}, \frac{1}{2a}\right).$$

Show that

$$c_n(h) = \begin{cases} \dfrac{\sin\frac{\pi}{a}(t-na)}{\frac{\pi}{a}(t-na)} & \text{if } t \neq na, \\ 1 & \text{if } t = na. \end{cases}$$

(c) Use the expression for the Fourier coefficients of a product (Exercise 5.13) to deduce Shannon's formula: For all $t \in \mathbb{R}$,

$$f(t) = \sum_{n=-\infty}^{+\infty} f(na) \frac{\sin \frac{\pi}{a}(t - na)}{\frac{\pi}{a}(t - na)}, \quad 0 < a \leq \frac{1}{2\lambda_c}.$$

Lesson 19

The Space $\mathscr{S}$ ($\mathbb{R}$)

We have seen in the last few lessons how it is necessary to restrict the choice of functions in $L^1(\mathbb{R})$ if we wish to use the differentiation formulas and define the inverse Fourier transform. In this lesson, we are going to introduce a subspace of $L^1(\mathbb{R})$ that is invariant under the Fourier transform, differentiation, and multiplication by polynomials.

19.1 Rapidly decreasing functions

19.1.1 Definition A function $f : \mathbb{R} \to \mathbb{C}$ is said to decay rapidly, or be rapidly decreasing, if for all $p \in \mathbb{N}$,

$$\lim_{|x|\to\infty} |x^p f(x)| = 0.$$

For example, the function $f(x) = e^{-|x|}$ decays rapidly. It is important to note that in spite of the name, this definition does not imply that the function is monotonic in a neighborhood of infinity ($f(x) = e^{-|x|}\sin x$ is also rapidly decreasing).

The following is a useful property about the integrability of rapidly decreasing functions.

19.1.2 Proposition *If f is locally integrable, $f \in L^1_{\text{loc}}(\mathbb{R})$, and rapidly decreasing, then $x^p f(x)$ is in $L^1(\mathbb{R})$ for all $p \in \mathbb{N}$.*

Proof. Since f decays rapidly, there is an $M > 0$ such that for all $|x| > M$ we have $|x^{p+2} f(x)| \leq 1$. Thus

$$\begin{aligned}\int_{\mathbb{R}} |x^p f(x)|\,dx &\leq \int_{|x|\leq M} |x^p f(x)|\,dx + \int_{|x|>M} \frac{1}{x^2}|x^{p+2} f(x)|\,dx \\ &\leq M^p \int_{|x|\leq M} |f(x)|\,dx + \int_{|x|>M} \frac{1}{x^2}\,dx < +\infty. \qquad \square\end{aligned}$$

This leads to an important property of the Fourier transform of a rapidly decreasing function; it generalizes Proposition 17.2.1(iii).

19.1.3 Proposition *If $f \in L^1(\mathbb{R})$ decays rapidly, then $\widehat{f}$ is infinitely differentiable.*

Proof. $x^p f(x)$ is in $L^1(\mathbb{R})$ for all $p \in \mathbb{N}$ by Proposition 19.1.2. This implies by Proposition 17.2.1(i) that $\widehat{f}$ is in C^∞. □

Conversely, what does $\widehat{f} \in C^\infty$ imply about f? A partial answer is given by the next result.

19.1.4 Proposition *Assume that f is in C^∞. If $f^{(k)}$ is in $L^1(\mathbb{R})$ for all $k \in \mathbb{N}$, then $\widehat{f}$ decays rapidly.*

Proof. $\widehat{f^{(k)}}(\xi) = (2i\pi\xi)^k \widehat{f}(\xi)$ for all $k \in \mathbb{N}$ by Proposition 17.2.1(ii). By the Riemann–Lebesgue theorem, $\lim_{|\xi|\to+\infty} |\xi|^k |\widehat{f}(\xi)| = 0$. □

Said another way, we have just proved the following results:

(i) The faster f decreases at infinity, the greater the regularity of $\widehat{f}$.

(ii) The more regular f is, the faster $\widehat{f}$ decays.

In particular, if $f \in C^\infty(\mathbb{R})$ and decreases rapidly, the same is true for $\widehat{f}$. Note the similarity of this result and Proposition 5.3.4 about the Fourier coefficients of a periodic function in $C^\infty(\mathbb{R})$.

19.2 The space $\mathscr{S}$ ($\mathbb{R}$)

19.2.1 Definition $\mathscr{S}$ ($\mathbb{R}$), or simply $\mathscr{S}$, denotes the vector space of functions $f : \mathbb{R} \to \mathbb{C}$ that have the following two properties:

(i) f is infinitely differentiable.

(ii) f and all of its derivatives decay rapidly.

The space $\mathscr{S}$ ($\mathbb{R}$) is called the Schwartz class of functions, and it is named for the French mathematician Laurent Schwartz.

19.2.2 Proposition *The space $\mathscr{S}$ has the following properties:*

(i) *$\mathscr{S}$ is invariant under multiplication by a polynomial.*

(ii) *$\mathscr{S}$ is invariant under derivation; that is, $f \in \mathscr{S} \Rightarrow f' \in \mathscr{S}$.*

(iii) *$\mathscr{S} \subset L^1(\mathbb{R})$.*

The proof is left as an exercise. The main result of this lesson is contained in the next theorem.

19.2.3 Theorem *The space $\mathscr{S}$ is invariant under the Fourier transform; that is, $f \in \mathscr{S} \Rightarrow \widehat{f} \in \mathscr{S}$.*

Proof. Assume $f \in \mathscr{S}$. Then f is in $L^1(\mathbb{R})$ and decays rapidly. Thus $\widehat{f} \in C^\infty$ by Proposition 19.1.3. Since $f^{(k)}$ is rapidly decreasing for all $k \in \mathbb{N}$, it is integrable for all k by Proposition 19.1.2. We deduce from Proposition 19.1.4 that $\widehat{f}$ is rapidly decreasing. We need to show that the derivatives of $\widehat{f}$ are all rapidly decreasing. Since $\big((-2i\pi x)^q f(x)\big)^{(p)}$ is integrable, we see from (17.5) and (17.6) that

$$\frac{1}{(2i\pi)^p}\mathscr{F}\left(\big((-2i\pi x)^q f(x)\big)^{(p)}\right)(\xi) = \xi^p \mathscr{F}\left((-2i\pi x)^q f(x)\right)(\xi) = \xi^p \widehat{f}^{(q)}(\xi). \tag{19.1}$$

The right-hand term is the Fourier transform of an integrable function, so by the Riemann–Lebesgue theorem $\lim_{|\xi|\to\infty} |\xi^p \widehat{f}^{(q)}(\xi)| = 0$. □

We will need the notion of *convergence in* $\mathscr{S}$ in later parts of the book.

19.2.4 Definition A sequence $(f_n)_{n\in\mathbb{N}}$ of elements in $\mathscr{S}$ tends to 0 as n tends to infinity if

$$\lim_{n\to\infty} \sup_{x\in\mathbb{R}} |x^p f_n^{(q)}(x)| = 0$$

for all p and q in $\mathbb{N}$. We write $f_n \to 0$ in $\mathscr{S}$, or simply $f_n \to 0$ if there is no chance for misunderstanding.

Clearly, $f_n \to f$ in $\mathscr{S}$ if and only if $f_n - f \to 0$ in $\mathscr{S}$. Definition 19.2.4 implies that the sequence f_n and the sequences of all derivatives $f_n^{(q)}$ converge uniformly on $\mathbb{R}$ (take $p = 0$). We will often make use of the following consequences of convergence in $\mathscr{S}$.

19.2.5 Proposition *If the sequence $(f_n)_{n\in\mathbb{N}}$ tends to 0 in $\mathscr{S}$, then*

(i) *$f_n' \to 0$ in $\mathscr{S}$ (continuity of derivation);*

(ii) *$Pf_n \to 0$ in $\mathscr{S}$ for all polynomials P;*

(iii) *$f_n \to 0$ in $L^1(\mathbb{R})$;*

(iv) *$\widehat{f} \to 0$ in $\mathscr{S}$ (continuity of the Fourier transform).*

Proof.

(i) This is part of the definition.

(ii) It is sufficient to prove this for $P(x) = x^k$; thus we must show that $x^p(x^k f_n(x))^{(q)}$ converges uniformly on $\mathbb{R}$. But this follows directly from the definition and from Leibniz's formula for the derivatives of a product.

(iii) Take $\varepsilon > 0$. Since $f_n \to 0$ in $\mathscr{S}$, there is an $N > 0$ such that for all $n \geq N$ and all $x \in \mathbb{R}$, $|(1 + x^2)f_n(x)| \leq \varepsilon$. Thus for all $n \geq N$, $\int |f_n(x)|\,dx \leq \varepsilon \int (1 + x^2)^{-1}\,dx = \varepsilon\pi$, which proves that $f_n \to 0$ in $L^1(\mathbb{R})$.

(iv) We use (19.1); thus $|\xi^p \widehat{f_n}^{(q)}(\xi)| = (2\pi)^{q-p} |\mathscr{F}\left((x^q f_n(x))^{(p)}\right)(\xi)|$. Let $g_n(x) = (x^q f_n(x))^{(p)}$. We know from Proposition 19.2.2 that g_n is in $\mathscr{S}$ and from (i) and (ii) that $g_n \to 0$ in $\mathscr{S}$; hence $\|g_n\|_1 \to 0$ by (iii). Since $|\widehat{g}_n(\xi)| \leq \|g_n\|_1$, (iv) is proved. □

19.3 The inverse Fourier transform on $\mathscr{S}$

We have seen (Theorem 18.1.1) that if f and $\widehat{f}$ are integrable, then at all points where f is continuous,

$$f(x) = \int_{\mathbb{R}} e^{2i\pi x\xi} \widehat{f}(\xi)\, d\xi. \tag{19.2}$$

If f is in $\mathscr{S}$, then $\widehat{f}$ is in $\mathscr{S}$ and hence is integrable. Since f is continuous everywhere, (19.2) is true for all $x \in \mathbb{R}$. In other notation,

$$f = \overline{\mathscr{F}}(\mathscr{F} f) \tag{19.3}$$

for all $f \in \mathscr{S}$. In the same way, $f = \mathscr{F}(\overline{\mathscr{F}} f)$. This means that $\mathscr{F}$ is a 1-to-1 mapping on $\mathscr{S}$, and its inverse is

$$\mathscr{F}^{-1} = \overline{\mathscr{F}}. \tag{19.4}$$

19.3.1 Theorem *The Fourier transform $\mathscr{F}$ is a linear 1-to-1 mapping from $\mathscr{S}$ onto $\mathscr{S}$ that is continuous in the sense of convergence on $\mathscr{S}$. The inverse mapping is $\mathscr{F}^{-1} = \overline{\mathscr{F}}$; in other words, the relations*

$$\widehat{f}(\xi) = \int_{\mathbb{R}} e^{-2i\pi\xi x} f(x)\, dx,$$
$$f(x) = \int_{\mathbb{R}} e^{+2i\pi x\xi} \widehat{f}(\xi)\, d\xi$$

are true for all $f \in \mathscr{S}$ and all $x, \xi \in \mathbb{R}$.

Proof. We have proved everything except the continuity of $\mathscr{F}$ and $\overline{\mathscr{F}}$. The continuity of $\mathscr{F}$ is given by Proposition 19.2.5(iv); the continuity of $\overline{\mathscr{F}}$ is proved in the same way. □

An often cited representative of the space $\mathscr{S}$ is a Gaussian, which is a function of the form

$$g(x) = \beta e^{-\alpha(x-m)^2}.$$

The function $g(x) = e^{-\pi x^2}$ plays a special role in analysis.

19.3.2 Proposition *The Fourier transform of $g(x) = e^{-\pi x^2}$ is the same function, $\widehat{g}(\xi) = e^{-\pi\xi^2}$.*

This result was established in Section 17.3.2. The function $g(x) = e^{-\pi x^2}$ is a fixed point of the Fourier transform.

19.4 Exercises

Exercise 19.1 Find an example of a function in $C^\infty(\mathbb{R})$ that decays rapidly but whose derivative does not decay rapidly.

***Exercise 19.2**

(a) Show that if f and g are in $\mathscr{S}(\mathbb{R})$, then the product fg belongs to $\mathscr{S}(\mathbb{R})$.

(b) Show that the mapping $f \mapsto fg$ is continuous from $\mathscr{S}(\mathbb{R})$ to $\mathscr{S}(\mathbb{R})$.

Exercise 19.3 Let $f : \mathbb{R} \to \mathbb{C}$ be measurable. We say that f is slowly increasing if there exist $C > 0$ and $N \in \mathbb{N}$ such that

$$|f(x)| \leq C(1+x^2)^N \quad \text{for all } x \in \mathbb{R}.$$

Suppose that $f \in C^\infty(\mathbb{R})$ and all of its derivatives are slowly increasing, and suppose that $g \in \mathscr{S}(\mathbb{R})$. Show that $fg \in \mathscr{S}(\mathbb{R})$.

Exercise 19.4 Assume that $g \in \mathscr{S}(\mathbb{R})$ and that $f = P/Q$, where P and Q are polynomials. Show that if Q has no real zeros, then $fg \in \mathscr{S}(\mathbb{R})$.

Exercise 19.5 Prove Proposition 19.2.2.

Exercise 19.6 Show that $\mathscr{S}(\mathbb{R}) \subset L^p(\mathbb{R})$, $p \geq 1$.
Hint: Write $f(x) = (1+x^2)^{-1/p}(1+x^2)^{1/p} f(x)$ and note that $(1+x^2)^{1/p} f(x)$ is bounded on $\mathbb{R}$.

***Exercise 19.7 (the density of $\mathscr{D}(\mathbb{R})$ in $\mathscr{S}(\mathbb{R})$)**
Assume that ϕ is in $\mathscr{S}(\mathbb{R})$ and that α is in $\mathscr{D}(\mathbb{R})$ with $\alpha = 1$ on $[-1,1]$. Define $\alpha_n(x) = \alpha(x/n)$ for $n \in \mathbb{N}^\star$. Show that the sequence $\phi_n = \alpha_n \phi$ is in $\mathscr{D}(\mathbb{R})$ and that it converges to ϕ in the topology of $\mathscr{S}(\mathbb{R})$.

Lesson 20

The Convolution of Functions

The convolution, like the Fourier transform, is one of the essential tools of signal processing. The results that we establish in this lesson will be restricted to functions. Our development will often rely on the theorems on integration established in Lesson 14. The limits of the notion of "function" and practical applications lead naturally to generalize the Fourier transform, convolution, and associated concepts to distributions. The study of distributions will begin with Lesson 26; the convolution for distributions will be developed in Lesson 32.

20.1 Definitions and examples

20.1.1 Definition The convolution of two functions f and g from $\mathbb{R}$ to $\mathbb{C}$ is the function $f * g$, if it exists, defined by

$$f * g(x) = \int_{\mathbb{R}} f(x-t)g(t)\,dt = \int_{\mathbb{R}} f(u)g(x-u)\,du.$$

If no assumptions are made about f and g, the convolution is clearly not defined. Take, for example, $f = g \equiv 1$. We give assumptions in Section 20.2 that imply the existence of $f * g$. But first we examine two examples that allow us to visualize some properties of the convolution.

20.1.2 Examples

(a) Let $f = g = \chi_{[0,1]}$. Then

$$\int_{\mathbb{R}} f(x-t)g(t)\,dt = \int_0^1 \chi_{[0,1]}(x-t)\,dt = \text{measure}\ \big([0,1] \cap [x-1, x]\big),$$

which is the "hat" function

$$f * g(x) = \begin{cases} 0 & \text{if } x \leq 0, \\ x & \text{if } 0 \leq x \leq 1, \\ 2 - x & \text{if } 1 \leq x \leq 2, \\ 0 & \text{if } x \geq 2. \end{cases}$$

This convolution is illustrated in Figure 20.1.

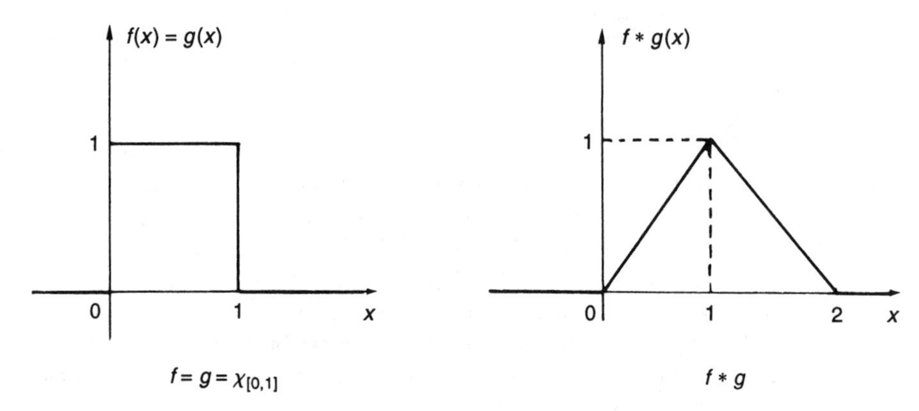

FIGURE 20.1. The convolution $f * g$ is continuous.

(b) Take $f \in L^1(\mathbb{R})$ and $g = \frac{1}{2h}\chi_{[-h,+h]}$ with $h > 0$. Then

$$f * g(x) = \frac{1}{2h} \int_{-h}^{+h} f(x - t)\, dt = \frac{1}{2h} \int_{x-h}^{x+h} f(u)\, du,$$

which is the average of f on the interval $[x - h, x + h]$. The continuity of $f * g$ is a direct consequence of Proposition 14.5.4.

These two examples illustrate an important property of the convolution: It regularizes a function by averaging. We will study this in detail in Lesson 21, but first we are going to present some conditions that imply the existence of the convolution. For this we will need the notion of the support of a function.

We have already given the definition of the support of a continuous function (Definition 15.1.5). We are now dealing with measurable functions that are defined almost everywhere, and it is necessary to proceed cautiously. For example, if we simply extend the previous definition of support to measurable functions, we have $\overline{\{x \in \mathbb{R} \mid \chi_{\mathbb{Q}}(x) \neq 0\}} = \mathbb{R}$ for the support of the characteristic function of the rationals. This definition is not what we want, for it clearly depends on the particular function we have taken as the representative of its class. We need a definition that gives the same result for all functions that are equal almost everywhere.

20.1.3 Definition (support of a measurable function)

Let $f : \mathbb{R} \to \mathbb{C}$ be a measurable function. Let θ_i, $i \in I$, be the family of open sets in $\mathbb{R}$ such that for all $i \in I$, $f = 0$ a.e. on θ_i. Let $\theta = \cup_{i\in I}\theta_i$ and define the support of f, $\mathrm{supp}(f)$, to be the closed set $\mathbb{R} \setminus \theta$, that is,

$$\mathrm{supp}(f) = \mathbb{R} \setminus \theta.$$

It is left as an exercise to verify that this definition extends Definition 15.1.5, that $f = 0$ a.e. on θ, and that $f = g$ a.e. implies $\mathrm{supp}(f) = \mathrm{supp}(g)$.

In the first example of Section 20.1.2, $\mathrm{supp}(f) = \mathrm{supp}(g) = [0, 1]$, and we saw that the convolution $f * g$ spreads these supports with the result that $\mathrm{supp}(f * g) = [0, 2]$. In general, we have the following result.

20.1.4 Lemma *Let f and g be two functions for which $f * g$ exists. Then*

$$\mathrm{supp}(f * g) \subset \overline{\mathrm{supp}(f) + \mathrm{supp}(g)}.$$

Proof. Let S denote $\mathbb{R}\setminus(\mathrm{supp}(f)+\mathrm{supp}(g))$ and let S^o denote the interior of S. Suppose that $x \in S$. Then for all $t \in \mathrm{supp}(f)$ we have $(x-t) \notin \mathrm{supp}(g)$, and consequently $\int_{\mathbb{R}} g(x-t)f(t)\,dt = 0$. Now let θ_{f*g} be the largest open set on which $f*g = 0$ a.e. We have just seen that $x \in S^o \Rightarrow x \in \theta_{f*g}$. Thus $x \in \mathbb{R} \setminus \theta_{f*g} = \mathrm{supp}(f * g)$ implies that $x \in \mathbb{R} \setminus S^o$. Since $\mathbb{R} \setminus S^o = \overline{\mathbb{R} \setminus S}$, this proves the result. □

20.2 Convolution in $L^1(\mathbb{R})$

We will establish the existence of the convolution for integrable functions.

20.2.1 Proposition *If f and g are in $L^1(\mathbb{R})$, then the following hold:*

(i) *$f * g$ is defined almost everywhere and $f * g$ belongs to $L^1(\mathbb{R})$.*

(ii) *The convolution is a continuous bilinear operator from $L^1(\mathbb{R}) \times L^1(\mathbb{R})$ to $L^1(\mathbb{R})$ with*

$$\|f * g\|_1 \le \|f\|_1\|g\|_1. \tag{20.1}$$

Proof.

(i) Since f and g are in $L^1(\mathbb{R})$, Fubini's theorem implies that the function $(y, z) \mapsto f(y)g(z)$ is in $L^1(\mathbb{R}^2)$. By making the change of variables $y = x - t$ and $z = t$, we have

$$\iint_{\mathbb{R}\times\mathbb{R}} f(y)g(z)\,dy\,dz = \iint_{\mathbb{R}\times\mathbb{R}} f(x-t)g(t)\,dx\,dt.$$

The function $x \mapsto \int_{\mathbb{R}} f(x-t)g(t)\,dt$ is thus defined almost everywhere and belongs to $L^1(\mathbb{R})$, again by Fubini's theorem.

(ii) To establish the inequality (20.1) we write

$$|f * g(x)| \leq \int_{\mathbb{R}} |f(x-t)||g(t)|\,dt = |f| * |g|(x).$$

Thus

$$\int_{\mathbb{R}} |f * g|(x)\,dx \leq \int_{\mathbb{R}} |f| * |g|(x)\,dx = \int_{\mathbb{R}} dx \int_{\mathbb{R}} |f(x-t)||g(t)|\,dt$$
$$= \int_{\mathbb{R}} |g(t)| \Big(\int_{\mathbb{R}} |f(x-t)|\,dx \Big)\,dt = \|g\|_1 \|f\|_1. \qquad \square$$

Can the hypotheses of this last result be weakened? If f and g are in $L^1_{\text{loc}}(\mathbb{R})$, the result is false (take $f = g \equiv 1$). However, we have the following result.

20.2.2 Proposition *Assume that $f \in L^1_{\text{loc}}(\mathbb{R})$ and that $g \in L^1(\mathbb{R})$.*

(i) *If* supp(g) *is bounded, then $f * g(x)$ exists a.e. and belongs to $L^1_{\text{loc}}(\mathbb{R})$.*

(ii) *If f is bounded, then $f * g(x)$ exists for all x and belongs $L^\infty(\mathbb{R})$.*

Proof.

(i) g is zero a.e. outside some interval $[-a, a]$. Take x in a finite interval $[\alpha, \beta]$. For all $t \in [-a, a]$ and all $x \in [\alpha, \beta]$,

$$f(x-t)g(t) = \chi_{[\alpha-a,\beta+a]}(x-t) f(x-t) g(t),$$

and thus

$$f * g(x) = \int_{-a}^{+a} f(x-t)g(t)\,dt = \big(\chi_{[\alpha-a,\beta+a]} f\big) * g(x).$$

$f * g$ coincides on $[\alpha, \beta]$ with the convolution of two functions in $L^1(\mathbb{R})$, so by Proposition 20.2.1(i) it is defined a.e. and is integrable. Thus $f * g$ is defined a.e. and is integrable on all compact sets.

(ii) If $f \in L^\infty(\mathbb{R})$, then

$$\Big| \int_{\mathbb{R}} f(u) g(x-u)\,du \Big| \leq \|f\|_\infty \int_{\mathbb{R}} |g(x-u)|, du = \|f\|_\infty \|g\|_1$$

for all x, and $\|f * g\|_\infty \leq \|f\|_\infty \|g\|_1$. $\square$

20.3 Convolution in $L^p(\mathbb{R})$

If p and q are two real positive numbers (perhaps $+\infty$) such that $\frac{1}{p} + \frac{1}{q} = 1$, we say that p and q are harmonic conjugates, or simply conjugates.

20.3.1 Proposition *Assume that $f \in L^p(\mathbb{R})$ and $g \in L^q(\mathbb{R})$ (p and q conjugates). Then the following hold:*

(i) *$f * g$ is defined everywhere and is continuous and bounded on $\mathbb{R}$.*

(ii) $\|f * g\|_\infty \le \|f\|_p \|g\|_q$. (20.2)

Proof. We will prove this result in two particularly important cases: $p = 1$, $q = +\infty$ and $p = 2$, $q = 2$.

First case: $p = 1$, $q = +\infty$.

Since we have already seen (Proposition 20.2.2(ii)) that $f * g$ is defined everywhere and bounded, we only need to prove continuity. For this we write

$$\begin{aligned}|f * g(x) - f * g(y)| &\le \int |f(x-t) - f(y-t)||g(t)|\,dt \\ &\le \|g\|_\infty \int |f(x-t) - f(y-t)|\,dt.\end{aligned}$$

We first establish the continuity when f is continuous with compact support. Thus let $(-a, a)$ be an open interval containing $\operatorname{supp}(f)$. For $|x - y|$ sufficiently small,

$$\begin{aligned}\int_{\mathbb{R}} |f(x-t) - f(y-t)|\,dt &= \int_{\mathbb{R}} |f(x-y+u) - f(u)|\,du \\ &= \int_{-a}^{+a} |f(x-y+u) - f(u)|\,du \\ &\le 2a \sup_{|u|\le a} |f(x-y+u) - f(u)|.\end{aligned}$$

Since f is uniformly continuous on $[-a, a]$ it follows that $f * g$ is continuous on $\mathbb{R}$, in fact, uniformly continuous on $\mathbb{R}$.

When $f \in L^1(\mathbb{R})$, we argue using the density of C_c^0 in $L^1(\mathbb{R})$ (Theorem 15.3.3). Thus let f_n be a sequence of continuous functions with compact support such that $\lim_{n\to\infty} \|f_n - f\|_1 = 0$. Adding and subtracting $f_n * g$ at x and y, we see that

$$\begin{aligned}|f * g(x) - f * g(y)| \le &|f * g(x) - f_n * g(x)| + |f_n * g(y) - f * g(y)| \\ &+ |f_n * g(x) - f_n * g(y)|,\end{aligned}$$

so

$$|f * g(x) - f * g(y)| \le 2\|g\|_\infty \|f - f_n\|_1 + |f_n * g(x) - f_n * g(y)|.$$

The first term on the right-hand side tends to 0 as n tends to infinity, and the second term is uniformly continuous for each fixed n; it follows directly that $f * g$ is uniformly continuous on $\mathbb{R}$.

Second case: $p = 2$, $q = 2$.

From Schwarz's inequality we have

$$|f * g(x)| \leq \int_{\mathbb{R}} |f(x-t)||g(t)|\,dt \leq \Big(\int_{\mathbb{R}} |f(x-t)|^2\,dt\Big)^{1/2}\Big(\int_{\mathbb{R}} |g(t)|^2\,dt\Big)^{1/2},$$

and hence $\|f*g\|_\infty \leq \|f\|_2\|g\|_2$. The continuity is established as in the first case using Schwarz's inequality and the density of $C_c^0(\mathbb{R})$ in $L^2(\mathbb{R})$ (Section 15.3.4).

For $p \neq 1, 2$ one uses Hölder's inequality (Lemma 15.2.4) and imitates the arguments given above. □

We next examine the convolution of a function $L^1(\mathbb{R})$ with a function of $L^2(\mathbb{R})$, which is a case not included in the last result.

20.3.2 Proposition *If $f \in L^1(\mathbb{R})$ and $g \in L^2(\mathbb{R})$, then the following hold:*

(i) *$f * g(x)$ exists almost everywhere.*

(ii) *$f * g$ is in $L^2(\mathbb{R})$, and*

$$\|f * g\|_2 \leq \|f\|_1\|g\|_2. \tag{20.3}$$

Proof.

(i) Write

$$|f(u)g(x-u)| = \Big(|f(u)||g(x-u)|^2\Big)^{1/2}\Big(|f(u)|\Big)^{1/2}. \tag{20.4}$$

Since $|f| \in L^1(\mathbb{R})$ and $|g|^2 \in L^1(\mathbb{R})$, the function $u \mapsto |f(u)||g(x-u)|^2$ is integrable for almost all x (Proposition 20.2.1(i)). The right-hand term of (20.4), being the product of two square integrable functions, is integrable. Thus $f * g(x)$ is defined for almost all x.

(ii) Using the Schwarz inequality and (20.4) we see that

$$\begin{aligned}|f * g(x)| &\leq \int_{\mathbb{R}} |f(u)||g(x-u)|\,du\\ &\leq \Big(\int_{\mathbb{R}} |f(u)||g(x-u)|^2\,du\Big)^{1/2}\Big(\int_{\mathbb{R}} |f(u)|\,du\Big)^{1/2},\end{aligned}$$

and thus

$$|f * g(x)|^2 \leq \Big(|f| * |g|^2(x)\Big)\|f\|_1.$$

Integrating both sides of the last inequality shows that

$$\begin{aligned}\int_{\mathbb{R}} |f * g(x)|^2\,dx &\leq \|f\|_1 \int_{\mathbb{R}} |f| * |g|^2(x)\,dx\\ &\leq \|f\|_1\|f\|_1\|g^2\|_1,\end{aligned}$$

and finally,

$$\|f * g\|_2 \leq \|f\|_1\|g\|_2.$$

□

20.3.3 Remark The last result can be generalized to the convolution $L^p(\mathbb{R}) * L^q(\mathbb{R})$ with $\frac{1}{p} + \frac{1}{q} - 1 = \frac{1}{r}$, where p, q, r are ≥ 1. For $f \in L^p(\mathbb{R})$ and $g \in L^q(\mathbb{R})$, $f * g$ is in $L^r(\mathbb{R})$ [Kho72]. In Proposition 20.3.2 we have $p = 1$, $q = 2$, and $r = 2$.

20.4 Convolution of functions with limited support

When one observes a signal, it exists from some time t_{i} to time t_{f}, with the possibility that $t_{\mathrm{i}} = -\infty$ and $t_{\mathrm{f}} = +\infty$. Signals whose support is limited on the left (lies to the right of some finite point) are of particular interest.

20.4.1 Definition

$C_+^0 = \{f \in C^0(\mathbb{R}) \mid \operatorname{supp}(f) \subset [a, +\infty] \text{ for some } a \in \mathbb{R}\}$.

$C_{\mathrm{pw}+} = \{f \text{ is piecewise continuous } \mid \operatorname{supp}(f) \subset [a, +\infty] \text{ for some } a \in \mathbb{R}\}$.

The function spaces C_-^0 and $C_{\mathrm{pw}-}$ are defined similarly for functions whose support is limited on the right.

Recall that $C_{\mathrm{c}}^0(\mathbb{R})$ denotes the continuous functions with bounded support and that C_{pw} denotes the functions that are piecewise continuous, that is, f is continuous except for a finite number of points $a_1, \ldots, a_k$ where $f(a_j^+)$ and $f(a_j^-)$ exist (see Section 5.2.1).

20.4.2 Proposition *If f and g are in $C_{\mathrm{c}}^0(\mathbb{R})$, the convolution $f * g$ exists and belongs to $C_{\mathrm{c}}^0(\mathbb{R})$.*

Proof. Consider this to be a case of the convolution $L^1(\mathbb{R}) * L^\infty(\mathbb{R})$ (Proposition 20.3.1): $f * g$ is defined, continuous, and bounded on $\mathbb{R}$. By Lemma 20.1.4,

$$\operatorname{supp}(f * g) \subset \overline{\operatorname{supp}(f) + \operatorname{supp}(g)}.$$

Thus $f * g$ has bounded support and belongs to $C_{\mathrm{c}}^0(\mathbb{R})$. □

20.4.3 Remark In the statement of this last result one can assume that f and g are in C_{pw} and have bounded support. The convolution $f * g$ is again in $C_{\mathrm{c}}^0(\mathbb{R})$.

20.4.4 Proposition *Given f and g in $C_{\mathrm{pw}+}$, the convolution $f * g$ exists and belongs to C_+^0.*

Proof. Suppose that $\operatorname{supp}(f) \subset [a, +\infty)$ and $\operatorname{supp}(g) \subset [b, +\infty)$. Then $f(x - t) = 0$ if $x - t < a$, and $g(t) = 0$ if $t < b$. Thus

$$f * g(x) = 0 \quad \text{if} \quad x < a + b. \tag{20.5}$$

If $x \geq a+b$ and $M > x$, then

$$f * g(x) = \int_b^{M-a} f(x-t)g(t)\,dt. \tag{20.6}$$

In (20.6), only the values of f on the interval $[2a+b-M, M-b]$ are used, so the convolution can be written

$$f * g(x) = \big(f\chi_{[2a+b-M,M-b]}\big) * \big(g\chi_{[b,M-a]}\big)(x). \tag{20.7}$$

Hence $f * g$ agrees on $[a+b, M)$ with the convolution of two functions in C_{pw} that have bounded support. $f * g$ is thus in C^0_+. □

20.5 Summary

$L^1 * L^1$	$\subset$	L^1
$L^1 * L^\infty$	$\subset$	$L^\infty \cap C^0$
$L^2 * L^2$	$\subset$	$L^\infty \cap C^0$
$L^2 * L^1$	$\subset$	L^2
$C_{\text{pw}+} * C_{\text{pw}+}$	$\subset$	C^0_+
$C^0_{\text{c}} * C^0_{\text{c}}$	$\subset$	C^0_{c}

20.6 Exercises

****Exercise 20.1** Let $f : \mathbb{R} \to \mathbb{C}$ be a measurable function. With the notation of Definition 20.1.3, $\operatorname{supp}(f) = \mathbb{R}\backslash\theta$. Show that $f = 0$ a.e. on θ.

Hint: The proof of this theoretical result is delicate. Write $\theta = \bigcup_{n=1}^{\infty} K_n$, where

$$K_n = \Big\{x \in \theta \mid \text{ distance}(x, \mathbb{R}\backslash\theta) \geq \frac{1}{n} \text{ and } |x| \leq n\Big\}.$$

Note that K_n is compact and hence is in the union of a finite number of the open sets θ_i.

Exercise 20.2 Let $f = \chi_{[0,1]}$. Show that $h = f * (f * f)$ makes sense and compute h. What is the regularity of h?

Exercise 20.3 Show that

$$\chi_{[-a,a]} * \sin x = 2 \sin a \, \sin x,$$
$$\chi_{[-a,a]} * \cos x = 2 \sin a \, \cos x.$$

Exercise 20.4 Compute $u * u$, where u denotes the Heaviside function.

Exercise 20.5 Suppose $f \in L^1(\mathbb{R})$ and $g \in L^p(\mathbb{R})$, $1 \leq p < +\infty$. Show that $f * g \in L^p(\mathbb{R})$ and that $\|f * g\|_p \leq \|f\|_1 \|g\|_p$.

Hint: The case $p = 1$ is done in Proposition 20.2.1, and the case $p = 2$ is done in Proposition 20.3.2. For $p \neq 1, 2$, imitate the proof in Proposition 20.3.2 by writing

$$|f(u)g(x-u)| = (|f(u)||g(x-u)|^p)^{\frac{1}{p}} |f(u)|^{1-\frac{1}{p}};$$

then use Hölder's inequality (Lemma 15.2.4).

Exercise 20.6 Show that the convolution of a slowly increasing function f with a rapidly decreasing function g is well-defined.

Hint: Write $|f(x-t)g(t)| \leq C\left(1 + (x-t)^2\right)^n \left(1 + t^2\right)^{-n-2}$.

***Exercise 20.7** Assume that f and g are in $L^2_{\mathrm{p}}(0, a)$ and that their periodic convolution is

$$f * g(x) = \int_0^a f(x-t)g(t)\,dt = \int_0^a f(t)g(x-t)\,dt.$$

(a) Show that $f * g$ exists and belongs to $L^\infty_{\mathrm{p}}(0, a) \cap C^0([0, a])$ and that

$$\|f * g\|_\infty \leq \|f\|_2 \|g\|_2.$$

(b) Show that $c_n(f * g) = a c_n(f) c_n(g)$.

Lesson 21

Convolution, Derivation, and Regularization

We saw in Lesson 20 conditions under which the convolution of two functions is well-defined. We turn now to several important properties of the convolution, some of which will be extended to distributions in Lesson 32. In the current lesson we focus on regularization.

21.1 Convolution and continuity

We have shown that $f * g$ is continuous on $\mathbb{R}$ when $f \in L^p(\mathbb{R})$, $g \in L^q(\mathbb{R})$, and $\frac{1}{p} + \frac{1}{q} = 1$ (Proposition 20.3.1). Here is a consequence of that result.

21.1.1 Proposition *Suppose that $f \in L^p(\mathbb{R})$ has bounded support and that g is in $L^q_{\text{loc}}(\mathbb{R})$ with $\frac{1}{p} + \frac{1}{q} = 1$. Then the convolution $f * g$ is defined and continuous for all $x \in \mathbb{R}$.*

Proof. f is zero a.e. outside some interval $[-a, a]$. Suppose x is in a bounded interval $[\alpha, \beta]$. $f * g$ agrees on $[\alpha, \beta]$ with $f * g\chi_{[\alpha - a, \beta - a]}$, and this reduces to a convolution $L^p * L^q$ as in Proposition 20.3.1. Thus $f * g$ is defined and continuous everywhere. □

21.1.2 Example The convolution of a function in $L^1(\mathbb{R})$ with a bounded function having compact support is continuous.

21.2 Convolution and derivation

The last result can be generalized: Convolution with a function of class C^p yields a function in $C^p(\mathbb{R})$.

21.2.1 Proposition *Let f be in $L^1(\mathbb{R})$ and let g be in $C^p(\mathbb{R})$. Assume that $g^{(k)}$ is bounded for $k = 0, 1, \ldots, p$. Then* (i) *$f * g \in C^p(\mathbb{R})$ and* (ii) *$(f * g)^{(k)} = f * g^{(k)}$ for $k = 1, 2, \ldots, p$.*

Proof. By applying Proposition 20.3.1 with $p = 1$ and $q = +\infty$ we see that $f * g^{(k)}$ is continuous for $k = 0, 1, \ldots, p$. The function $x \mapsto f(t)g(x-t)$ is p-times differentiable, and for $k = 0, 1, \ldots, p$,

$$|f(t)g^{(k)}(x-t)| \leq M_k |f(t)|,$$

where $M_k = \sup_{y \in \mathbb{R}} |g^{(k)}(y)|$. Since $f \in L^1(\mathbb{R})$, we can differentiate under the integral sign (Proposition 14.2.2); hence

$$(f * g)^{(k)}(x) = \int_{\mathbb{R}} f(t) g^{(k)}(x-t)\, dt = f * g^{(k)}(x). \qquad \square$$

21.2.2 Remark $(f * g)^{(k)}$ is bounded on $\mathbb{R}$ for $k = 0, 1, \ldots, p$ because $(f * g)^{(k)} = f * g^{(k)}$ is a convolution of the type $L^1 * L^\infty$.

21.3 Convolution and regularization

21.3.1 Definition A sequence of functions ρ_n in $\mathscr{D}(\mathbb{R})$ (Definition 15.1.7) is called a regularizing sequence if it satisfies the following conditions:

(i) $\rho_n(x) \geq 0$ for all $x \in \mathbb{R}$.

(ii) $\displaystyle\int_{\mathbb{R}} \rho_n(x)\, dx = 1$.

(iii) The support of ρ_n is in $[-\varepsilon_n, \varepsilon_n]$, $\varepsilon_n > 0$, and $\lim\limits_{n \to \infty} \varepsilon_n = 0$.

To see that such a sequence exists, take $\rho \in \mathscr{D}(\mathbb{R})$ defined by

$$\rho(x) = \begin{cases} \dfrac{1}{c}\, e^{-\frac{1}{1-x^2}} & \text{if } |x| \leq 1, \\ 0 & \text{if } |x| > 1, \end{cases}$$

with

$$c = \int_{-1}^{1} e^{-\frac{1}{1-x^2}}\, dx,$$

and let $\rho_n(x) = n\rho(nx)$. In practice, regularizing sequences are used without defining them explicitly. As we will see, the details are not important; one uses only properties (i), (ii), and (iii).

21.3.2 Definition If $f \in L^1(\mathbb{R})$, the functions $f * \rho_n$ are called regularizations of f.

It is clear from the properties of ρ_n and Proposition 21.2.1 that $f * \rho_n$ is in $C^\infty(\mathbb{R})$. But what is the relation between f and its regularizations?

21.3.3 Theorem (density of $\mathscr{D}(\mathbb{R})$ in $L^1(\mathbb{R})$) *Let f be a function in $L^1(\mathbb{R})$. For $\varepsilon > 0$ there exists g_ε in $\mathscr{D}(\mathbb{R})$ such that $\|f - g_\varepsilon\|_1 \leq \varepsilon$.*

Proof. First choose f_ε in $C^0_c(\mathbb{R})$ such that $\|f - f_\varepsilon\|_1 \leq \varepsilon/2$ (Theorem 15.3.3). Assume that $\operatorname{supp}(f_\varepsilon) \subset [a, b]$. Now consider the regularizations g_n of f_ε, namely, $g_n = f_\varepsilon * \rho_n$. Let $K = [a-1, b+1]$. Then $\operatorname{supp}(g_n) \subset K$ for sufficiently large n (Lemma 20.1.4). Since $\rho_n \in C^\infty(\mathbb{R})$, g_n is in $\mathscr{D}(\mathbb{R})$.

We wish to estimate $\|f_\varepsilon - g_n\|_1$. For sufficiently large n,

$$\int_{\mathbb{R}} |f_\varepsilon(x) - g_n(x)|\, dx \leq (b - a + 2) \sup_{x \in K} |f_\varepsilon(x) - g_n(x)|. \tag{21.1}$$

Since $\int_{\mathbb{R}} \rho_n(t)\, dt = 1$, we can write

$$f_\varepsilon(x) - g_n(x) = \int_{\mathbb{R}} (f_\varepsilon(x) - f_\varepsilon(x-t))\rho_n(t)\, dt.$$

Thus

$$|f_\varepsilon(x) - g_n(x)| \leq \sup_{|t| \leq \varepsilon_n} |f_\varepsilon(x) - f_\varepsilon(x-t)|,$$

and

$$\sup_{x \in K} |f_\varepsilon(x) - g_n(x)| \leq \sup_{\substack{x \in K \\ |t| \leq \varepsilon_n}} |f_\varepsilon(x) - f_\varepsilon(x-t)|. \tag{21.2}$$

f_ε is uniformly continuous, so the right-hand side of (21.2) tends to 0 as $n \to +\infty$. Returning to (21.1) we see that

$$\lim_{n \to \infty} \|f_\varepsilon - g_n\|_1 = 0. \tag{21.3}$$

In particular, there is an N such that $n > N$ implies $\|f_\varepsilon - g_n\|_1 \leq \varepsilon/2$. Thus for all sufficiently large n, $\|f - g_n\|_1 \leq \|f - f_\varepsilon\|_1 + \|f_\varepsilon - g_n\|_1 \leq \varepsilon$, which proves that $\mathscr{D}(\mathbb{R})$ is dense in $L^1(\mathbb{R})$. □

21.3.4 Remark The proof shows that if f is continuous, the sequence $f * \rho_n$ tends to f uniformly on all compact sets.

21.3.5 Remark One can prove in the same way that $\mathscr{D}(\mathbb{R})$ is dense in $L^p(\mathbb{R})$, $1 < p < \infty$. Then, since $\mathscr{D}(\mathbb{R}) \subset \mathscr{S}(\mathbb{R})$, $\mathscr{S}(\mathbb{R})$ is dense in $L^p(\mathbb{R})$.

21.3.6 Proposition *If $f \in L^1(\mathbb{R})$ and ρ_n is a regularizing sequence, then $\lim_{n \to \infty} \|f - f * \rho_n\|_1 = 0$.*

Proof. Take $\varepsilon > 0$. By Theorem 21.3.3 there is a g_ε in $\mathscr{D}(\mathbb{R})$ such that $\|f - g_\varepsilon\|_1 \leq \varepsilon/4$. From (20.1) we see that

$$\|f * \rho_n - g_\varepsilon * \rho_n\|_1 \leq \|f - g_\varepsilon\|_1 \|\rho_n\|_1 = \|f - g_\varepsilon\|_1,$$

and hence that

$$\begin{aligned}\|f - f * \rho_n\|_1 &\leq \|f - g_\varepsilon\|_1 + \|g_\varepsilon - g_\varepsilon * \rho_n\|_1 + \|g_\varepsilon * \rho_n - f * \rho_n\|_1 \\ &\leq 2\|f - g_\varepsilon\|_1 + \|g_\varepsilon - g_\varepsilon * \rho_n\|_1.\end{aligned}$$

We know from (21.3) that

$$\lim_{n\to\infty} \|g_\varepsilon - g_\varepsilon * \rho_n\|_1 = 0.$$

Thus there is an N such that for all $n > N$, $\|g_\varepsilon - g_\varepsilon * \rho_n\|_1 \leq \varepsilon/2$, in which case $\|f - f * \rho_n\|_1 \leq \varepsilon$. This proves that $f * \rho_n$ tends to f in $L^1(\mathbb{R})$. □

21.3.7 Remark A similar argument can be used to show that the regularizations $f * \rho_n$ of a function $f \in L^p(\mathbb{R})$, $1 < p < \infty$, tend to f in $L^p(\mathbb{R})$. (See Exercise 21.4.)

21.4 The convolution $\mathscr{S}(\mathbb{R}) * \mathscr{S}(\mathbb{R})$.

Since $\mathscr{S}(\mathbb{R})$ is in $L^1(\mathbb{R})$, we know that the convolution of two functions in $\mathscr{S}(\mathbb{R})$ is in $L^1(\mathbb{R})$. There is a better result.

21.4.1 Proposition *Assume that f and g are in $\mathscr{S}(\mathbb{R})$. Then the following hold:*

(i) *$f * g$ is in $\mathscr{S}(\mathbb{R})$.*

(ii) *The convolution is a continuous operator from $\mathscr{S}(\mathbb{R}) \times \mathscr{S}(\mathbb{R})$ to $\mathscr{S}(\mathbb{R})$.*

Proof.

(i) $f*g \in C^\infty(\mathbb{R})$ (Proposition 21.2.1). We look at the behavior at infinity. First,

$$\lim_{|x|\to\infty} f * g(x) = \lim_{|x|\to\infty} \int_{\mathbb{R}} f(x-t)g(t)\,dt = 0$$

by dominated convergence: $f \in \mathscr{S}$ and $|f(x-t)g(t)| \leq \|f\|_\infty |g(t)|$, which is integrable. To study $\lim_{|x|\to\infty} x^p (f * g)^{(q)}(x)$ we use the formula

$$x^p (f * g)^{(q)}(x) = \sum_{j=0}^{p} \beta_j (x^{p-j} f) * (x^j g^{(q)}),$$

where the β_j are binomial coefficients. Thus $x^p(f * g)^{(q)}$ is written as a sum of convolutions of elements in $\mathscr{S}(\mathbb{R})$, which is invariant under differentiation and multiplication by polynomials (Proposition 19.2.2). Thus, by what we have just shown for $f * g$, $\lim_{|x|\to\infty} x^p (f * g)^{(q)}(x) = 0$.

(ii) To prove continuity, consider two sequences f_n and g_m in $\mathscr{S}(\mathbb{R})$ that converge in $\mathscr{S}$ to f and g. Adding and subtracting $f_n * g$, we have

$$\|f_n * g_m - f * g\|_\infty \leq \|(f_n - f) * g\|_\infty + \|f_n * (g_m - g)\|_\infty.$$

Using (20.2), this becomes

$$\|f_n * g_m - f * g\|_\infty \leq \|f_n - f\|_\infty \|g\|_1 + \|f_n\|_1 \|g_m - g\|_\infty,$$

and it follows that $f_n * g_m$ converges to $f * g$. For expressions of the form $x^p (f_n * g_m)^{(q)}(x)$ we use the decomposition given in (i). □

21.4.2 Remark Note that we did not use the regularity of f in the last proof. By modifying this proof we can show that $g \mapsto f * g$ is continuous from $\mathscr{S}(\mathbb{R})$ to $\mathscr{S}(\mathbb{R})$ when $g \in \mathscr{S}(\mathbb{R})$ and $f \in L^1_{\text{loc}}(\mathbb{R})$ decreases rapidly.

21.5 Exercises

Exercise 21.1 Assume that f is in $L^1(\mathbb{R})$ and $g(x) = e^{2i\pi x}$. Compute $f * g$.

Exercise 21.2 Suppose $f \in C^0(\mathbb{R})$ and $h > 0$. Show that $g = \dfrac{1}{2h}\chi_{[-h,h]} * f$ is in $C^1(\mathbb{R})$ and compute g'.

***Exercise 21.3** Show that $\mathscr{D}(\mathbb{R})$ is dense in $L^p(\mathbb{R})$, $1 \leq p < +\infty$. Deduce from this that $\mathscr{S}(\mathbb{R})$ is dense in $L^p(\mathbb{R})$, $1 \leq p < +\infty$.

Exercise 21.4 Take $f \in L^p(\mathbb{R})$, $1 \leq p < +\infty$. Use Exercise 20.5 to show that $\lim_{n \to +\infty} \|f - f * \rho_n\|_p = 0$.

***Exercise 21.5** Let I be an open interval in $\mathbb{R}$. Show that $\mathscr{D}(I)$ is dense in $L^p(I)$, $1 \leq p < +\infty$.

***Exercise 21.6** Suppose $f \in L^1_{\text{loc}}(\mathbb{R})$ is such that

$$\int_{\mathbb{R}} f(t)\phi(t)\,dt = 0 \quad \text{for all } \phi \in \mathscr{D}(\mathbb{R}).$$

(a) Show that the regularizations $\rho_n * f$ are zero.

(b) Take $a > 0$ and $b = a + 1$. Show that for all $|x| \leq a$,

$$\rho_n * f(x) = \rho_n * (\chi_{[-b,b]} f)(x) = 0,$$

and from this deduce that

$$\int_{-a}^{a} |f(x)|\,dx \leq \|\chi_{[-b,b]} f - \rho_n * \chi_{[-b,b]} f\|_1.$$

(c) Conclude that $f = 0$ a.e. on $\mathbb{R}$.

***Exercise 21.7** Suppose f and g are in $L^2(\mathbb{R})$. Prove that $\lim_{|x|\to\infty} f*g(x) = 0$.

Exercise 21.8 If f and g are in $\mathscr{S}(\mathbb{R})$, show that

$$x^p(f*g)^{(q)}(x) = \sum_{j=0}^{p} \binom{p}{j} (x^{p-j}f) * (x^j g^{(q)})(x).$$

Lesson 22

The Fourier Transform on $L^2(\mathbb{R})$

In signal processing, $L^2(\mathbb{R})$ models the space of signals that are functions of a continuous variable (usually time) and that have finite energy. Until now, the Fourier transform has been defined only for integrable functions, and $L^2(\mathbb{R})$ is not included in $L^1(\mathbb{R})$. The purpose of this lesson is to extend the Fourier transform to $L^2(\mathbb{R})$; we will do this using results that have been established for $\mathscr{S}(\mathbb{R})$.

22.1 Extension of the Fourier transform

We proved in Section 19.2 that $\mathscr{F}$ is a continuous linear operator from $\mathscr{S}$ to $\mathscr{S}$. We are going to extend $\mathscr{F}$ to $L^2(\mathbb{R})$ using the following density result.

22.1.1 Proposition (density of $\mathscr{S}$ in $L^2(\mathbb{R})$)

$\mathscr{S}$ is a dense linear subspace of $L^2(\mathbb{R})$.

Proof. We need to show that $\mathscr{S} \subset L^2(\mathbb{R})$. For $f \in \mathscr{S}$, there exists an $A > 0$ such that $|(1+x^2)f(x)| \le A$ for all $x \in \mathbb{R}$. Hence

$$\int_{\mathbb{R}} |f(x)|^2\, dx \le A^2 \int_{\mathbb{R}} \frac{dx}{(1+x^2)^2} < +\infty.$$

The density follows from the density of $\mathscr{D}(\mathbb{R})$ in $L^2(\mathbb{R})$ (Section 21.3.5). □

The Fourier transform is an isometry from $\mathscr{S}$ to $\mathscr{S}$ in the L^2 norm.

22.1.2 Proposition (The Plancherel–Parseval equality)

For f and g in $\mathscr{S}$,

(i)
$$\int_{\mathbb{R}} \widehat{f}(\xi)\overline{\widehat{g}}(\xi)\, d\xi = \int_{\mathbb{R}} f(x)\overline{g}(x)\, dx,$$

(ii)
$$\int_{\mathbb{R}} |\widehat{f}(\xi)|^2 d\xi = \int_{\mathbb{R}} |f(x)|^2 dx.$$

Proof. The first equation follows directly from (17.4): Let $h(\xi) = \overline{\widehat{g}}(\xi)$. From (17.4),

$$\int_{\mathbb{R}} \widehat{f}(\xi)h(\xi)\,d\xi = \int_{\mathbb{R}} f(x)\widehat{h}(x)\,dx.$$

But $\overline{\widehat{g}}(\xi) = \displaystyle\int_{\mathbb{R}} e^{2i\pi\xi x}\overline{g}(x)\,dx = \overline{\mathscr{F}}\,\overline{g}(\xi)$. Thus $\widehat{h} = \overline{g}$, which proves (i). The second relation is derived from (i) by taking $f = g$. □

$\mathscr{F}$ is extended to $L^2(\mathbb{R})$ using the density of $\mathscr{S}$ in $L^2(\mathbb{R})$ and the fact that $L^2(\mathbb{R})$ is complete. This will be an application of the following result.

22.1.3 Proposition *Let E and F be two normed vector spaces. Assume that F is complete and that G is a dense linear subspace of E. If A is a continuous linear operator from G to F, then there exists has a unique continuous linear extension of A, denoted by $\widetilde{A}$, from E to F. Furthermore, the norm of $\widetilde{A}$ is equal to the norm of A.*

Proof. Let f be an element of E. Since G is dense in E, there is a sequence f_n in G such that $\lim_{n\to\infty} \|f - f_n\| = 0$. Being convergent, f_n is a Cauchy sequence, and since A is continuous,

$$\|Af_n - Af_m\| \le \|A\|\|f_n - f_m\|.$$

This shows that Af_n is a Cauchy sequence in F. Since F is complete, Af_n converges to some element g in F. It is easy to show that g does not depend on the sequence f_n that converges to f. Thus by letting $\widetilde{A}f = g$, $\widetilde{A}$ is well-defined on F.

$\widetilde{A}$ is linear by definition, and

$$\|\widetilde{A}f\| = \|g\| = \lim_{n\to\infty} \|Af_n\| \le \lim_{n\to\infty} \|A\| \cdot \|f_n\| = \|A\| \cdot \|f\|.$$

Thus $\|\widetilde{A}\| \le \|A\|$. Since $\widetilde{A}f = Af$ for all $f \in G$, we have

$$\|\widetilde{A}\| = \sup_{\substack{f\in E\\ f\neq 0}} \frac{\|\widetilde{A}f\|}{\|f\|} \ge \sup_{\substack{f\in G\\ f\neq 0}} \frac{\|Af\|}{\|f\|} = \|A\|,$$

and consequently $\|\widetilde{A}\| = \|A\|$. Finally, G being dense in E, it is clear that $\widetilde{A}$ is unique. □

$\mathscr{F}$ is an isometry on $\mathscr{S}$ in the L^2 norm. By applying the last result with $E = F = L^2(\mathbb{R})$ and $G = \mathscr{S}$ we have the next theorem.

22.1.4 Theorem *The Fourier transform $\mathscr{F}$ and its inverse $\overline{\mathscr{F}}$ extend uniquely to isometries on $L^2(\mathbb{R})$. Using the same notation for these extensions, we have the following results for all f and g in $L^2(\mathbb{R})$:*

(i) $\mathscr{F}\ \overline{\mathscr{F}}\ f = \overline{\mathscr{F}}\ \mathscr{F}\ f = f$ a.e.

(ii) $\displaystyle\int_{\mathbb{R}} f(x)\overline{g}(x)\,dx = \int_{\mathbb{R}} \mathscr{F}\ f(\xi)\overline{\mathscr{F}\ g}(\xi)\,d\xi.$

(iii) $\|f\|_2 = \|\mathscr{F}\ f\|_2.$

Proof. Having extended $\mathscr{F}$ from $\mathscr{S}$ to L^2, the other results follow by using the density of $\mathscr{S}$ in $L^2(\mathbb{R})$ and taking limits. □

We now examine some properties of this extension. The first result is that (17.4) is true in $L^2(\mathbb{R})$.

22.1.5 Proposition *If f and g are in $L^2(\mathbb{R})$, $\mathscr{F}\ f\cdot g$ and $f\cdot\mathscr{F}\ g$ are in $L^1(\mathbb{R})$, and*

$$\int_{\mathbb{R}} \mathscr{F}\ f(t)g(t)\,dt = \int_{\mathbb{R}} f(u)\mathscr{F}\ g(u)\,du. \tag{22.1}$$

Proof. We have just seen that $\mathscr{F}\ f$ is in $L^2(\mathbb{R})$, so that the product of $\mathscr{F}\ f$ and g is in $L^1(\mathbb{R})$. The same is true for f and $\mathscr{F}\ g$. Let f_n and g_n be sequences in $\mathscr{S}$ that tend to f and g respectively. Since $\mathscr{F}\ f_n = \widehat{f_n}$ and $\mathscr{F}\ g_n = \widehat{g}_n$, and since $\mathscr{S}\subset L^1(\mathbb{R})$, it follows from (17.4) that

$$\int_{\mathbb{R}} \mathscr{F}\ f_n(t)g_n(t)\,dt = \int_{\mathbb{R}} f_n(u)\mathscr{F}\ g_n(u)\,du.$$

Equation (22.1) follows by passing to the limit. □

22.1.6 Proposition *The Fourier transform defined on $L^1(\mathbb{R})$ and the one obtained by extension to $L^2(\mathbb{R})$ coincide on $L^1(\mathbb{R})\cap L^2(\mathbb{R})$. If $f\in L^2(\mathbb{R})$, then $\mathscr{F}\ f$ is the limit in $L^2(\mathbb{R})$ of the sequence g_n defined by*

$$g_n(\xi) = \int_{-n}^{n} e^{-2i\pi\xi x} f(x)\,dx.$$

Proof. Denote the Fourier transform on $L^1(\mathbb{R})$ by $\widehat{f}$ and that on $L^2(\mathbb{R})$ by $\mathscr{F}\ f$, as we have been doing. Take $f\in L^1(\mathbb{R})\cap L^2(\mathbb{R})$ and $\psi\in\mathscr{S}(\mathbb{R})$. Applying (17.4) and (22.1) we have

$$\int_{\mathbb{R}} \psi\widehat{f} = \int_{\mathbb{R}} \widehat{\psi} f = \int_{\mathbb{R}} \mathscr{F}\ \psi f = \int_{\mathbb{R}} \psi\mathscr{F}\ f.$$

Thus $\displaystyle\int_{\mathbb{R}} (\widehat{f} - \mathscr{F}\ f)\psi = 0$ for all $\psi\in\mathscr{S}(\mathbb{R})$. Since $\widehat{f} - \mathscr{F}\ f$ is in $L^1_{\text{loc}}(\mathbb{R})$, we conclude (Exercise 21.6) that $\widehat{f} = \mathscr{F}\ f$ a.e.

Let $f_n = f\chi_{[-n,n]}$. By dominated convergence (Theorem 14.1.1), we know that $\lim_{n\to\infty}\|f_n - f\|_2 = 0$. Since $f_n\in L^1(\mathbb{R})\cap L^2(\mathbb{R})$, we have $g_n = \widehat{f_n} = \mathscr{F}\ f_n$, and $\lim_{n\to\infty}\|\mathscr{F}\ f - g_n\|_2 = 0$ by the continuity of $\mathscr{F}$. □

22.1.7 Remark If $f \in L^2(\mathbb{R})$, then $\overline{\mathscr{F}} f$ is the limit in $L^2(\mathbb{R})$ of the sequence h_n defined by $h_n(\xi) = \int_{-n}^{n} e^{2i\pi\xi x} f(x)\, dx$.

22.1.8 Remark We will continue to denote the Fourier transform by $\widehat{f}$ or $\mathscr{F} f$. The meaning of these notations is now clear, depending on whether $f \in L^1(\mathbb{R})$ or $f \in L^2(\mathbb{R})$.

22.2 Application to the computation of certain Fourier transforms

When we know that the Fourier transform $\widehat{f}$ of a function $f \in L^1(\mathbb{R})$ is in $L^1(\mathbb{R})$, we can compute $\widehat{\widehat{f}}$ and obtain new transforms. This is the way we obtained the table in Section 18.2.2 from that in Section 17.3.3. However, for $f_\varepsilon(x) = e^{-\varepsilon a x} u(\varepsilon x)$ with $\varepsilon = \pm 1$ and $\mathrm{Re}(a) > 0$, $\widehat{f_\varepsilon}(\xi) = \dfrac{\varepsilon}{\varepsilon a + 2i\pi\xi}$, which is not in $L^1(\mathbb{R})$. It is, however, in $L^2(\mathbb{R})$, and in this case we can compute $\mathscr{F}(\widehat{f_\varepsilon})$.

22.2.1 Proposition

(i) *If* $f \in L^2(\mathbb{R})$, *then* $\mathscr{F}\mathscr{F} f = f_\sigma$ a.e.

(ii) *If* $f \in L^1(\mathbb{R}) \cap L^2(\mathbb{R})$, *then* $\mathscr{F}\widehat{f} = f_\sigma$ a.e.

Proof. To prove (i), we first show that $\mathscr{F} f = \overline{\mathscr{F}} f_\sigma$. Thus take a sequence f_n in $\mathscr{S}(\mathbb{R})$ such that $\lim_{n\to\infty} \|f - f_n\|_2 = 0$. We have $\mathscr{F} f_n = \overline{\mathscr{F}}(f_n)_\sigma$ by Proposition 18.2.1, and in the limit, $\mathscr{F} f = \overline{\mathscr{F}} f_\sigma$. If f is also in $L^1(\mathbb{R})$, then $\widehat{f} = \mathscr{F} f$, and this implies (ii). □

We are now able to compute the Fourier transform of f_ε.

22.2.2 The completion of Section 18.2.2

(i) $a \in \mathbb{C}$, $\mathrm{Re}(a) > 0$.

$$\frac{1}{a + 2i\pi x} \overset{\mathscr{F}}{\longmapsto} e^{a\xi} u(-\xi)$$

$$\frac{1}{a - 2i\pi x} \overset{\mathscr{F}}{\longmapsto} e^{-a\xi} u(\xi)$$

(ii)

$$\frac{\sin x}{x} \overset{\mathscr{F}}{\longmapsto} \pi\chi_{[-\frac{1}{2\pi}, \frac{1}{2\pi}]}(\xi)$$

With these results and those of Lessons 17 and 18, we can compute the Fourier transform of any rational function $P(x)/Q(x)$ by decomposing it into partial fractions.

22.3 The uncertainty principle

The purpose of this section is to develop the relation that exists between the localization of a signal and the localization of its spectrum.

Given a function $f : \mathbb{R} \to \mathbb{C}$ such that f, xf, and $\xi\widehat{f}$ are in $\mathrm{L}^2(\mathbb{R})$, we introduce the following definitions and notation:

$$\sigma_f^2 = \int_{\mathbb{R}} x^2 |f(x)|^2 \, dx \qquad \text{(energy dispersion of } f \text{ in time).}$$

$$\sigma_{\widehat{f}}^2 = \int_{\mathbb{R}} \xi^2 |\widehat{f}(\xi)|^2 \, d\xi \qquad \text{(energy dispersion in frequency).}$$

$$E_f = \int_{\mathbb{R}} |f(x)|^2 \, dx \qquad \text{(energy of } f\text{).}$$

The value Δt, defined by

$$\Delta t^2 = \frac{\sigma_f^2}{E_f},$$

is called the *effective duration* of the signal f; $\Delta\lambda$, defined by

$$\Delta\lambda^2 = \frac{\sigma_{\widehat{f}}^2}{E_f},$$

is called the *effective bandwidth*. The uncertainty principle is a relation between Δt and $\Delta\lambda$ that says that one cannot arbitrarily localize a signal in both time and frequency. This relation is

$$\Delta t \cdot \Delta\lambda \geq \frac{1}{4\pi}, \tag{22.2}$$

which is the content of the next result.

22.3.1 Proposition *Let $f : \mathbb{R} \to \mathbb{C}$ be a function in $C^1(\mathbb{R})$ such that f, f' and xf are in $L^2(\mathbb{R})$. Then*

$$\sigma_f \cdot \sigma_{\widehat{f}} \geq \frac{E_f}{4\pi}. \tag{22.3}$$

Proof. We assume the following two results (see Exercises 22.6 and 22.7):

(i) $\lim\limits_{|x| \to \infty} x|f(x)|^2 = 0$.

(ii) $\widehat{f'}(\xi) = 2i\pi\xi\widehat{f}(\xi)$.

The second formula will be proved in the more general context of tempered distributions (Proposition 31.2.4). Also, note that f being differentiable almost everywhere does not imply (ii) (take $f(x) = \chi_{[-1,1]}(x)$).

Using (ii) and Theorem 22.1.4(iii) we see that

$$\sigma_{\widehat{f}}^2 = \frac{1}{4\pi^2} \int_{\mathbb{R}} |\widehat{f'}(\xi)|^2 \, d\xi = \frac{1}{4\pi^2} \int_{\mathbb{R}} |f'(x)|^2 \, dx.$$

On the other hand, $(f\overline{f})' = f'\overline{f} + f\overline{f}'$, and

$$\begin{aligned}\left|\int_{\mathbb{R}} x(f(x)\overline{f}(x))'\,dx\right| &\leq \int_{\mathbb{R}} |xf'(x)\overline{f}(x)|\,dx + \int_{\mathbb{R}} |xf(x)\overline{f}'(x)|\,dx \\ &\leq \left(\int_{\mathbb{R}} |x\overline{f}(x)|^2 dx\right)^{1/2}\left(\int_{\mathbb{R}} |f'(x)|^2 dx\right)^{1/2} \\ &\quad + \left(\int_{\mathbb{R}} |xf(x)|^2 dx\right)^{1/2}\left(\int_{\mathbb{R}} |\overline{f}'(x)|^2 dx\right)^{1/2} \\ &= 2\left(\int_{\mathbb{R}} |f'(x)|^2 dx\right)^{1/2}\left(\int_{\mathbb{R}} |xf(x)|^2 dx\right)^{1/2} \\ &= 4\pi\sigma_{\widehat{f}} \cdot \sigma_f.\end{aligned}$$

But

$$\int_{\mathbb{R}} x\Big(f(x)\overline{f}(x)\Big)'\,dx = \Big[x|f(x)|^2\Big]_{-\infty}^{\infty} - \int_{\mathbb{R}} |f(x)|^2\,dx = -E_f,$$

since $\lim_{|x|\to\infty} x|f(x)|^2 = 0$. Thus $\sigma_f \cdot \sigma_{\widehat{f}} \geq E_f/(4\pi)$. □

The next proposition shows that a Gaussian signal has the minimum effective duration for a given effective bandwidth.

22.3.2 Proposition *Let the effective bandwidth $\Delta\lambda$ be fixed. Then the signal*

$$f(t) = \alpha e^{-(2\pi\Delta\lambda)^2 t^2}$$

minimizes the effective duration.

Proof. In the proof of Proposition 22.3.1 we used the Schwarz inequality to obtain $E_f \leq 4\pi\sigma_f \cdot \sigma_{\widehat{f}}$. One has equality in the real case when tf and f' are proportional. This implies that f is of the form $f(t) = \alpha e^{-ct^2}$, where $c > 0$ because $f \in L^2(\mathbb{R})$. We know that $\widehat{f}(\lambda) = \alpha\sqrt{\frac{\pi}{c}}e^{-\frac{\pi^2}{c}\lambda^2}$ (Section 17.3.4), so that

$$\Delta\lambda^2 = \frac{\int_{\mathbb{R}} \lambda^2|\widehat{f}(\lambda)|^2 d\lambda}{\int_{\mathbb{R}} |\widehat{f}(\lambda)|^2 d\lambda} = \frac{\int_{\mathbb{R}} \lambda^2 e^{-\frac{2\pi^2\lambda^2}{c}}\,d\lambda}{\int_{\mathbb{R}} e^{-\frac{2\pi^2\lambda^2}{c}}\,d\lambda} = \frac{\frac{c}{4\pi^2}\int_{\mathbb{R}} e^{-\frac{2\pi^2\lambda^2}{c}}\,d\lambda}{\int_{\mathbb{R}} e^{-\frac{2\pi^2\lambda^2}{c}}\,d\lambda} = \frac{c}{4\pi^2}.$$

Hence $c = 4\pi^2\Delta\lambda^2$. □

22.4 Exercises

*Exercise 22.1

(a) Let a and b be two real numbers with $a < b$. Compute the Fourier transform of $f(\xi) = \dfrac{\sin \pi(b-a)\xi}{\pi\xi} e^{-i\pi(a+b)\xi}$. Compute $\mathscr{F} f$ when $a = -b = -\dfrac{1}{2\pi}$.

(b) Compute $\mathscr{F} f$ using Proposition 22.1.6(ii).

Exercise 22.2 Let $f(x) = \dfrac{1}{a^2+x^2}$ with $a \in \mathbb{R}$, $a \neq 0$. Compute the Fourier transform of f two different ways: by direct computation (see Section 18.2.2) and by decomposing f into partial fractions and using the Fourier transform on $L^2(\mathbb{R})$ (see Section 22.2.2).

Exercise 22.3 Let $f(x) = \dfrac{x}{a^2+x^2}$ with $a \in \mathbb{R}$, $a \neq 0$. Show that $f \notin L^1(\mathbb{R})$. Compute the Fourier transform of f.

Exercise 22.4 In Exercise 18.5 on Shannon's formula, show that the assumptions on f imply that $f \in L^2(\mathbb{R})$ and that $\sum_{n=-\infty}^{+\infty} |f(na)|^2 < +\infty$.

Exercise 22.5 Evaluate $\displaystyle\int_{\mathbb{R}} \frac{\sin^2 x}{x^2}\,dx$ and $\displaystyle\int_{\mathbb{R}} \frac{dx}{(1+x^2)^2}\,dx$.

Hint: Use Theorem 22.1.4(iii). The results are π and $\dfrac{\pi}{2}$.

Exercise 22.6 Take $f \in C^1(\mathbb{R}) \cap L^2(\mathbb{R})$ and suppose in addition that f' and xf are in $L^2(\mathbb{R})$.

(a) Show that $x|f|^2$ and $(x|f|^2)'$ are in $L^1(\mathbb{R})$.

(b) Use this (and [Bre83] p. 130) to prove that $\displaystyle\lim_{|x|\to+\infty} x|f(x)|^2 = 0$.

**Exercise 22.7 Assume that $f \in C^1(\mathbb{R}) \cap L^2(\mathbb{R})$ with $f' \in L^2(\mathbb{R})$. We wish to establish the formula

$$\mathscr{F} f'(\xi) = 2i\pi\xi \mathscr{F} f(\xi) \quad \text{a.e.}$$

(a) Let $h_n(\xi) = \displaystyle\int_{-n}^{n} e^{2i\pi\xi x} f'(x)\,dx$. Show that h_n converges to $\mathscr{F} f'$ in $L^2(\mathbb{R})$. From this deduce the existence of a strictly increasing sequence $(n_k)_{k\in\mathbb{N}}$ such that h_{n_k} converges a.e. to $\mathscr{F} f'$ (use [Bre83] p. 58).

(b) By integrating by parts and using [Bre83] p. 130, show that there is a strictly increasing sequence (k_j) such that $h_{n_{k_j}}$ converges a.e. to $2i\pi\xi\mathscr{F} f$.

Lesson 23

Convolution and the Fourier Transform

The Fourier transform has the remarkable property that it interchanges convolution and multiplication. Formally, we have these relations:

$$\widehat{f * g}(\xi) = \widehat{f}(\xi) \cdot \widehat{g}(\xi),$$
$$\widehat{f \cdot g}(\xi) = \widehat{f} * \widehat{g}(\xi).$$

We will establish conditions under which these formulas are valid.

23.1 Convolution and the Fourier transform in $L^1(\mathbb{R})$

First we are going to complete a result about the Fourier transform. We saw (Theorem 18.1.1) that $\overline{\mathscr{F}}\,\widehat{f}(t) = f(t)$ at every point t where f is continuous when f and $\widehat{f}$ are in $L^1(\mathbb{R})$. In particular, if $f \in \mathscr{S}(\mathbb{R})$, then $\overline{\mathscr{F}}\,\widehat{f}(t) = f(t)$ for all $t \in \mathbb{R}$. We use the density of $\mathscr{S}(\mathbb{R})$ in $L^1(\mathbb{R})$ to prove the next result.

23.1.1 Proposition *If f and $\widehat{f}$ are in $L^1(\mathbb{R})$, then $\overline{\mathscr{F}}\,\widehat{f} = f$ a.e.*

Proof. Since $\mathscr{S}(\mathbb{R})$ is dense in $L^1(\mathbb{R})$, there exists a sequence f_n in $\mathscr{S}(\mathbb{R})$ such that $\lim_{n\to\infty} \|f - f_n\|_1 = 0$. As we have noted, $\overline{\mathscr{F}}\,\widehat{f_n}(t) = f_n(t)$ for all $n \in \mathbb{N}$ and all $t \in \mathbb{R}$. We are going to show that $\int_{\mathbb{R}} (f(t) - \overline{\mathscr{F}}\,\widehat{f}(t))\varphi(t)\,dt = 0$ for all $\varphi \in \mathscr{S}(\mathbb{R})$, and this implies (by Exercise 21.6 or by (27.6)) that $\overline{\mathscr{F}}\,\widehat{f} = f$ a.e.

Since $\widehat{f_n}$ is in $\mathscr{S}(\mathbb{R}) \subset L^1(\mathbb{R})$, we have by (17.4)

$$\int_{\mathbb{R}} f_n(t)\varphi(t)\,dt = \int_{\mathbb{R}} \overline{\mathscr{F}}\,\widehat{f_n}(t)\varphi(t)\,dt = \int_{\mathbb{R}} \widehat{f_n}(u)\overline{\mathscr{F}}\,\varphi(u)\,du.$$

It is clear that $\lim_{n\to\infty} \int_{\mathbb{R}} f_n(t)\varphi(t)\,dt = \int_{\mathbb{R}} f(t)\varphi(t)\,dt$, and with (17.3) we have $\lim_{n\to\infty} \|\widehat{f}_n - \widehat{f}\|_\infty = 0$. Thus

$$\lim_{n\to\infty} \int_{\mathbb{R}} \widehat{f}_n(u)\overline{\mathscr{F}}\varphi(u)\,du = \int_{\mathbb{R}} \widehat{f}(u)\overline{\mathscr{F}}\varphi(u)\,du.$$

Now, $\widehat{f} \in L^1(\mathbb{R})$ and $\mathscr{F}\varphi \in \mathscr{S}(\mathbb{R})$, so by (17.4),

$$\int_{\mathbb{R}} \widehat{f}(u)\overline{\mathscr{F}}\varphi(u)\,du = \int_{\mathbb{R}} \overline{\mathscr{F}}\widehat{f}(t)\varphi(t)\,dt.$$

Finally, for all $\varphi \in \mathscr{S}(\mathbb{R})$,

$$\int_{\mathbb{R}} f(t)\varphi(t)\,dt = \int_{\mathbb{R}} \overline{\mathscr{F}}\widehat{f}(t)\varphi(t)\,dt,$$

and this proves the result. □

This result implies that if f and $\widehat{f}$ are in $L^1(\mathbb{R})$, then f is continuous, or more precisely, the equivalence class to which f belongs contains a continuous representative, namely, $\overline{\mathscr{F}}\widehat{f}$.

23.1.2 Proposition *Given f and g in $L^1(\mathbb{R})$, we have*

(i) $\widehat{f * g}(\xi) = \widehat{f}(\xi) \cdot \widehat{g}(\xi)$ *for all* $\xi \in \mathbb{R}$.

(ii) *If in addition $\widehat{f}$ and $\widehat{g}$ are in $L^1(\mathbb{R})$, then*

$$\widehat{f \cdot g}(\xi) = \widehat{f} * \widehat{g}(\xi) \quad \textit{for all} \quad \xi \in \mathbb{R}.$$

Proof.

(i) $f * g$ is in $L^1(\mathbb{R})$ by §20.2.1. The computation of $\widehat{f * g}(\xi)$ is a direct application of Fubini's theorem:

$$\begin{aligned}\int_{\mathbb{R}} e^{-2i\pi\xi x} f * g(x)\,dx &= \int_{\mathbb{R}} e^{-2i\pi\xi x}\Big(\int_{\mathbb{R}} f(x-t)g(t)\,dt\Big)\,dx \\ &= \int_{\mathbb{R}} g(t)\Big(\int_{\mathbb{R}} e^{-2i\pi\xi x} f(x-t)\,dx\Big)\,dt \\ &= \int_{\mathbb{R}} g(t)e^{-2i\pi\xi t}\widehat{f}(\xi)\,dt = \widehat{g}(\xi)\cdot\widehat{f}(\xi).\end{aligned}$$

(ii) Note that (i) is true for $\overline{\mathscr{F}}$ by changing i to $-i$. Since $\widehat{f}$ and $\widehat{g}$ are in $L^1(\mathbb{R})$, we can apply (i) and Proposition 23.1.1:

$$\overline{\mathscr{F}}(\widehat{f} * \widehat{g})(x) = \overline{\mathscr{F}}\widehat{f}(x)\cdot\overline{\mathscr{F}}\widehat{g}(x) = f(x)g(x) \quad \text{a.e.}$$

Note that fg is in $L^1(\mathbb{R})$ because both f and g are in $L^1(\mathbb{R}) \cap C(\mathbb{R})$. Taking the Fourier transform of both sides of the last equation shows that $\widehat{f} * \widehat{g}(\xi) = \widehat{f \cdot g}(\xi)$ for all $\xi \in \mathbb{R}$. □

This result is particularly important for functions in $\mathscr{S}(\mathbb{R})$.

23.1.3 Proposition *If f and g are in $\mathscr{S}(\mathbb{R})$, then*

(i) $\widehat{f * g} = \widehat{f} \cdot \widehat{g}$;

(ii) $\widehat{f \cdot g} = \widehat{f} * \widehat{g}$.

Proof. Proposition 23.1.2 applies directly because $\mathscr{S}$ is invariant under the Fourier transform. □

23.2 Convolution and the Fourier transform in $L^2(\mathbb{R})$

We extended the Fourier transform from $\mathscr{S}$ to $L^2(\mathbb{R})$ in Lesson 22. The convolution is a continuous operator from $L^2(\mathbb{R}) \times L^2(\mathbb{R})$ to $L^\infty \cap C^0$ (Proposition 20.3.1).

23.2.1 Proposition *Given f and g in $L^2(\mathbb{R})$, we have*

(i) $f * g(t) = \overline{\mathscr{F}}(\widehat{f} \cdot \widehat{g})(t)$ *for all t in $\mathbb{R}$;*

(ii) $\widehat{f \cdot g} = \widehat{f} * \widehat{g}$ *for all t in $\mathbb{R}$.*

Proof.

(i) We establish the result using the density of $\mathscr{S}$ in $L^2(\mathbb{R})$ and applying Proposition 23.1.2(i). Thus let f_n and g_n be two sequences in $\mathscr{S}$ such that

$$\lim_{n\to\infty} \|f - f_n\|_2 = 0 \quad \text{and} \quad \lim_{n\to\infty} \|g - g_n\|_2 = 0.$$

We see that $f_n * g_n = \overline{\mathscr{F}}(\widehat{f_n} \cdot \widehat{g_n})$ by taking the inverse Fourier transform of both sides of Proposition 23.1.2(i). On the other hand,

$$\begin{aligned}\|\widehat{f} \cdot \widehat{g} - \widehat{f_n} \cdot \widehat{g_n}\|_1 &\le \|\widehat{f} - \widehat{f_n}\|_2 \|\widehat{g}\|_2 + \|\widehat{f_n}\|_2 \|\widehat{g} - \widehat{g_n}\|_2 \\ &= \|f - f_n\|_2 \|g\|_2 + \|f_n\|_2 \|g - g_n\|_2\end{aligned}$$

by Theorem 22.1.4(iii), and hence $\lim_{n\to\infty} \|\widehat{f} \cdot \widehat{g} - \widehat{f_n} \cdot \widehat{g_n}\|_1 = 0$. By applying the Riemann–Lebesgue theorem (Theorem 17.1.3) to the inverse Fourier transform, we see that $\overline{\mathscr{F}}(\widehat{f_n} \cdot \widehat{g_n})$ tends to $\overline{\mathscr{F}}(\widehat{f} \cdot \widehat{g})$ uniformly on $\mathbb{R}$.

The last step is to determine the limit of $f_n * g_n$. From (20.2) we have

$$\|f * g - f_n * g_n\|_\infty \le \|f - f_n\|_2 \|g\|_2 + \|f_n\|_2 \|g - g_n\|_2;$$

thus $f_n * g_n$ converges uniformly to $f * g$, which is continuous. We conclude that

$$f * g(t) = \overline{\mathscr{F}}(\widehat{f} \cdot \widehat{g})(t)$$

for all t in $\mathbb{R}$.

(ii) The proof is similar to that of (i) and is left as an exercise. □

23.2.2 Remark With reference to the last proposition, note that the formula $\widehat{f * g} = \widehat{f} \cdot \widehat{g}$ does not make sense a priori, since $f * g$ is only in $L^\infty(\mathbb{R})$. This formula is true whenever $f * g$ is in $L^1(\mathbb{R})$.

When $f \in L^2(\mathbb{R})$ and $g \in L^1(\mathbb{R})$, the convolution and the Fourier transform are well defined, and we have the next result.

23.2.3 Proposition *If $f \in L^2(\mathbb{R})$ and $g \in L^1(\mathbb{R})$, then $\widehat{f} \cdot \widehat{g}$ is in $L^2(\mathbb{R})$ and $f * g = \overline{\mathscr{F}}(\widehat{f} \cdot \widehat{g})$, with equality in $L^2(\mathbb{R})$.*

Proof. We proceed as in Proposition 23.2.1. Take two sequences f_n and g_n in $\mathscr{S}$ such that $\lim_{n\to\infty} \|f - f_n\|_2 = 0$ and $\lim_{n\to\infty} \|g - g_n\|_1 = 0$. We know that $\overline{\mathscr{F}}(\widehat{f_n} \cdot \widehat{g_n}) = f_n * g_n$; we first study the convergence of $\widehat{f_n} \cdot \widehat{g_n}$.

Since $\widehat{f}$ is in $L^2(\mathbb{R})$ and $\widehat{g}$ is in $L^\infty(\mathbb{R})$, $\widehat{f} \cdot \widehat{g}$ is in $L^2(\mathbb{R})$ and

$$\begin{aligned}\|\widehat{f} \cdot \widehat{g} - \widehat{f_n} \cdot \widehat{g_n}\|_2 &\leq \|\widehat{f} - \widehat{f_n}\|_2 \|\widehat{g}\|_\infty + \|\widehat{f_n}\|_2 \|\widehat{g_n} - \widehat{g}\|_\infty \\ &= \|f - f_n\|_2 \|\widehat{g}\|_\infty + \|f_n\|_2 \|\widehat{g_n} - \widehat{g}\|_\infty.\end{aligned}$$

Since $\lim_{n\to\infty} \|g - g_n\|_1 = 0$ implies $\lim_{n\to\infty} \|\widehat{g} - \widehat{g_n}\|_\infty = 0$, $\widehat{f_n} \cdot \widehat{g_n}$ converges to $\widehat{f} \cdot \widehat{g}$ in $L^2(\mathbb{R})$. Consequently, $\overline{\mathscr{F}}(\widehat{f_n} \cdot \widehat{g_n})$ tends to $\overline{\mathscr{F}}(\widehat{f} \cdot \widehat{g})$ in $L^2(\mathbb{R})$. Finally, we must examine the convergence of $f_n * g_n$. The convolution is continuous from $L^2(\mathbb{R}) * L^1(\mathbb{R})$ to $L^2(\mathbb{R})$ (Proposition 20.3.2). Hence $f_n * g_n$ converges to $f * g$ in $L^2(\mathbb{R})$, and we have $f * g = \overline{\mathscr{F}}(\widehat{f} \cdot \widehat{g})$ in $L^2(\mathbb{R})$. □

23.3 Convolution and the Fourier transform: Summary

23.3.1 The Fourier transform on $L^1(\mathbb{R})$

The Fourier transform of a function in $L^1(\mathbb{R})$ is denoted by $\widehat{f}$ or $\mathscr{F} f$.

Riemann–Lebesgue theorem:

$$\mathscr{F} : L^1(\mathbb{R}) \to L^\infty(\mathbb{R}) \cap C^0(\mathbb{R}),$$
$$\lim_{|x|\to\infty} \mathscr{F} f(x) = 0.$$

Exchange formula: $\displaystyle\int_{\mathbb{R}} f(t)\widehat{g}(t)\,dt = \int_{\mathbb{R}} \widehat{f}(u) g(u)\,du.$

Derivation formulas: $\widehat{f}^{(k)} = [(-2i\pi x)^k f]^\wedge; \quad \widehat{f^{(k)}} = (2i\pi\xi)^k \widehat{f}.$

Translation formulas: $\tau_a \widehat{f} = [e^{2i\pi a x} f]^\wedge; \quad \widehat{\tau_a f} = e^{-2i\pi\xi a} \widehat{f}.$

Properties:

$$\begin{array}{llll} f & \text{even} & \Longrightarrow & \widehat{f} \ \text{even}, \\ f & \text{odd} & \Longrightarrow & \widehat{f} \ \text{odd}, \\ f & \text{real, even} & \Longrightarrow & \widehat{f} \ \text{real, even}, \\ f & \text{real, odd} & \Longrightarrow & \widehat{f} \ \text{imaginary, odd}. \end{array}$$

Fourier transforms:

(i) $a \in \mathbb{C}$, $\mathrm{Re}(a) > 0$, $\varepsilon = \pm 1$, $k = 0, 1, 2, \ldots$;

$$\frac{x^k}{k!} e^{-\varepsilon a x} u(\varepsilon x) \xrightarrow{\mathscr{F}} \frac{\varepsilon}{(\varepsilon a + 2i\pi\xi)^{k+1}},$$

$$e^{-a|x|} \xrightarrow{\mathscr{F}} \frac{2a}{a^2 + 4\pi^2\xi^2},$$

$$\mathrm{sign}(x) e^{-a|x|} \xrightarrow{\mathscr{F}} \frac{-4i\pi\xi}{a^2 + 4\pi^2\xi^2}.$$

(ii) $a \in \mathbb{R}$, $a > 0$;

$$e^{-ax^2} \xrightarrow{\mathscr{F}} \sqrt{\frac{\pi}{a}} e^{-\frac{\pi^2}{a}\xi^2}.$$

23.3.2 The inverse Fourier transform

The operator $\overline{\mathscr{F}}$ is the conjugate of $\mathscr{F}$, and $\mathscr{F}^{-1} = \overline{\mathscr{F}}$. If f and $\widehat{f}$ are in $L^1(\mathbb{R})$, then

$$\overline{\mathscr{F}}\, \widehat{f}(t) = f(t) \quad \text{a.e.},$$
$$\mathscr{F}\, \widehat{f} = f_\sigma.$$

This last formula leads one to find new Fourier transforms (see the table in Section 18.2.2).

The space $\mathscr{S}(\mathbb{R})$ of functions in C^∞ that decay rapidly is dense in $L^p(\mathbb{R})$ for $1 \le p < +\infty$.

$\mathscr{F}$ is linear, 1-to-1 onto, and bicontinuous from $\mathscr{S}(\mathbb{R})$ to $\mathscr{S}(\mathbb{R})$.

23.3.3 The Fourier transform on $L^2(\mathbb{R})$

$\mathscr{F}$ is an isometry from $L^2(\mathbb{R})$ onto $L^2(\mathbb{R})$: $\|f\|_2 = \|\mathscr{F} f\|_2$. In particular, $\mathscr{F}$ preserves the scalar product in $L^2(\mathbb{R})$:

$$\int_{\mathbb{R}} f(x)\overline{g}(x)\, dx = \int_{\mathbb{R}} \widehat{f}(\xi)\overline{\widehat{g}}(\xi)\, d\xi.$$

23.3.4 Convolution

f	g	$f * g$	Continuity
L^1	L^1	L^1	$\lVert f * g\rVert_1 \leq \lVert f\rVert_1 \lVert g\rVert_1$
L^1	L^∞	$L^\infty \cap C^0$	$\lVert f * g\rVert_\infty \leq \lVert f\rVert_1 \lVert g\rVert_\infty$
L^2	L^2	$L^\infty \cap C^0$	$\lVert f * g\rVert_\infty \leq \lVert f\rVert_2 \lVert g\rVert_2$
L^2	L^1	L^2	$\lVert f * g\rVert_2 \leq \lVert f\rVert_2 \lVert g\rVert_1$
$\mathscr{S}$	$\mathscr{S}$	$\mathscr{S}$	

Regularization: If $f \in L^p(\mathbb{R})$, $1 \leq p < +\infty$, then

$$\lim_{n \to \infty} \lVert \rho_n * f - f \rVert_p = 0.$$

23.3.5 Convolution and the Fourier transform

$$\left.\begin{array}{l} f \in L^1(\mathbb{R}) \\ g \in L^1(\mathbb{R}) \end{array}\right\} \implies \widehat{f * g}(\xi) = \widehat{f}(\xi) \cdot \widehat{g}(\xi) \qquad \text{for all } \xi \in \mathbb{R},$$

$$\left.\begin{array}{l} f, \widehat{f} \in L^1(\mathbb{R}) \\ g, \widehat{g} \in L^1(\mathbb{R}) \end{array}\right\} \implies \widehat{f \cdot g}(\xi) = \widehat{f} * \widehat{g}(\xi) \qquad \text{for all } \xi \in \mathbb{R},$$

$$\left.\begin{array}{l} f \in L^2(\mathbb{R}) \\ g \in L^2(\mathbb{R}) \end{array}\right\} \implies \left\{\begin{array}{l} f * g(t) = \overline{\mathscr{F}}(\widehat{f} \cdot \widehat{g})(t) \\ \widehat{f \cdot g}(t) = \widehat{f} * \widehat{g}(t) \end{array}\right. \qquad \text{for all } t \in \mathbb{R},$$

$$\left.\begin{array}{l} f \in L^2(\mathbb{R}) \\ g \in L^1(\mathbb{R}) \end{array}\right\} \implies f * g(t) = \overline{\mathscr{F}}(\widehat{f} \cdot \widehat{g})(t) \qquad \text{for a.e. } t \in \mathbb{R}.$$

23.4 Exercises

Exercise 23.1 Compute $\widehat{f * f}$ for $f = \chi_{[0,1]}$.

Exercise 23.2 Compute $f * f$ when $f(t) = \dfrac{\sin 2\pi\lambda t}{\pi t}$ with $\lambda > 0$.

Exercise 23.3 Compute the Fourier transform of

$$g(x) = \frac{\sin^2 x}{x^2}$$

and deduce that

$$\int_{\mathbb{R}} \frac{\sin^2 x}{x^2}\, dx = \pi.$$

Exercise 23.4 Let $g(x) = e^{-\pi x^2}$. Compute $g * g$.

Exercise 23.5 Let $f_a(x) = \dfrac{2a}{a^2 + 4\pi^2 x^2}$ with $a \in \mathbb{C}$ and $\mathrm{Re}(a) > 0$. If $b \in \mathbb{C}$ and $\mathrm{Re}(b) > 0$, compute $f_a * f_b$.

Hint: Use the Fourier transform to show that $f_a * f_b = f_{a+b}$.

Exercise 23.6 If f and g are in $L^2(\mathbb{R})$, show that $f * g$ is in $C^0(\mathbb{R})$ and that

$$\lim_{|x| \to \infty} f * g(x) = 0.$$

Chapter VII

Analog Filters

Lesson 24

Applications to Analog Filters Governed by a Differential Equation

The tools we have just developed (convolution and the Fourier transform for functions) are going to be used to study analog filters that are governed by a linear differential equation with constant coefficients,

$$\sum_{k=0}^{q} b_k g^{(k)} = \sum_{j=0}^{p} a_j f^{(j)}, \quad a_p \cdot b_q \neq 0, \tag{24.1}$$

where f is the input and $g = A(f)$ is the output. Other conditions must be given to eliminate ambiguity among the possible solutions of (24.1).

24.1 The case where the input and output are in $\mathscr{S}$

This case is very special. The input has no reason to be so regular, but we will see that this is a step toward more general cases.

We assume that $f \in \mathscr{S}$ and look for a solution g in $\mathscr{S}$. If such a g exists, we can take the Fourier transform of both sides of (24.1). Thus

$$\sum_{k=0}^{q} b_k (2i\pi\lambda)^k \widehat{g}(\lambda) = \sum_{j=0}^{p} a_j (2i\pi\lambda)^j \widehat{f}(\lambda). \tag{24.2}$$

Consider the two polynomials

$$P(x) = \sum_{j=0}^{p} a_j x^j \quad \text{and} \quad Q(x) = \sum_{k=0}^{q} b_k x^k$$

and assume that the rational function $P(x)/Q(x)$ has no poles on the imaginary axis. Then $P(2i\pi\lambda)/Q(2i\pi\lambda)$ has no poles for real λ, and (24.2) is equivalent to

$$\widehat{g}(\lambda) = \frac{P(2i\pi\lambda)}{Q(2i\pi\lambda)} \widehat{f}(\lambda). \tag{24.3}$$

This equality completely determines g in $\mathscr{S}$, if it exists, and thus proves the uniqueness of a solution of (24.1) in $\mathscr{S}$. The existence of a solution also follows from (24.3), since the function

$$G(\lambda) = \frac{P(2i\pi\lambda)}{Q(2i\pi\lambda)}\widehat{f}(\lambda)$$

is in $\mathscr{S}$ whenever f is in $\mathscr{S}$. By applying Theorem 19.3.1, we see that $g = \mathscr{F}^{-1}(G)$ is a solution of (24.1) in $\mathscr{S}$.

24.1.1 Proposition *If $P(x)/Q(x)$ has no poles on the imaginary axis and if f is in $\mathscr{S}$, then (24.1) has a unique solution $g \in \mathscr{S}$. In this case, the system*

$$A : \mathscr{S} \to \mathscr{S},$$
$$f \mapsto g$$

is a filter.

Proof. We have proved the first part of the result and thus need only to show that A is a filter on $\mathscr{S}$. The linearity and invariance present no difficulty. To prove continuity in the topology of $\mathscr{S}$, suppose that a sequence f_n tends to 0 in $\mathscr{S}$. Then $\widehat{f_n}$ tends to 0 in $\mathscr{S}$, as does $\widehat{g}_n$ given by (24.3). Thus g_n tends to 0 by Theorem 19.3.1. □

The differential equation (24.1) has a unique solution without initial conditions being specified. This is because we require the solution g to be in $\mathscr{S}$, which means that g and all of its derivatives vanish at infinity.

We assume in what follows that P/Q has no poles on the imaginary axis. Also, note that $P \not\equiv 0$, since we assume that $a_p \neq 0$.

24.1.2 The output expressed as a convolution ($p < q$)

If we assume that $\deg P < \deg Q$, then the transfer function

$$H(\lambda) = \frac{P(2i\pi\lambda)}{Q(2i\pi\lambda)} \tag{24.4}$$

is in $L^2(\mathbb{R}) \cap L^\infty(\mathbb{R})$. By decomposing this rational function into partial fractions, we see from Sections 18.2.2 and 22.2.2 that it has an inverse Fourier transform $h = \mathscr{F}^{-1}H$ that is bounded, rapidly decreasing, continuous except perhaps at the origin, and satisfies (24.3),

$$\widehat{g} = \widehat{h} \cdot \widehat{f},$$

which by Proposition 23.2.1(i) implies that

$$g = h * f. \tag{24.5}$$

This is the same kind of formula that we obtained in Section 2.4 for the RC filter. The response is the convolution of the input with a fixed function h that is called the *impulse response*. Note that if $\deg P \geq \deg Q$, the computations we have just made no longer make sense.

24.2 Generalized solutions of the differential equation

The formula $g = h * f$, obtained when f is in $\mathscr{S}$, makes sense in the following more general cases.

24.2.1 If f is in $L^1(\mathbb{R})$, then g is in $L^1(\mathbb{R}) \cap L^2(\mathbb{R}) \cap L^\infty(\mathbb{R})$ (Propositions 20.2.1, 20.3.1, and 20.3.1) and

$$\begin{aligned} \|g\|_1 &\leq \|h\|_1 \ \|f\|_1, \\ \|g\|_2 &\leq \|h\|_2 \ \|f\|_1, \\ \|g\|_\infty &\leq \|h\|_\infty \|f\|_1. \end{aligned} \tag{24.6}$$

24.2.2 If f is in $L^2(\mathbb{R})$, then g is in $L^2(\mathbb{R})$, it is bounded and continuous (Proposition 20.3.1), it tends to 0 at infinity (Proposition 23.2.1(i)), and

$$\begin{aligned} \|g\|_2 &\leq \|h\|_1 \|f\|_2, \\ \|g\|_\infty &\leq \|h\|_2 \|f\|_2. \end{aligned} \tag{24.7}$$

24.2.3 If f is in $L^\infty(\mathbb{R})$, then g is also bounded and (proposition 20.3.1)

$$\|g\|_\infty \leq \|h\|_1 \|f\|_\infty. \tag{24.8}$$

The system A defined in Proposition 24.1.1 in continuous from $L^\infty(\mathbb{R})$ to $L^\infty(\mathbb{R})$, and thus it is a filter. Similarly, (24.6) and (24.7) show that A is continuous from $L^1(\mathbb{R})$ to $L^p(\mathbb{R})$ ($p = 1, 2, \infty$), and from $L^2(\mathbb{R})$ to $L^q(\mathbb{R})$ ($q = 2, \infty$).

24.2.4 Definition The response of a filter to the unit step function is called the step response of the filter. This response, h_1, is well-defined as a generalized solution of (24.1). It is bounded by (24.8) and is given by

$$h_1(t) = h * u(t) = \int_{-\infty}^{t} h(s)\, ds. \tag{24.9}$$

24.3 The impulse response when $\deg P < \deg Q$

The impulse response $h = \mathscr{F}^{-1} H$ is computed by decomposing H into partial fractions. The poles of P/Q are assumed to lie off the imaginary axis. There are two cases to consider: P/Q has only simple poles or P/Q has multiple poles.

24.3.1 The case where $P(x)/Q(x)$ has only simple poles

In this case, H can be decomposed in the form

$$H(\lambda) = \sum_{k=0}^{q} \frac{\beta_k}{2i\pi\lambda - z_k}, \tag{24.10}$$

where $z_1, \dots, z_q$ are the poles. From Section 22.2.2, read for $\mathscr{F}^{-1}$, we conclude that

$$h(t) = \Big(\sum_{k \in K_-} \beta_k e^{z_k t} \Big) u(t) - \Big(\sum_{k \in K_+} \beta_k e^{z_k t} \Big) u(-t), \tag{24.11}$$

where we have defined

$$K_- = \big\{k \in \{1, 2, \dots, q\} \mid \mathrm{Re}(z_k) < 0\big\},$$

$$K_+ = \big\{k \in \{1, 2, \dots, q\} \mid \mathrm{Re}(z_k) > 0\big\}.$$

24.3.2 The case where $P(x)/Q(x)$ has multiple poles

Let $z_1, z_2, \dots, z_l$ the poles and let $m_1, m_2, \dots, m_l$ be their multiplicities. Then we can write H as

$$H(\lambda) = \sum_{k=1}^{l} \sum_{m=1}^{m_k} \frac{\beta_{k,m}}{(2i\pi\lambda - z_k)^m}. \tag{24.12}$$

By using the results in Section 17.3.4, we see that

$$h(t) = \Big(\sum_{k \in K_-} P_k(t) e^{z_k t} \Big) u(t) - \Big(\sum_{k \in K_+} P_k(t) e^{z_k t} \Big) u(-t), \tag{24.13}$$

where

$$P_k(t) = \sum_{m=1}^{m_k} \beta_{k,m} \frac{t^{m-1}}{(m-1)!}.$$

24.3.3 The case of purely imaginary poles

What we have done so far does not allow us to treat an equation like

$$g'' + \omega^2 g = f,$$

where $P(x)/Q(x) = 1/(x^2 + \omega^2)$ has two poles are on the imaginary axis. In this case h is a sinusoid and the Fourier transform of H (when H is considered to be a function) is no longer defined. This problem will be resolved in Section 35.2.3 in the context of distributions.

24.3.4 The case where $\deg P = \deg Q$

Take for example the equation

$$g'' - \omega^2 g = f''.$$

Again, what we have done so far does not apply. Nevertheless, we can still manage to solve the equation. Changing the unknown function to $g_0 = g - f$ lowers the order of the right-hand side:

$$g_0'' - \omega^2 g_0 = \omega^2 f.$$

Then we have $g_0 = h_0 * f$ and $g = f + h_0 * f$. This is no longer a convolution like (24.5), but it will serve the same purpose. On the other hand, it is clear that we can obtain g as

$$g = h_1 * f' \quad \text{or} \quad g = h_2 * f''.$$

In the general case, we change the unknown function to $g_0 = g + \lambda f$ and find that

$$\sum_{k=0}^{q} b_k g_0^{(k)} = \sum_{k=0}^{q} (a_k - \lambda b_k) f^{(k)}.$$

Taking $\lambda = a_q/b_q$ reduces the degree of the right-hand side and brings us back to the case $p < q$. We can then write

$$g = \lambda f + h_0 * f. \tag{24.14}$$

(Note that it is possible that $h_0 \equiv 0$; this happens when $P(x) = \lambda Q(x)$ for all x (see Exercise 24.2).) The representation (24.14) leads to estimates like those given in Section 24.2. In Section 35.2 we will give an expression for g as a convolution without the condition $p < q$, but in this case h will be a distribution.

24.3.5 Summary

When P/Q has no poles on the imaginary axis and $\deg P \leq \deg Q$, a unique generalized solution of (24.1) can be defined under the sole condition that $f \in L^1(\mathbb{R}) \cup L^2(\mathbb{R}) \cup L^\infty(\mathbb{R})$. $A(f) = g$ is a filter that we will call the *generalized filter* A associated with (24.1). The output g is given by $g = h * f$ or possibly by a formula like (24.14).

24.4 Stability

24.4.1 Definition An analog system $A : X \to Y$ is said to be stable if there exists an $M > 0$ such that $\|Af\|_\infty \leq M\|f\|_\infty$ for all $f \in L^\infty(\mathbb{R}) \cap X$.

By (24.8), the generalized filter A is stable when $\deg P < \deg Q$. If $\deg P = \deg Q$, the system is still stable from what we have seen in Section 24.3.4.

24.4.2 Theorem *The generalized filter governed by equation (24.1), whose output g is defined by (24.5) or (24.14), is stable when $\deg P \leq \deg Q$ and the poles of $P(x)/Q(x)$ are not on the imaginary axis.*

$$\left.\begin{array}{l}\deg P \leq \deg Q \text{ and } P/Q \text{ has no} \\ \text{poles on the imaginary axis.}\end{array}\right\} \implies \left\{\begin{array}{l}\text{The generalized filter} \\ A \text{ is stable.}\end{array}\right.$$

24.5 Realizable systems

24.5.1 Definition A system is said to be realizable (or causal) if the equality of two input signals for $t < t_0$ implies the equality of the two output signals for $t < t_0$ (see Section 2.1.2).

For a filter, which is by definition linear and invariant, this condition becomes the following: For all $t_0 \in \mathbb{R}$,

$$f(t) = 0 \text{ for } t < t_0 \quad \implies \quad Af(t) = 0 \text{ for } t < t_0.$$

We will see that the realizability of the filter defined in Section 24.3.5 depends simply on its impulse response or on the position of the poles.

Assume that $\deg P \leq \deg Q$.

$$\begin{array}{c}\text{The generalized filter} \\ A \text{ is realizable.}\end{array} \iff \operatorname{supp}(h) \subset [0, +\infty).$$

If $\operatorname{supp}(h) \subset [0, +\infty)$, the output

$$g(t) = \int_0^{+\infty} h(s) f(t - s)\, ds$$

is 0 for $t < t_0$ when $f(t) = 0$ for $t < t_0$. We prove the other direction by contradiction. Thus suppose that there is a $t_1 < 0$ such that $h(t_1) > 0$. Since h is continuous at t_1, there is an interval (a, b) such that $b < 0$ and $a < t_1 < b$ implies that $h(t) > 0$. For the causal input signal

$$f(t) = \chi_{[0,b-a]}(t),$$

we have an output signal

$$g(t) = \int_{t-b+a}^{t} h(s)\,ds$$

with $g(b) > 0$. This contradicts the fact that A is causal. This is the proof when $\deg P < \deg Q$. In case $\deg P = \deg Q$, one uses the trick introduced in Section 24.3.4.

From formulas (24.11) and (24.13) we see that $\mathrm{supp}(h) \subset [0, +\infty)$ if and only if K_+ is empty. Thus if $\deg P \leq \deg Q$, we have the following result:

$$\left.\begin{array}{l}\text{The generalized filter}\\ A \text{ is realizable.}\end{array}\right\} \iff \left\{\begin{array}{l}\text{The poles of P/Q are located to}\\ \text{the left of the imaginary axis.}\end{array}\right.$$

24.5.2 Theorem *For the generalized filter defined in Section 24.3.5 with* $\deg P \leq \deg Q$ *to be realizable, it is necessary and sufficient that all the poles of* P/Q *have strictly negative real parts.*

For $\deg P = \deg Q$, the property results from the fact that the output can be written as $g = \lambda f + h_0 * f$, $\lambda \in \mathbb{C}$. In summary, if $\deg P \leq \deg Q$, we have the following result:

$$\left.\begin{array}{l}\text{The real parts of all the}\\ \text{poles of } P/Q \text{ are negative.}\end{array}\right\} \iff \left\{\begin{array}{l}\text{The generalized filter } A\\ \text{is realizable and stable.}\end{array}\right.$$

24.6 Gain and response time

The gain of a filter of the type described in Section 24.3.5 is defined to be the constant

$$K = H(0).$$

From (24.9) we see that

$$K = \widehat{h}(0) = \lim_{t \to +\infty} h_1(t),$$

which is the ratio between the asymptotic value of the step response and the height of the input step function. The response time is defined to be the time it takes the step response to reach and maintain a certain percentage of its limit, in general 95%:

$$t_r = \min\left\{ t \;\middle|\; \left|\frac{h_1(t) - K}{K}\right| \leq \frac{5}{100} \text{ for all } t > t_r \right\}.$$

24.7 The Routh criterion

The stability of a system depends on the location of the roots of the characteristic equation $Q(x) = 0$ in the complex plane. We note that it is not necessary to compute the roots of this equation to determine whether all their real parts are negative. It is possible to use the *Routh criterion*: The roots of the equation

$$a_0 x^p + a_1 x^{p-1} + \cdots + a_{p-1} x + a_p = 0$$

with real coefficients will all have strictly negative real parts if and only if the elements of the first column of the following array all have the same sign:

$$\begin{bmatrix} a_0 & a_2 & a_4 & a_6 & \cdots \\ a_1 & a_3 & a_5 & a_7 & \cdots \\ b_1^1 & b_2^1 & b_3^1 & b_4^1 & \cdots \\ b_1^2 & b_2^2 & b_3^2 & b_4^2 & \cdots \\ \vdots & \vdots & \vdots & \vdots & \end{bmatrix}$$

with

$$b_k^1 = \frac{a_1 a_{2k} - a_0 a_{2k+1}}{a_1},$$

$$b_k^2 = \frac{b_1^1 a_{2k+1} - a_1 b_{k+1}^1}{b_1^1},$$

etc., for $k = 1, 2, \ldots$.

EXAMPLES

(a) $Q(x) = x^4 + 3x^3 + 6x^2 + 9x + 12$.

The Routh matrix is

$$\begin{bmatrix} 1 & 6 & 12 \\ 3 & 9 & 0 \\ 3 & 12 & \\ -3 & & \end{bmatrix},$$

and thus the real parts of the roots are not all negative.

(b) $Q(x) = x^3 + (2k+1)x^2 + (k+1)^2 x + k^2 + 1 = 0$.

The Routh matrix is

$$\begin{bmatrix} 1 & (k+1)^2 & 0 \\ 2k+1 & k^2+1 & 0 \\ \dfrac{2k(k^2+2k+2)}{2k+1} & 0 & \\ k^2+1 & & \end{bmatrix}.$$

For the elements in the first column all to have the same sign, we must have $2k+1>0$ and $2k>0$. Thus the real parts of the roots of Q are strictly negative if and only if $k>0$.

24.8 Exercises

Exercise 24.1 Compute explicitly the output g of the generalized filter defined by

$$g'-ag=f, \quad a>0,$$

and show that it is stable. Is it realizable? Compute the step response.

Exercise 24.2 Let $a,b\in\mathbb{R}$. We wish to study the differential equation

$$g'-ag=f'-bf.$$

In which cases ($a=b$ and $a\neq b$) can one define a generalized filter? Discuss stability and causality.

Exercise 24.3 Consider the generalized filter determined by

$$g''+2ag'+bg=f$$

given $a,b\in\mathbb{R}$.

(a) Determine the regions of the (a,b)-plane where the poles of Q are not on the imaginary axis.

(b) Determine the regions corresponding to a realizable filter.

(c) Show that the filter is unstable if $b=0$.

Exercise 24.4 Does (24.5) define a function when f is slowly increasing?
Hint: See Exercise 20.6.

Exercise 24.5 Compute the transfer function and the impulse response of the generalized filter

$$g'''+g=f''+f.$$

Is the filter stable? Realizable?

Lesson 25

Examples of Analog Filters

25.1 Revisiting the RC filter

The RC filter was studied in Section 2.4. The equation is

$$RCg' + g = f,$$

and

$$\frac{P(x)}{Q(x)} = \frac{1}{1 + RCx}, \qquad z_1 = -\frac{1}{RC}.$$

The filter is stable and realizable (fortunately!). Formula (24.11) shows that

$$h(t) = \frac{1}{RC} e^{-\frac{t}{RC}} u(t)$$

and

$$g(t) = \frac{1}{RC} \int_{-\infty}^{t} e^{-\frac{t-s}{RC}} f(s)\, ds.$$

By taking $f = u$, we obtain the step response

$$h_1(t) = (1 - e^{-\frac{t}{RC}}) u(t).$$

The gain is $K = 1$. At the times $t = RC$ and $t = 3RC$,

$$h_1(RC) = 1 - e^{-1} \approx 0.63,$$
$$h_1(3RC) = 1 - e^{-3} \approx 0.95.$$

The response time is $t_r = 3RC$. The number RC is called the *time constant* of the filter, or *RC-constant*; it provides a good characterization of the time it takes the filter to respond to an abrupt change in the input. In this sense, it characterizes the system's dynamics. The impulse response and step response are illustrated, respectively, in Figures 25.1 and 25.2

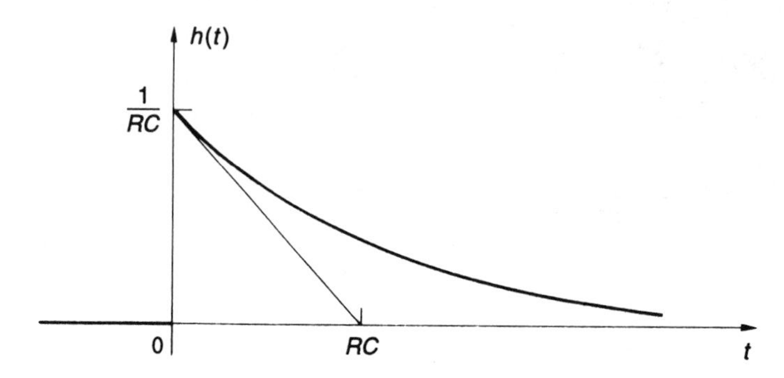

FIGURE 25.1. Impulse response of the *RC* filter.

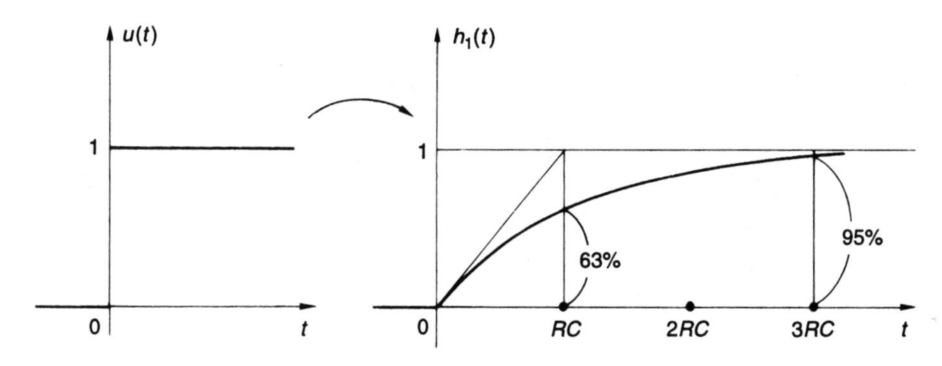

FIGURE 25.2. Step response of the *RC* filter.

25.2 The *RLC* circuit

If v is the voltage across the capacitance and f is the applied voltage, by Ohm's law,

$$LCv'' + RCv' + v = f,$$

which defines a second-order filter (Figure 25.3).

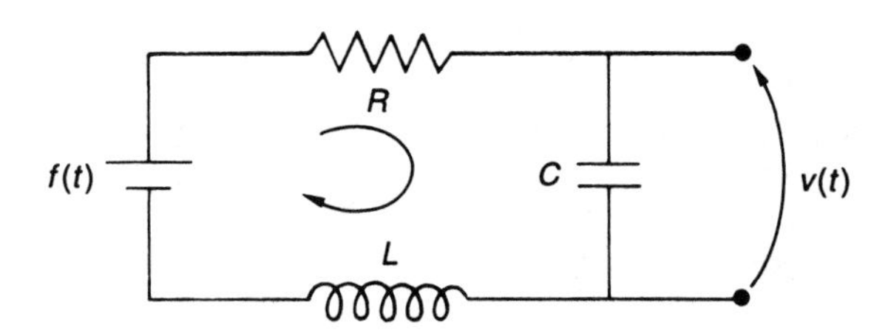

FIGURE 25.3. *RLC* circuit.

Thus

$$\frac{P(x)}{Q(x)} = \frac{1}{LCx^2 + RCx + 1},$$

and there are three cases to consider that depend on the sign of

$$\Delta = C^2\Big(R^2 - 4\frac{L}{C}\Big).$$

First case: $\Delta < 0$ $\Big(R < 2\sqrt{\frac{L}{C}}\Big)$.

Let

$$\omega = \sqrt{4\frac{L}{C} - R^2}, \quad \alpha = \frac{R}{2L}, \quad \beta = \frac{\omega}{2L}.$$

The two poles are complex conjugates and have negative real parts:

$$z = -\alpha + i\beta \quad \text{and} \quad \overline{z} = -\alpha - i\beta.$$

The partial fraction representation of H is

$$H(\lambda) = \frac{1}{i\omega C}\left[\frac{1}{2i\pi\lambda - z} - \frac{1}{2i\pi\lambda - \overline{z}}\right],$$

and we have the representation of h from (24.11) (Figure 25.4):

$$h(t) = \frac{2}{\omega C}\, e^{-\frac{R}{2L}t} \sin\Big(\frac{\omega}{2L}t\Big) \cdot u(t). \tag{25.1}$$

The response to the input f is thus

$$v(t) = \frac{2}{\omega C}\int_{-\infty}^{t} e^{-\alpha(t-x)} \sin\beta(t-x) f(x)\, dx$$

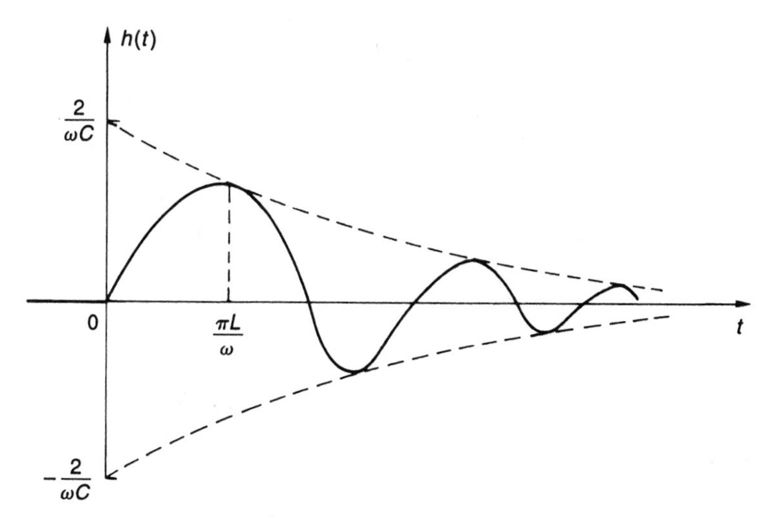

FIGURE 25.4. Impulse response of the RLC circuit (R small).

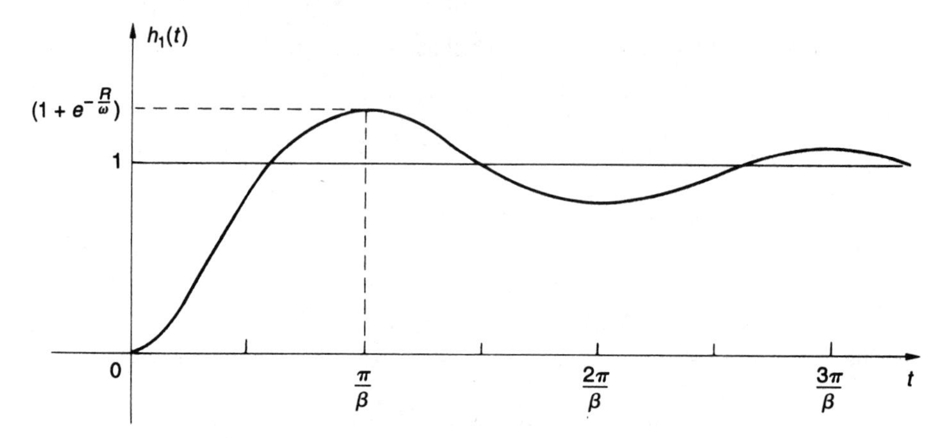

FIGURE 25.5. Step response of the RLC circuit (R small).

and the step response is

$$h_1(t) = \frac{2}{\omega C}\left(\int_0^t e^{-\alpha x}\sin\beta x\,dx\right)u(t).$$

This integral is evaluated by integrating by parts two times:

$$h_1(t) = \left[1 - e^{-\alpha t}(\cos\beta t + \frac{\alpha}{\beta}\sin\beta t)\right]u(t).$$

The step response oscillates around the limit value $K = 1$ (Figure 25.5).

Second case: $\Delta = 0$ $\left(R = 2\sqrt{\frac{L}{C}}\right)$.

In this case,

$$\frac{P(x)}{Q(x)} = \frac{1}{LC\left(x + \frac{R}{2L}\right)^2}$$

has a double real negative pole:

$$z = -\frac{R}{2L}.$$

The impulse response is

$$h(t) = \frac{1}{LC}te^{-\frac{R}{2L}t}u(t), \tag{25.2}$$

and

$$v(t) = \frac{1}{LC}\int_{-\infty}^t (t-s)e^{-\frac{R}{2L}(t-s)}f(s)\,ds.$$

The step response is

$$h_1(t) = \left[1 - \left(1 + \frac{R}{2L}t\right)e^{-\frac{R}{2L}t}\right]u(t).$$

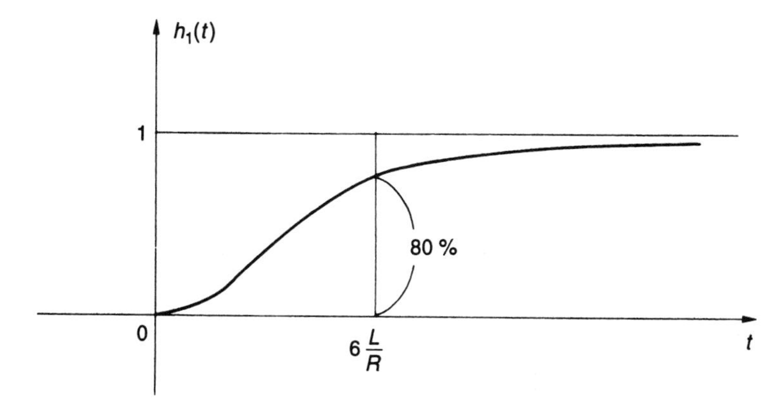

FIGURE 25.6. Step response of the RLC circuit in the critical case.

This response no longer oscillates around its asymptote (Figure 25.6).

Third case: $\Delta > 0$ $\left(R > 2\sqrt{\frac{L}{C}}\right)$.

Here we have

$$H(\lambda) = \frac{1}{LC(2i\pi\lambda - z_1)(2i\pi\lambda - z_2)},$$

and $P(x)/Q(x)$ has two real negative poles:

$$z_1 = -\frac{R+\omega}{2L}, \quad z_2 = -\frac{R-\omega}{2L}, \quad \text{where} \quad \omega = \sqrt{R^2 - 4\frac{L}{C}}.$$

H is decomposed as

$$H(\lambda) = -\frac{1}{\omega C}\left[\frac{1}{2i\pi\lambda - z_1} - \frac{1}{2i\pi\lambda - z_2}\right],$$

and

$$h(t) = \frac{-1}{\omega C}\left[e^{z_1 t} - e^{z_2 t}\right] u(t). \tag{25.3}$$

The step response is (Figure 25.7)

$$h_1(t) = \left[1 + \frac{2L}{C\omega(R+\omega)}e^{z_1 t} - \frac{2L}{C\omega(R-\omega)}e^{z_2 t}\right] u(t).$$

The response is slower than in the critical case $\Delta = 0$. The gain is 1 in all three cases. The RLC filter is stable and realizable.

25.3 Another second-order filter: $-\frac{1}{\omega}g''+g=f$

In this example,

$$\frac{P(x)}{Q(x)} = \frac{-\omega^2}{x^2 - \omega^2}, \quad \omega > 0,$$

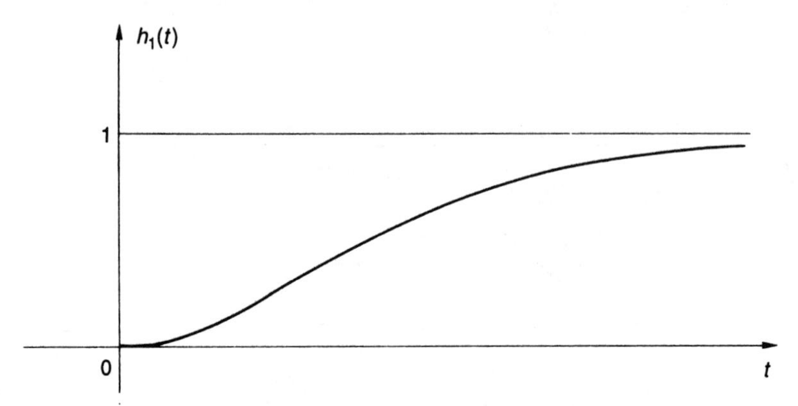

FIGURE 25.7. Step response of the RLC filter (R large).

so that

$$H(\lambda) = \frac{\omega^2}{4\pi^2\lambda^2 + \omega^2}.$$

From Section 18.2.2, the impulse response is (Figure 25.8)

$$h(t) = \frac{1}{2}\omega e^{-\omega|t|}.$$

Thus the output is

$$g(t) = \frac{1}{2}\omega \int_{\mathbb{R}} e^{-\omega|t-s|} f(s)\, ds,$$

and the step response is (Figure 25.9)

$$h_1(t) = \begin{cases} \frac{1}{2}e^{\omega t} & \text{if } t \le 0, \\ \frac{1}{2}(2 - e^{-\omega t}) & \text{if } t \ge 0. \end{cases}$$

The filter is stable but not realizable.

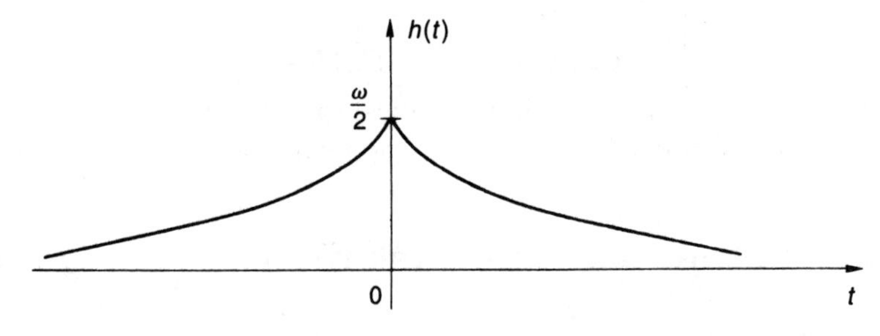

FIGURE 25.8. Impulse response of the filter $-\frac{1}{\omega^2}g'' + g = f$.

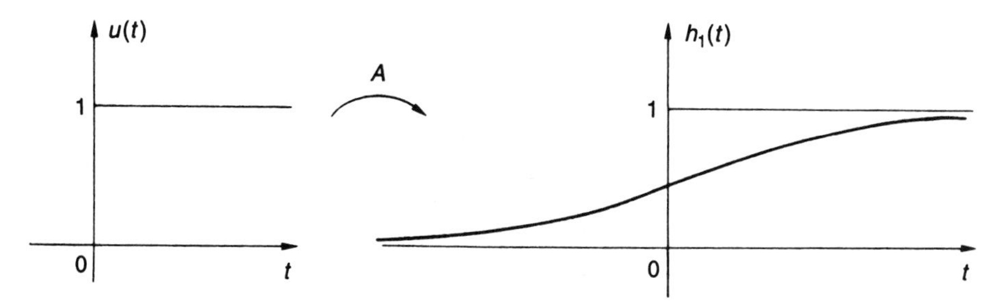

FIGURE 25.9. Step response of the filter $-\frac{1}{\omega^2}g'' + g = f$.

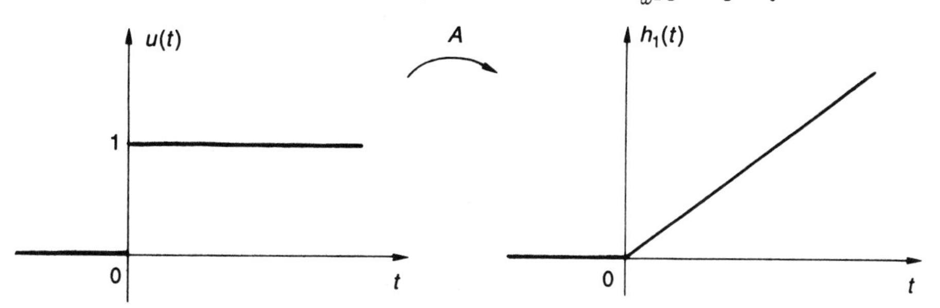

FIGURE 25.10. Step response of the integrator.

25.4 Integrator and differentiator filters

25.4.1 The integrator $g' = f$

In this case, we have

$$\frac{P(x)}{Q(x)} = \frac{1}{x}$$

There is one pole at the origin, and the results of Lesson 24 do not apply. If f is in $\mathscr{S}$, one cannot in general find a g in $\mathscr{S}$. It is easy to study this directly: g is a primitive of f, and if we limit the search to causal signals, then g is determined by having to be causal. In this case,

$$g(t) = \int_{-\infty}^{t} f(s)\, ds,$$

which can be written in terms of the Heaviside function:

$$g = u * f.$$

This is a convolution system whose impulse response is the unit step function. The step response is defined and is the ramp $h_1(t) = tu(t)$ (see Figure 5.10). The gain is infinite; the system is unstable but realizable if limited to causal signals.

25.4.2 The differentiator $g = f'$

Here we have
$$\frac{P(x)}{Q(x)} = x \quad \text{and} \quad \deg P > \deg Q.$$

This filter is clearly realizable but unstable. Neither the impulse response nor the step response can be defined with the tools developed so far. These will be defined later in the context of distributions.

25.5 The ideal low-pass filter

It is customary to describe a filter by the way it modifies the frequencies of the input signal. This is just to say that a filter is described by its transfer function H, since the frequencies of the input and output are related by
$$\widehat{g}(\lambda) = H(\lambda)\widehat{f}\,(\lambda). \tag{25.4}$$

The ideal low-pass filter does not change the frequencies λ for $|\lambda| < \lambda_c$ (λ_c is the cutoff frequency) and completely suppresses the others. Thus the transfer function of the ideal filter is
$$H(\lambda) = \begin{cases} 1 & \text{if} \quad |\lambda| < \lambda_c, \\ 0 & \text{otherwise.} \end{cases}$$

From Section 22.2.2, the h in $L^2(\mathbb{R})$ for which $\widehat{h} = H$ is
$$h(t) = \frac{\sin 2\pi\lambda_c t}{\pi t}.$$

If we consider only input signals with finite energy, then f, h, and H are in $L^2(\mathbb{R})$, and (25.4) can be expressed as (Proposition 23.2.1(i))
$$g = h * f.$$

We know that g is continuous, bounded, and zero at infinity. The right-hand side of (25.4) is in $L^2(\mathbb{R})$ because $\widehat{f}$ is in $L^2(\mathbb{R})$. Thus $\widehat{g}$ and g are in $L^2(\mathbb{R})$. The important issue here is the form of h; it tells us that *the ideal low-pass filter is not realizable.* This is indeed troublesome, but not at all surprising. Faced with the impossibility of having an ideal low-pass filter, the best we can expect is to find realizable filters whose transfer functions approximate that of the ideal filter. In general, the transfer functions of these "real filters" will have a bell-shaped amplitude (see Figure 2.1) and unbounded support. These ideas are illustrated in Figure 25.11.

The better $|H(\lambda)|$ approximates the centered rectangular window, the better will be the performance of the realizable filter. In Section 2.4 we saw that the RC filter acts as a crude low-pass filter. We will see in the next section that the Butterworth filters provide better realizable approximations to the ideal low-pass filter.

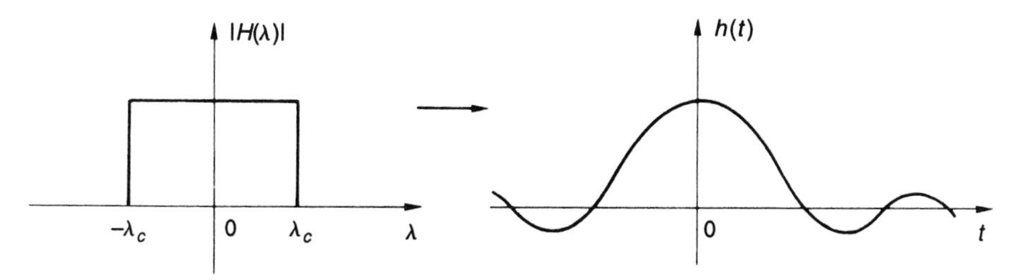

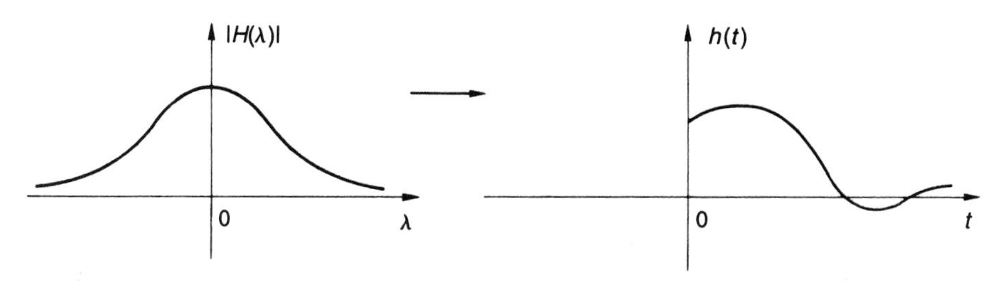

FIGURE 25.11. A real low-pass filter can only attenuate higher frequencies.

25.6 The Butterworth filters

The Butterworth filters are the filters whose energy spectra have the form

$$|H(\lambda)|^2 = \frac{1}{1 + \left(\dfrac{\lambda}{\lambda_c}\right)^{2n}}, \quad \lambda_c > 0. \tag{25.5}$$

For $n = 1$ we have the RC filter with $\lambda_c = 1/(2\pi RC)$. The motivation for increasing n is to produce a cleaner cutoff around λ_c.

As n increases, frequencies in the pass band $|\lambda| < \lambda_c$ are less attenuated, and frequencies in the attenuation band $|\lambda| > \lambda_c$ are more suppressed (see Figure 25.12). Since we have some freedom to choose the phase (only the modulus has been given), we will determine $H(\lambda)$ to obtain a stable and realizable filter. If we require h to be real, then

$$|H(\lambda)|^2 = H(\lambda)\overline{H(\lambda)} = H(\lambda)H(-\lambda). \tag{25.6}$$

The poles of $|H(\lambda)|^2$ are the complex numbers

$$p_k = \lambda_c e^{i\frac{\pi}{2n}(2k+1)}, \quad k = 0, 1, \dots, 2n-1,$$

and they occur in conjugate pairs. We want $H(\lambda)$ to be a rational function

$$H(\lambda) = \frac{P(2i\pi\lambda)}{Q(2i\pi\lambda)} = F(2i\pi\lambda),$$

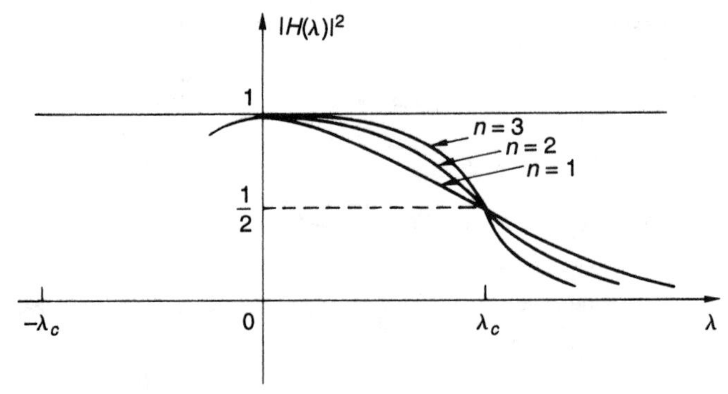

FIGURE 25.12. Energy spectra of the Butterworth filters.

and furthermore, we want the poles

$$z_k = \frac{p_k}{2i\pi}$$

of F to lie to the left of the imaginary axis. This means that the poles p_k must lie above the real axis. Thus, for the poles of $H(\lambda)$ we select those p_k whose imaginary parts are positive. The remaining p_k (the conjugates of the ones selected) are the poles of $H(-\lambda)$. Here are two examples.

Case $n = 2$:

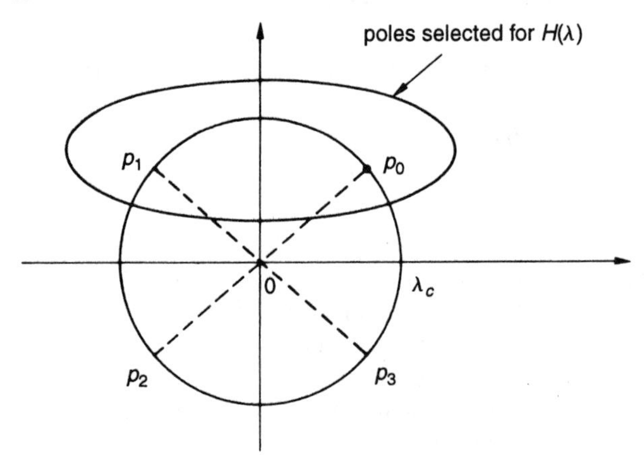

FIGURE 25.13. Butterworth filter of order 2.

In this case (Figure 25.13),

$$H(\lambda) = \frac{p_0 p_1}{(\lambda - p_0)(\lambda - p_1)}.$$

Case $n = 3$:

Here we have (Figure 25.14)

$$H(\lambda) = \frac{-p_0 p_1 p_2}{(\lambda - p_0)(\lambda - p_1)(\lambda - p_2)}.$$

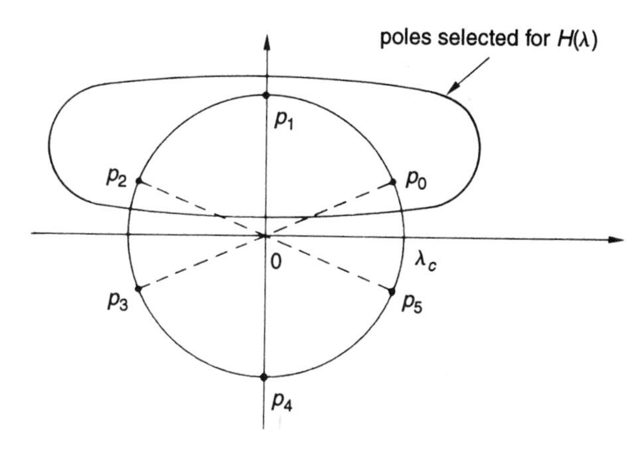

FIGURE 25.14. Butterworth filter of order 3.

We will compute the impulse response for the case $n = 2$. Thus

$$p_0 = p = \frac{\lambda_c}{\sqrt{2}}(1+i) \quad \text{and} \quad p_1 = -\overline{p}.$$

If we let $a = \pi\lambda_c\sqrt{2}$, we have

$$H(\lambda) = -\frac{2i\pi \mid p \mid^2}{p+\overline{p}}\left[\frac{1}{2i\pi\lambda+\alpha} - \frac{1}{2i\pi\lambda+\overline{\alpha}}\right],$$

where $\alpha = a(1-i)$. Referring to Section 22.2.2,

$$h(t) = -ia(e^{-\alpha t} - e^{-\overline{\alpha}t})u(t) = 2ae^{-at}\sin at \cdot u(t).$$

This impulse response has the same form as that of the *RLC* circuit, which is equation (25.1).

25.7 The general approximation problem

There are many ways to approximate the ideal low-pass filter with stable, realizable filters. The Butterworth filters belong to the class of polynomial filters ($P(x) = 1$). The *Chebyshev filters* are also in this class. These are obtained by letting

$$|H(\lambda)|^2 = \frac{1}{1+a^2T_n^2(\lambda)},$$

where $T_n(\lambda)$ is the Chebyshev polynomial of degree n and a is a parameter that determines the amplitude of the oscillations in the pass band. We also mention the *elliptic filters*: $|H(\lambda)|^2$ has the same form as above, but $T_n(\lambda)$ is replaced by a rational function. For an account of this we refer to [BL80].

The general approximation problem, given the frequency specifications, amounts to looking for a rational function that falls within a predetermined template (see Figure 25.15).

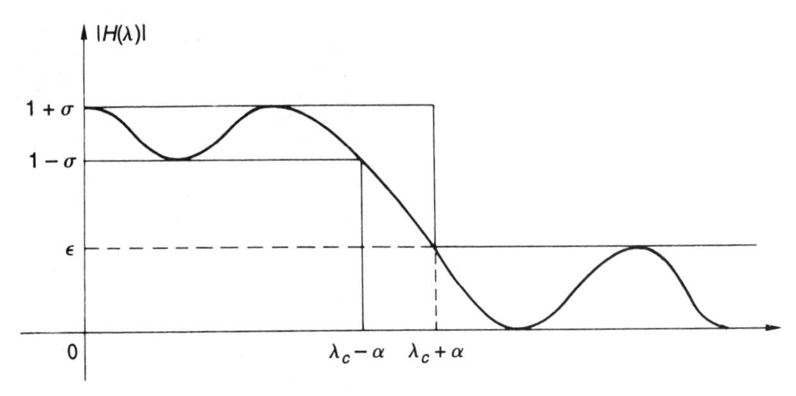

FIGURE 25.15. Approximation of the ideal low-pass filter with given frequency specifications.

25.8 Exercises

Exercise 25.1 Show that it is possible to choose the constants R, L, and C such that the RLC circuit is a Butterworth filter of order 2.

Hint: Take $R = \sqrt{2L/C}$ and compute $|H(\lambda)|^2$ as in Section 25.2, First case. One finds that

$$|H(\lambda)|^2 = \frac{1}{1+\left(\dfrac{\lambda}{\lambda_C}\right)^4} \quad \text{with} \quad \lambda_C = \frac{1}{2\pi\sqrt{LC}}.$$

Exercise 25.2 Discuss the stability of the generalized system

$$g^{(4)} + 6g^{(3)} + 11g'' + 6g' + kg = f$$

as a function of k.

Exercise 25.3 Consider the following electric filter

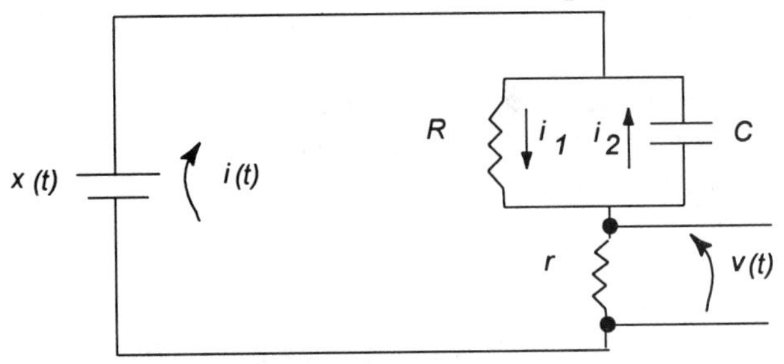

where x is the input and where the voltage v across the resistance r is the output.

(a) Show that x and v are related by $RrCv' + (R+r)v = rx + RrCx'$.

(b) Compute the transfer function and the step response.

(c) Assume that r is small with respect to R. What is the role of this filter?

Chapter VIII

Distributions

Lesson 26

Where Functions Prove to Be Inadequate

We are going to take a turn here that will lead to a new environment in which signals are no longer modeled solely by functions. The two themes for this heuristic introduction are impulse and derivation.

26.1 The impulse in physics

Intuitively, an impulse is a very strong signal having a very short duration. It is like a sharp "right to the jaw," or, less personally, like the collision of two solid bodies, one large and one small. The acceleration experienced by the smaller one is short and intense, and its velocity appears to be discontinuous, since it changes rapidly from one value to another. We first consider a simple example.

26.1.1 The notion of a point mass

We restrict the example to a one-dimensional mass distribution. Thus imagine a unit mass distributed on the x-axis between the values $-h$ and h with a density $d_h(x)$ (see Figure 26.1). This density function has the following properties:

(i) $d_h(x) \geq 0$ for all $x \in \mathbb{R}$.

(ii) $d_h(x) = 0$ if $|x| > h$.

(iii) $\displaystyle\int_{\mathbb{R}} d_h(x)\,dx = 1$, the total mass.

If we imagine that this constant mass is compressed into the point $x = 0$, that is, if we let h tend to 0, then we have, in the limit, what is called a point mass at the origin. This is like the situation where a physical body is observed from so far away that it seems to have no dimension and appears as a point.

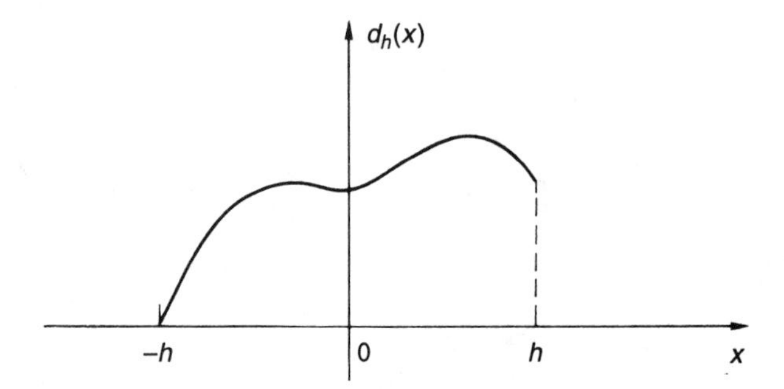

FIGURE 26.1. Density distribution of a mass.

But what happens mathematically to the density d_h as h tends to zero? If we assume that some sort of limit density $d(x)$ exists, then we would like it to satisfy the following conditions:

(i) $d(x) \geq 0$ for all $x \in \mathbb{R}$.

(ii) $d(x) = 0$ if $x \neq 0$.

(iii) $\displaystyle\int_{\mathbb{R}} d_h(x)\,dx = 1.$

One has the idea that at $x = 0$ the value $d(0)$ is infinite. The situation is similar to that of a point charge carried by an elementary particle.

26.1.2 A collision between two solid bodies

Let us try to see what happens in mechanics when a force becomes more and more intense and brief. Let S be a solid body of mass m at rest on a surface where it can slide without friction; think of a hockey puck about to be hit. Between the instants $t = -h$ and $t = h$ the stick applies a force f_h whose graph has, for example, the shape shown in Figure 26.2.

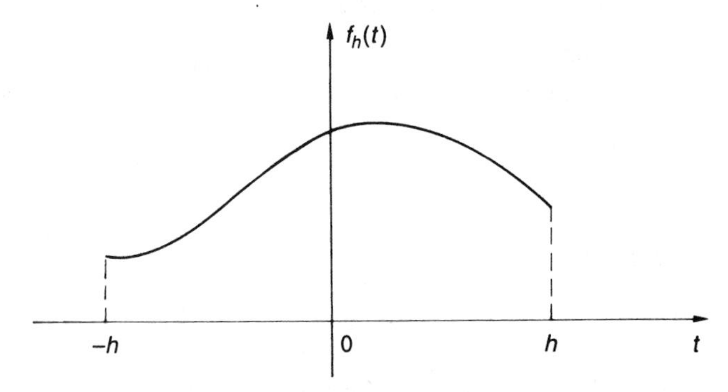

FIGURE 26.2. Force applied to a puck.

We imagine that the duration of the force becomes shorter and shorter ($h \to 0$) while always imparting the same energy E_f to S. These applied forces become more and more intense, and in the limit we have an instantaneous shock at time $t = 0$. If v_h and $\gamma_h = v'_h$ are respectively the velocity and the acceleration of S, its kinetic energy at time t is

$$E(t) = \frac{1}{2} m v_h^2(t),$$

which is constant and equal to E_f after time $t = h$:

$$E_f = \frac{1}{2} m v_h^2(h).$$

Thus $v_h(h)$ is a constant as a function of h, and we have

$$v_h(h) = \int_{-h}^{h} \gamma_h(t)\, dt = C, \quad \text{a constant.}$$

Newton's second law, $f_h = m\gamma_h$, implies that

$$\int_{-h}^{h} f_h(t)\, dt = C.$$

By taking $C = 1$, we see that the forces f_h satisfy the following three conditions:

(i) $f_h(t) \geq 0$ for all $t \in \mathbb{R}$.

(ii) $f_h(t) = 0$ if $|t| \geq h$.

(iii) $\displaystyle\int_{\mathbb{R}} f_h(t)\, dt = 1$.

At the limit, we will have a shock $f(t)$ that has the following properties:

(i) $f(t) \geq 0$ for all $t \in \mathbb{R}$.

(ii) $f(x) = 0$ if $t \neq 0$.

(iii) $\displaystyle\int_{\mathbb{R}} f(t))\, dt = 1$.

26.2 Uncontrolled skid on impact

From what we have just seen, the unit impulse at the origin will be an ideal signal, which we denote for the moment by imp(t), that satisfies the three conditions (i), (ii), and (iii). Unfortunately, even with the Lebesgue integral, these three conditions are incompatible for a function: The integral of a function that is zero almost everywhere is necessarily zero.

What is to be done?

People working in mechanics and theoretical particle physics around 1920–1930 (notably P. Dirac) were deterred by neither the question nor the contradiction. They had a useful tool, even if it was not conceptually satisfying. They used the imp(t) "function"—it was, in fact, called $\delta(t)$, but we change the name temporarily for clarity—which, desirable or not, was thought of as satisfying the conditions

$$\text{imp}(t) = \begin{cases} 0 & \text{if } t \neq 0, \\ +\infty & \text{if } t = 0, \end{cases}$$

$$\int_{\mathbb{R}} \text{imp}(t)\, dt = 1,$$

and whose graphic representation is shown Figure 26.3.

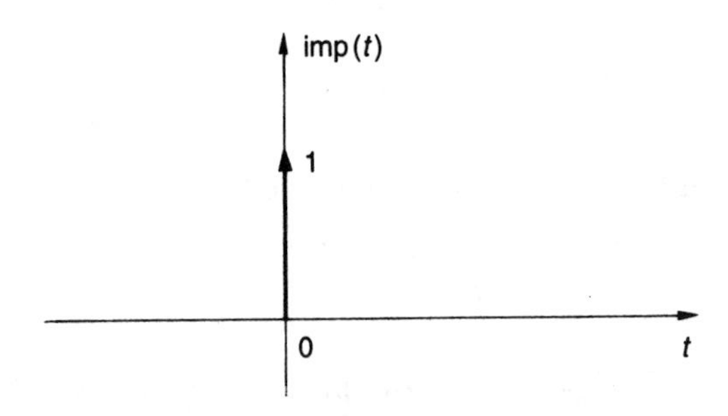

FIGURE 26.3. Unit impulse at the origin.

If we set aside rigor but respect the usual properties of functions, it is easy to exploit the formalism of (26.1). Take, for example, the computation of the integral

$$I = \int_{\mathbb{R}} \text{imp}(t) f(t)\, dt.$$

Assuming that f is differentiable, we integrate by parts by letting

$$\int_{-\infty}^{x} \text{imp}(t)\, dt = u(x) = \begin{cases} 0 & \text{if } x < 0, \\ 1 & \text{if } x > 0, \end{cases} \tag{26.1}$$

which is quite natural in view of (26.1). ($u(0)$ is not defined, but this is not important.) Here is the evaluation of I:

$$I = [u(x)f(x)]_{-\infty}^{+\infty} - \int_{\mathbb{R}} u(x) f'(x)\, dx;$$

$$I = f(+\infty) - \int_{0}^{+\infty} f'(x)\, dx = f(+\infty) - f(+\infty) + f(0);$$

$$I = f(0).$$

This relation makes sense even if f is only continuous at the origin, and it is thus possible to make practical use of integrals containing the impulse function by letting

$$\int_{\mathbb{R}} \operatorname{imp}(t) f(t)\, dt = f(0)$$

for all continuous functions.

In passing, we have "deduced" from (26.2) that the impulse is the derivative of Heaviside's unit step function, and everything is working out quite nicely. All the better, since we will see that this new derivation is far more satisfying than the old one.

26.3 A new-look derivation

With the example of the unit step, we are faced with two derivatives: the usual derivative, which is zero except at the origin, where it does not exist, and a derivation denoted by D that leads to the formula (see Figure 26.4)

$$\mathrm{D}u = \operatorname{imp}.$$

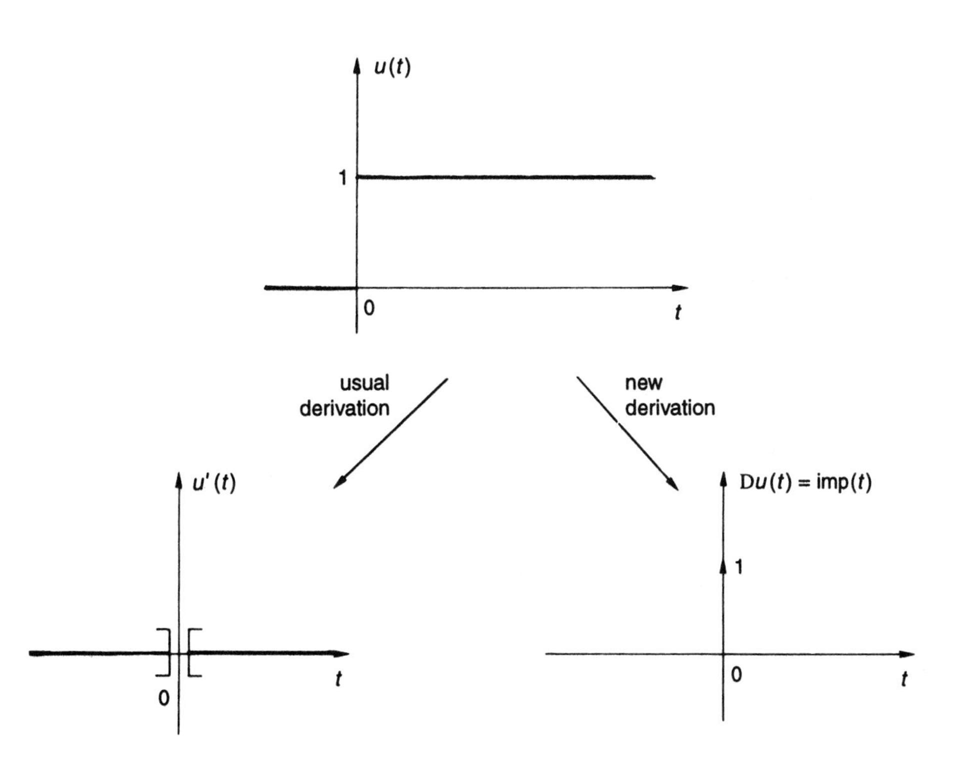

FIGURE 26.4. The two kinds of derivation.

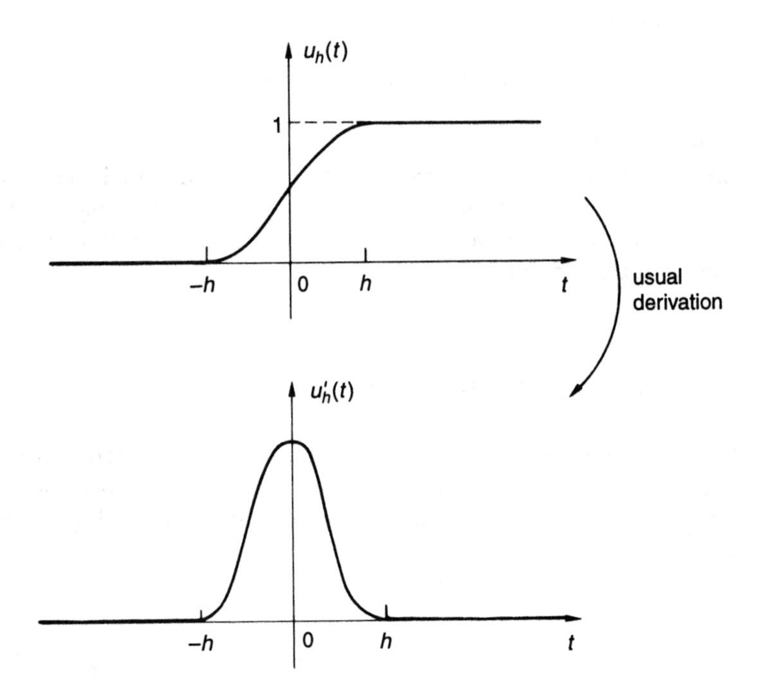

FIGURE 26.5. Establishment of an electric current.

From the modeling point of view, the unit step represents, for example, the instantaneous establishment of a constant electric current. We consider this phenomenon from a microscopic point of view, without going to the level of electrons, where the model would necessarily be discrete. Physically, there is no discontinuity at $t = 0$, but rather the continuous and very rapid (of order 10^{-7} seconds) establishment of the current. A more precise model would thus be a function $u_h(t)$ like the one shown in Figure 26.5.

For convenience, we put the time origin at the center of the transition phase. The usual derivative $u_h'(t)$ must satisfy the following conditions:

(i) $u_h'(t) \geq 0$ for all $t \in \mathbb{R}$.

(ii) $u_h'(t) = 0$ if $|t| > h$.

(iii) $\displaystyle\int_{\mathbb{R}} u_h'(t)\, dt = 1.$

Thus the functions u_h' have all the characteristics of approximations of the unit impulse. On the other hand, it is clear that taking the usual derivative and passing to the limit, wipes out the event that occurs at time $t = 0$. The information disappears. Thus it is more in keeping with the physical phenomenon to say "u is differentiable, and its derivative is the impulse" than to say "u is differentiable except at the origin, and its derivative is the zero function."

26.4 The birth of a new theory

It was necessary to wait until 1947 for the creation, by Laurent Schwartz, of a complete mathematical theory of these new objects. This is the *theory of distributions.* Since then, it has become an almost indispensable tool in theoretical physics and signal processing. Originally motivated by the study of partial differential equations, distribution theory has had an impact on most areas of mathematical analysis.

Distributions generalize the notion of function. We have already seen with the Lebesgue integral one way that functions needed to be generalized: Starting with an ordinary process that allows a well-defined value $f(x)$ to be associated with each x, we arrived at equivalence classes of functions that are equal almost everywhere, where the value of f a given point is no longer significant.

The new generalization will include the impulse as well an many other "generalized functions." The theory of distributions also contains the new derivation, which is called "derivation in the sense of distributions." This is a global concept, whereas the usual derivation applies only to "differentiable functions." Historically, it was around 1937 that the Soviet mathematician S.L. Sobolev first introduced the idea of a generalized derivative. Roughly, g is the generalized derivative of f if

$$\int_{\mathbb{R}} g(x)\varphi(x)\,dx = -\int_{\mathbb{R}} f(x)\varphi'(x)\,dx \tag{26.2}$$

for all regular functions φ that have bounded support. This point of view is taken in distribution theory, which was officially born in 1947 with the publication of Schwartz's first article in the Annals of the Fourier Institute at Grenoble. Little by little the idea that all continuous functions were differentiable spread throughout the mathematics community, to the general amazement of all! While the theory at first seemed rather esoteric and complicated (probably because of its heavy use of topology and the Lebesgue integral), mathematicians quickly realized that its actual use was much simpler than the theory: One could work formally and quickly without worrying about whether functions satisfied certain conditions, such as differentiability. A distribution is always differentiable, and in fact infinitely differentiable. A series divergent in the usual sense will often be convergent in the sense of distributions. One important property that we have already verified for the Heaviside function is the continuity of derivation:

$$f_n \to f \quad \Longrightarrow \quad \mathrm{D}f_n \to \mathrm{D}f,$$

where the limits are taken in the sense of distributions. Finally, the Fourier transform, an indispensable tool in so many areas, was until the advent of distributions defined only for integrable and square-integrable functions. One was not able to speak, for example, of the Fourier transform of the

Heaviside function. This restriction will be removed; in addition, we will see the theories of Fourier series and of the Fourier integral unified.

We are going to present only the elementary aspects of distribution theory. Our goal is to develop the tools needed for the principal applications of Fourier analysis.

Lesson 27

What Is a Distribution?

27.1 The basic idea

The basic idea for generalizing the notion of function in the context of distributions is to regard a function as an operator T_f (called a functional) acting by integration on functions themselves:

$$T_f(\varphi) = \int_{-\infty}^{+\infty} f(x)\varphi(x)\, dx. \tag{27.1}$$

This idea is analogous to that of identifying a real number a with the linear function

$$x \longmapsto ax.$$

It is this concept that allows one to go from the notion of derivative to that of differential.

Clearly, the integral in (27.1) does not always exist. If we want it to exist for rather general functions f, it is necessary to impose severe restrictions on the function φ, which is called a *test function*. We first require that φ vanish outside a bounded interval so that there will be no problem with convergence at infinity. Thus all test functions φ have bounded support.

To generalize derivation (or the derivative), we examine what happens when f is continuously differentiable. The functional associated with f' is

$$T_{f'}(\varphi) = \int_{-\infty}^{+\infty} f'(x)\varphi(x)\, dx,$$

and integration by parts shows that

$$T_{f'}(\varphi) = -\int_{-\infty}^{+\infty} f(x)\varphi'(x)\, dx. \tag{27.2}$$

This formula has the advantage that the derivative of f no longer appears, and thus the derived functional can be defined even though the function f is not differentiable. However, the test functions φ must be differentiable; indeed, they must be infinitely differentiable if we wish to iterate the operation.

27.2 The space $\mathscr{D}(\mathbb{R})$ of test functions

One is led naturally to require that test functions be infinitely differentiable and have bounded support. The space of these functions is denoted by $\mathscr{D}(\mathbb{R})$ or simply $\mathscr{D}$ (recall Definition 15.1.7).

$$\varphi \in \mathscr{D} \iff \begin{cases} \text{(i) } \varphi \text{ vanishes outside a bounded} \\ \quad\ \text{interval (which depends on } \varphi). \\ \text{(ii) } \varphi \text{ is infinitely differentiable} \\ \quad\ \text{in the usual sense on } \mathbb{R}. \end{cases}$$

In other words,

$$\mathscr{D}(\mathbb{R}) = \{\varphi : \mathbb{R} \to \mathbb{C} \mid \varphi \in C^\infty(\mathbb{R}),\ \text{supp}(\varphi) \text{ is bounded.}\}$$

It is not immediately obvious how to construct such a function or whether such functions exist, except for $\varphi = 0$. The "usual" functions (polynomials, rational functions, trigonometric functions, etc.) satisfy (ii) but not (i). In addition, the elementary infinitely differentiable functions are analytic. Thus if they are zero on some interval, no matter how small, they are identically zero. This is almost the opposite of what we want. To fix this situation, we can try to define φ a piece at a time:

$$\varphi(x) = \begin{cases} \theta(x) & \text{if } \ x \in (a,b), \\ 0 & \text{if } \ x \notin (a,b), \end{cases}$$

where θ is an elementary function. The difficulty here is that all of the derivatives of θ must vanish to the left of a and to the right of b. There exists, however, at least one explicit example that is always presented in this context. It is the following function, which we have already seen in Lessons 15 and 21:

$$\theta(x) = \begin{cases} \exp\Big(-\dfrac{1}{1-x^2}\Big) & \text{if } \ |x| < 1, \\ 0 & \text{if } \ |x| \geq 1. \end{cases} \tag{27.3}$$

One can verify, with a little patience, that this function is infinitely differentiable and that the derivatives all vanish at $x = \pm 1$. (Show that the nth derivative of θ is of the form $F_n(x)\theta(x)$, where F_n is a rational function.) It follows that $\mathscr{D} \neq \{0\}$, but this is still a modest result.

By translation and change of scale, we can construct a function in $\mathscr{D}$ whose support is an arbitrary bounded interval (a, b). In this way we find an infinite number of test functions whose supports are disjoint, and this shows that $\mathscr{D}$ is an infinite-dimensional vector space. In practice, we never need to use the explicit expression for a test function.

For those who have skipped over Lesson 21, we mention that it was shown there that $\mathscr{D}$ is dense in $L^1(\mathbb{R})$ and, more generally, in $L^p(\mathbb{R})$:

$$\overline{\mathscr{D}(\mathbb{R})} = L^p(\mathbb{R}), \quad 1 \leq p < +\infty.$$

27.3 The definition of a distribution

In formula (27.1), the functional T_f is linear on $\mathscr{D}$. We will add a continuity condition.

27.3.1 Definition A distribution is any mapping

$$T : \mathscr{D} \to \mathbb{C}$$

that is linear and continuous.

The variable of a distribution is a test function, and $T(\varphi)$ is a complex number. T is said to be a continuous linear functional (or linear form) on $\mathscr{D}$. The value of T at φ will be denoted in either of two ways:

$$T(\varphi) \quad \text{or} \quad \langle T, \varphi \rangle.$$

This second notation brings to mind a close relative of $T(\varphi)$, namely, the scalar product in $L^2(\mathbb{R})$ expressed by (27.1).

The continuity of T means the following:

$$\text{If } \varphi_n \to \varphi \text{ in } \mathscr{D}, \text{ then } T(\varphi_n) \to T(\varphi).$$

We still must specify the meaning of "$\varphi_n \to \varphi$ in $\mathscr{D}$"; thus we need to define a topology (or at least the concept of convergence) for $\mathscr{D}$. Here, and elsewhere in the book, we settle for a direct definition of convergence—the notion of a limit of a sequence—and avoid discussing the underlying topology. We will define what $\varphi_n \to 0$ means; by linearity, $\varphi_n \to \varphi$ will mean that $\varphi_n - \varphi \to 0$.

27.3.2 Definition (limits in $\mathscr{D}$) A sequence of elements $(\varphi_n)_{n \in \mathbb{N}}$ of $\mathscr{D}$ tends to 0 in $\mathscr{D}$ if the following hold:

(i) The supports of all the φ_n are contained in a fixed compact interval.

(ii) (φ_n) as well as all of the derived sequences tend to 0 uniformly on $\mathbb{R}$ as $n \to +\infty$.

There does not exist a distance function, much less a norm, on $\mathscr{D}$ that gives this notion of convergence. There is, however, a well-defined topology on $\mathscr{D}$ (see [Sch65b]). It is sufficient for our purposes to have the notion of convergence.

The following useful property follows directly from the definition.

27.3.3 Proposition *If a sequence of elements (φ_n) in $\mathscr{D}$ tends to 0 in $\mathscr{D}$, then the sequence (φ_n') is in $\mathscr{D}$ and tends to 0 in $\mathscr{D}$.*

It is clear that the set of distributions has the structure of a complex vector space with the obvious addition of two linear forms and multiplication by scalars. This space is called the *topological dual* of $\mathscr{D}$ or the *space of continuous linear functionals* on $\mathscr{D}$. We denote this dual space by $\mathscr{D}'(\mathbb{R})$ or simply $\mathscr{D}'$. This conforms with historical and customary notation.

27.3.4 Examples

(a) Point distributions: Let a be a real number and let δ_a be the mapping defined on $\mathscr{D}$ by

$$\delta_a(\varphi) = \varphi(a).$$

It is clear that δ_a is linear and continuous on $\mathscr{D}$ (convergence in $\mathscr{D}$ is uniform and hence pointwise). Thus δ_a is a distribution. For $a = 0$ we write simply δ.

(b) Let $(\lambda_n)_{n\in\mathbb{Z}}$ be a complex sequence and let $a > 0$. We write

$$T = \sum_{n=-\infty}^{+\infty} \lambda_n \delta_{na}$$

for the linear functional on $\mathscr{D}$ defined by

$$T(\varphi) = \sum_{n=-\infty}^{+\infty} \lambda_n \varphi(na).$$

This sum is, in fact, finite for each φ; hence there is no problem about convergence. Furthermore, T is continuous: If $\varphi_p \to 0$ and if $[A, B]$ is a bounded interval containing the supports of all the φ_p, then

$$T(\varphi_p) = \sum_{A \le na \le B} \lambda_n \varphi_p(na)$$

is a finite sum that tends to 0 as $p \to +\infty$. Thus T is a distribution. For $\lambda_n = 1$, we have *Dirac's comb*, which will be used often in the sequel.

27.4 Distributions as generalized functions

We wish to show that many functions can be identified as distributions via the relation (27.1). The integral in this formula is well-defined whenever f is locally integrable. The integrability of $f\varphi$ is a consequence of the inequality

$$\int_{\mathbb{R}} |f(x)||\varphi(x)|\,dx \leq \|\varphi\|_\infty \int_a^b |f(x)|\,dx, \tag{27.4}$$

where $\operatorname{supp}(\varphi) \subset (a, b)$.

27.4.1 Proposition *If f is locally integrable on $\mathbb{R}$, then the functional T_f defined by*

$$T_f(\varphi) = \int_{\mathbb{R}} f(x)\varphi(x)\,dx \tag{27.5}$$

is a distribution.

Proof. T_f is clearly linear. The continuity on $\mathscr{D}$ follows directly from (27.4): Let (a, b) be an interval containing the supports of the elements φ_n of a sequence in $\mathscr{D}$ that tends to zero. □

It is clear from (27.5) that if f and g are locally integrable and equal almost everywhere, they define the same distribution T_f on $\mathscr{D}$; that is,

$$f = g \text{ a.e.} \quad \Longrightarrow \quad T_f = T_g.$$

We are going to investigate the converse of this implication.

27.4.2 A question of identification

The mapping in Proposition 27.4.1 of $L^1_{\text{loc}}(\mathbb{R})$ into $\mathscr{D}'(\mathbb{R})$ given by $f \mapsto T_f$ is well-defined. The identification of $L^1_{\text{loc}}(\mathbb{R})$ with its image in $\mathscr{D}'$ is possible if this mapping, which is linear, is also 1-to-1; that is, if

$$T_f = 0 \quad \Longrightarrow \quad f = 0 \text{ a.e.}$$

(See Figure 27.1.) The proof of this property was given as an exercise (Exercise 21.6). Thus for all f in $L^1_{\text{loc}}(\mathbb{R})$

$$T_f = 0 \quad \Longleftrightarrow \quad f = 0 \text{ a.e.} \tag{27.6}$$

The distribution T_f is thus identified uniquely with the locally integrable function f. From now on we will make the identification $f \leftrightarrow T_f$ using any of several notations:

$$\langle T_f, \varphi\rangle = T_f(\varphi) = \langle f, \varphi\rangle = \int_{\mathbb{R}} f\varphi.$$

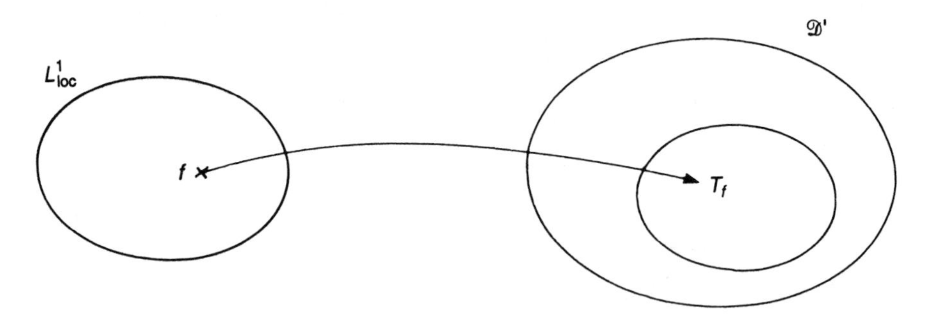

FIGURE 27.1. Embedding of $L^1_{\text{loc}}(\mathbb{R})$ in $\mathscr{D}'$.

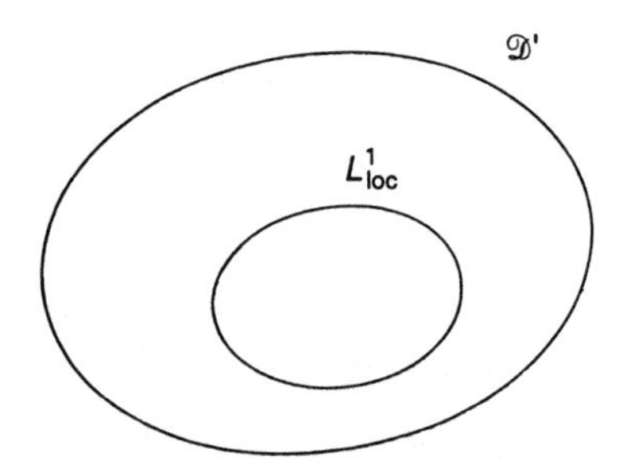

FIGURE 27.2. Identification of L^1_{loc} as a subspace of $\mathscr{D}'$.

The distributions T_f associated with locally integrable functions are said to be *regular*. With this identification we have the situation illustrated in Figure 27.2.

It is based on this embedding that distributions are called *generalized functions*. More precisely, distributions generalize locally integrable functions. Here are two examples:

The unit step function (Heaviside's function) u is identified with the distribution

$$T_u(\varphi) = \int_0^{+\infty} \varphi(x)\,dx.$$

The constant function $f = K$, $K \in \mathbb{C}$, is identified with the constant distribution of the form

$$T_f(\varphi) = K \int_{\mathbb{R}} \varphi(x)\,dx.$$

Unfortunately, a function as simple as $f(x) = x^{-1}$, which is not locally integrable around the origin, cannot at this point be considered a distribution. We will see in Section 28.5 how this situation can be remedied.

27.4.3 There are nonregular distributions

A simple example of a nonregular distribution is given by the point distributions

$$\delta_a(\varphi) = \varphi(a).$$

We will prove this for $a = 0$; that is, we will show that there does not exist a locally integrable function f such that

$$\int_{\mathbb{R}} f\varphi = \varphi(0)$$

for all $\varphi \in \mathscr{D}$. We argue indirectly. Thus, assume that such an f exists and let $\rho \in \mathscr{D}$ be the function defined by (27.3). Then we have

$$\int_{\mathbb{R}} f(x)\rho(nx)\,dx = \rho(0)$$

for all $n \in \mathbb{N}$. This implies that

$$1 = |\rho(0)| \leq \int_{-\frac{1}{n}}^{\frac{1}{n}} |f(x)||\rho(nx)|\,dx \leq \int_{-\frac{1}{n}}^{\frac{1}{n}} |f(x)|\,dx.$$

As $n \to \infty$, the right-hand integral tends to 0; this contradiction proves that f does not exist.

27.5 Exercises

Exercise 27.1 Are the following functionals distributions?

(a) $T(\varphi) = |\varphi(0)|$.

(b) $T(\varphi) = a, \ a \in \mathbb{C}$.

(c) $T(\varphi) = \sum_{n=0}^{+\infty} \varphi^{(n)}(0)$.

(d) $T(\varphi) = \int_{\mathbb{R}} |x|^\alpha \varphi(x)\,dx, \ \alpha \in \mathbb{R}$.

Exercise 27.2 Suppose $\varphi \in \mathscr{D}(\mathbb{R})$ and $n \in \mathbb{N}^*$. Show that φ can be written in the form

$$\varphi(x) = \sum_{k=0}^{n-1} \frac{x^k}{k!}\varphi^{(k)}(0) + x^n \psi_n(x)$$

with $\psi_n \in C^\infty(\mathbb{R})$.

Hint: Use Taylor's formula with integral remainder.

Exercise 27.3

(a) Take $y_j \in \mathbb{C}$, $0 \le j \le n$, and $a \in \mathbb{R}$. We wish to find $\varphi \in \mathscr{D}(\mathbb{R})$ such that

$$\varphi^{(j)}(a) = y_j, \quad 0 \le j \le n. \tag{I}$$

- Let $\theta_a = k\chi_{[a-1,a+1]} * \theta$, where $k \in \mathbb{R}$ and θ is defined by (27.3). Show that $\theta_a \in \mathscr{D}(\mathbb{R})$ and that $\theta_a^{(j)}(a) = 0$ for all $j \ge 1$. Choose k such that $\theta_a(a) = 1$.
- Find a polynomial P of degree n such that $\varphi = P\theta_a$ satisfies (I).

(b) Let $(y_j)_{j\in\mathbb{N}}$ be a sequence that is dominated by a geometric sequence:

$$|y_j| \le AB^j$$

for some $A, B > 0$. For $a \in \mathbb{R}$, show that there exists $\varphi \in \mathscr{D}(\mathbb{R})$ such that $\varphi^{(j)}(a) = y_j$ for all $j \in \mathbb{N}$.

Exercise 27.4 (truncation) For $f \in C^\infty(\mathbb{R})$, show that there exists $\varphi \in \mathscr{D}(\mathbb{R})$ that agrees with f on $[-1, 1]$.

Exercise 27.5 Assume that $f \in C^\infty(\mathbb{R}\backslash 0)$ satisfies the following condition:

There is a $p \in \mathbb{N}$ such that for all $n \in \mathbb{N}$, $x^p f^{(n)}(x) \to 0$ when $x \to 0$.

(a) Give an example of such a function.

(b) Show that there exists $g \in C^\infty(\mathbb{R})$ that agrees with f outside $[-1, 1]$.

Exercise 27.6 Suppose $g \in C^\infty(\mathbb{R})$, $a \in \mathbb{R}$, and $n \in \mathbb{N}$. Prove the following equivalence:

$$g\delta_a^{(n)} = 0 \iff g^{(j)}(a) = 0, \; j = 0, \ldots, n.$$

Lesson 28

Elementary Operations on Distributions

In this lesson we intend to extend to distributions several concepts regularly applied to functions such as parity and periodicity, as well as the fundamental notion of derivation. Indeed, it is in the area of differential equations (ordinary and partial) that the use of distributions has been most fruitful. We will illustrate a general method for extending to distributions certain concepts that are known for functions. The difficulty comes from the fact that the argument of a distribution is a function and not a real variable.

28.1 Even, odd, and periodic distributions

The reflection f_σ of a function f is defined by

$$f_\sigma(x) = f(-x).$$

The function f is said to be even if $f_\sigma = f$; it is said to be odd if $f_\sigma = -f$. We will try to define the reflection T_σ of a distribution T. For a regular distribution T_f, the identification with the function f imposes the relation

$$(T_f)_\sigma = T_{f_\sigma}.$$

This means that

$$(T_f)_\sigma(\varphi) = \int_{\mathbb{R}} f_\sigma(t)\varphi(t)\,dt = \int_{\mathbb{R}} f(t)\varphi_\sigma(t)\,dt$$

for all $\varphi \in \mathscr{D}$, or written differently, that

$$(T_f)_\sigma(\varphi) = T_f(\varphi_\sigma).$$

The last formula is our guide for extending "reflection" to all distributions.

28.1.1 Definition The reflection T_σ of a distribution T is defined by

$$T_\sigma(\varphi) = T(\varphi_\sigma) \tag{28.1}$$

for all $\varphi \in \mathscr{D}$. The distribution T is said to be even if $T_\sigma = T$ and odd if $T_\sigma = -T$.

It is easy to show that T_σ thus defined is indeed a distribution.

EXAMPLE: δ is an even distribution.

28.1.2 Periodic distributions

Here again we note what happens in the case of functions. A function f has real period a, $a \neq 0$, if

$$f(x-a) = f(x)$$

for all $x \in \mathbb{R}$. In terms of the translation operator (Section 2.1.3) this is

$$\tau_a f = f.$$

Thus it is sufficient to define the translation operator for distributions to establish the desired definition. For a regular distribution T_f, its identification with f forces the relation

$$\tau_a T_f = T_{\tau_a f}.$$

As before, this expands to

$$\langle \tau_a T_f, \varphi \rangle = \langle T_{\tau_a f}, \varphi \rangle = \int f(x-a)\varphi(x)\,dx = \int f(x)\varphi(x+a)\,dx,$$

which means that

$$\langle \tau_a T_f, \varphi \rangle = \langle T_f, \tau_{-a}\varphi \rangle$$

for all $\varphi \in \mathscr{D}$. The last formula can be extended to all distributions.

28.1.3 Definition The translate $\tau_a T$ of a distribution T is defined by

$$\langle \tau_a T, \varphi \rangle = \langle T, \tau_{-a}\varphi \rangle \tag{28.2}$$

for all $\varphi \in \mathscr{D}$. A distribution T is said to be periodic with period $a \neq 0$ if

$$\tau_a T = T.$$

EXAMPLES:

(i) $\tau_a \delta_b = \delta_{a+b}$.

(ii) Dirac's comb, $T(\varphi) = \sum_{n=\infty}^{+\infty} \varphi(na)$, is periodic with period a.

28.2 Support of a distribution

It is again a matter of extending to distributions a concept that has been defined for functions (see Definition 15.1.5). If φ is continuous, then

$$\operatorname{supp}(\varphi) = \overline{\{x \in \mathbb{R} \mid \varphi(x) \neq 0\}}.$$

28.2.1 Definition A distribution is said to be null (be zero, vanish) on an open set Ω if $T(\varphi) = 0$ for all test functions φ for which $\operatorname{supp}(\varphi) \subset \Omega$.

For example, δ is null on $(0, +\infty)$. It is possible to prove the following equivalence for a regular distribution T_f:

$$T_f \text{ is null on } \Omega \iff f(x) = 0 \text{ for a.e. } x \in \Omega. \tag{28.3}$$

We will assume (28.3). This is slightly more general than the condition discussed in Section 27.4.2, which was that

$$T_f = 0 \iff f(x) = 0 \text{ for a.e. } x \in \mathbb{R}.$$

28.2.2 Definition The support of a distribution T, $\operatorname{supp}(T)$, is defined to be the complement of the largest open set on which T is null.

This is equivalent to saying that $\operatorname{supp}(T) = \bigcup_i \Omega_i$, where the union is taken over all open sets Ω_i on which T is null. Clearly, $\operatorname{supp}(T)$ is a closed subset of $\mathbb{R}$.

EXAMPLE: $\operatorname{supp}\Big(\sum_{i=1}^n \lambda_i \delta_{a_i}\Big) \subset \{a_1, a_2, \dots, a_n\}$, with equality in case none of the λ_i are zero. It is sufficient to show that T is null on the complement of $\{a_1, a_2, \dots, a_n\}$. Thus take $\varphi \in \mathscr{D}$ with $\operatorname{supp}(\varphi) \cap \{a_1, \dots, a_n\} = \emptyset$; clearly,

$$T(\varphi) = \sum_{i=1}^n \lambda_i \varphi(a_i) = 0.$$

To prove equality, suppose there is an index k such that a_k is not in $\operatorname{supp}(T)$. Then there is a neighborhood I of a_k that contains none of the other a_j and on which T is null. Let φ be a test function such that

$$\operatorname{supp}(\varphi) \subset I \quad \text{and} \quad \varphi(a_k) = 1.$$

We have

$$T(\varphi) = 0 = \lambda_k \varphi(a_k) = \lambda_k,$$

which proves the result.

28.2.3 Proposition *If T_f is a regular distribution, then*

$$\operatorname{supp}(T_f) = \operatorname{supp}(f).$$

Proof. This follows directly from Definition 20.1.3 of the support of a measurable function and (28.3). □

28.2.4 Definition The space of distributions whose supports lie to the right of some finite point is denoted by $\mathscr{D}'_+$:

$$T \in \mathscr{D}'_+ \iff \operatorname{supp}(T) \subset [t_0, +\infty) \text{ for some } t_0 \in \mathbb{R}.$$

This space, which corresponds to the space of causal signals, will be used often in the sequel.

28.3 The product of a distribution and a function

One should not conclude from what we have done so far that all operations on functions extend naturally to distributions. We do not know how, in general, to define the product of two distributions. If we could do this, and if we wish this operation to be consistent with the identification $f \mapsto T_f$, we would have the relation

$$T_f \cdot T_g = T_{fg}.$$

But T_{fg} does not necessarily exist, since f and g can be locally integrable without fg being locally integrable (take $f(x) = g(x) = |x|^{-1/2}$). There is no problem if one of the functions, say g, is continuous. In this case,

$$\langle T_{fg}, \varphi \rangle = \int_{\mathbb{R}} fg\varphi = \langle T_f, g\varphi \rangle$$

for all $\varphi \in \mathscr{D}$. This suggests the formula

$$\langle gT, \varphi \rangle = \langle T, g\varphi \rangle.$$

For the expression on the right to be well-defined, we must have $g\varphi \in \mathscr{D}$ for all φ, and this implies that g must be infinitely differentiable.

28.3.1 Definition The product of a distribution T by an infinitely differentiable function g, denoted by gT, is defined by

$$\langle gT, \varphi \rangle = \langle T, g\varphi \rangle, \quad \text{for all } \varphi \in \mathscr{D}. \tag{28.4}$$

It is easy to see that gT is a well-defined distribution.

EXAMPLE: Take $T = \delta_a$. Then

$$\langle g\delta_a, \varphi \rangle = \langle \delta_a, g\varphi \rangle = g(a)\varphi(a) = \langle g(a)\delta_a, \varphi \rangle,$$

and hence

$$g\delta_a = g(a)\delta_a. \tag{28.5}$$

As a particular case,

$$x\delta = 0.$$

28.4 The derivative of a distribution

From what we have seen in Section 27.1, we should define the derivative T' of the distribution T by the formula

$$\langle T', \varphi \rangle = -\langle T, \varphi' \rangle, \tag{28.6}$$

which we deduce directly from (27.2). We need to show that T' is a distribution. It is clearly a linear mapping defined on $\mathscr{D}$, since φ' is in $\mathscr{D}$ for all $\varphi \in \mathscr{D}$. It is also continuous: If $\varphi_n \to 0$ in $\mathscr{D}$, then $\varphi_n' \to 0$ in $\mathscr{D}$ (Proposition 27.3.3), and hence $T'(\varphi_n) \to 0$.

Since T' is a distribution, it has a derivative $T'' : \mathscr{D} \to \mathbb{C}$ given by

$$\langle T'', \varphi \rangle = \langle T, \varphi'' \rangle,$$

and so on.

28.4.1 Proposition *Each distribution T is infinitely differentiable, and the nth derivative of T satisfies the relation*

$$\langle T^{(n)}, \varphi \rangle = (-1)^n \langle T, \varphi^{(n)} \rangle$$

for all $\varphi \in \mathscr{D}$.

28.4.2 Examples

(a) The derivative of the point distribution δ_a is given by

$$\delta_a'(\varphi) = -\varphi'(a). \tag{28.7}$$

(b) For the derivative of the unit step u we have

$$\langle T_u, \varphi \rangle = \int_0^{+\infty} \varphi(x)\, dx,$$

and hence

$$\langle T_u', \varphi \rangle = -\langle T_u, \varphi' \rangle = -\int_0^{+\infty} \varphi'(x)\, dx = \varphi(0) = \langle \delta, \varphi \rangle.$$

Thus we see that distribution theory leads us to the derivative of Heaviside's unit step being the point distribution at the origin. It was our heuristic derivation of this result that prompted us to use the point distribution as a model for the unit impulse. We will see in the next lesson a more direct (mathematical) reason for modeling the unit impulse at $x = 0$ by δ.

28.4.3 The usual derivative and the derivative in the sense of distributions

If f is a locally integrable function, then the associated distribution T_f has a derivative, which is the distribution T'_f. We call this distribution *the derivative of f in the sense of distributions.* When f is absolutely continuous on all compact intervals $[a,b]$, integration by parts (Theorem 14.5.6) shows that

$$\langle T'_f, \varphi\rangle = -\int_{\mathbb{R}} f\varphi' = \int_{\mathbb{R}} f'\varphi = \langle T_f, \varphi'\rangle,$$

and hence that

$$T'_f = T_{f'}.$$

The identification of a locally integrable function with its associated distribution leads to the following result: *For a function that is absolutely continuous on all compact intervals $[a,b]$, the derivative of f in the sense of distributions agrees a.e. with the usual, or ordinary, derivative.* For example,

$$f(x) = |x|$$

and

$$f'(x) = \operatorname{sign}(x) \quad \text{for a.e. } x.$$

In the general case of a locally integrable function f, this derivative—whether or not it is a function—will be denoted by f' when there is no ambiguity. The derivative in the sense of distributions is expressed by the relation

$$\langle f', \varphi\rangle = -\langle f, \varphi'\rangle \quad \text{for all } \varphi \in \mathscr{D}. \tag{28.8}$$

For example,

$$u' = \delta. \tag{28.9}$$

28.4.4 The derivative in the sense of distributions of a discontinuous function

As an example, we take a function f that is continuously differentiable on the intervals $(-\infty, a)$, (a,b), and $(b,+\infty)$ and that has finite left and right limits at a and b (see Figure 28.1). Taking the derivative of f in the sense of distributions, we have

$$\langle T'_f, \varphi\rangle = -\int_{-\infty}^{+\infty} f\varphi' = -\int_{\infty}^{a} f(x)\varphi'(x)\,dx - \int_{a}^{b} f(x)\varphi'(x)\,dx$$
$$- \int_{b}^{+\infty} f(x)\varphi'(x)\,dx.$$

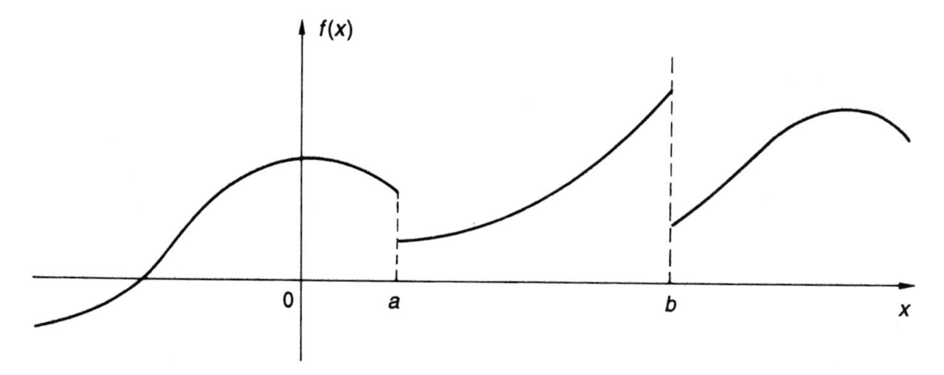

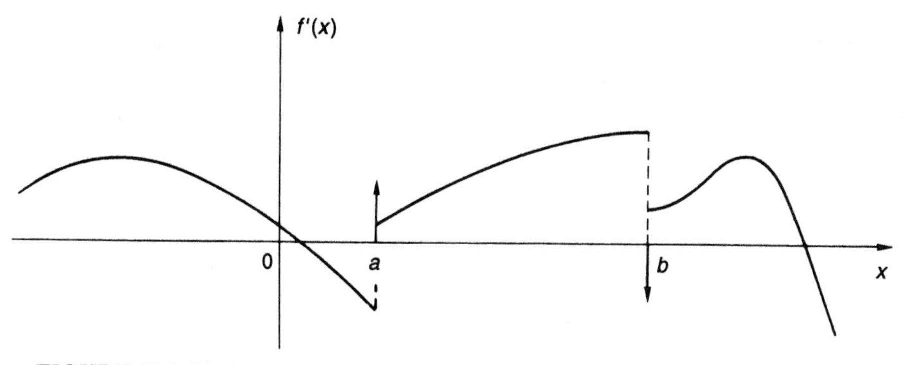

FIGURE 28.1. Derivative in the sense of distributions of a discontinuous function.

We integrate by parts on each of the intervals; this shows that

$$\langle T_f', \varphi\rangle = -f(a-)\varphi(a) + \int_{-\infty}^{a} f'(x)\varphi(x)\,dx - f(b-)\varphi(b) + f(a+)\varphi(a)$$
$$+ \int_a^b f'(x)\varphi(x)\,dx + f(b+)\varphi(b) + \int_b^{+\infty} f'(x)\varphi(x)\,dx.$$

This can be written

$$\langle T_f', \varphi\rangle = \lambda\varphi(a) + \mu\varphi(b) + \langle T_{f'}, \varphi\rangle$$

with

$$\lambda = f(a+) - f(a-),$$
$$\mu = f(b+) - f(b-).$$

Written in different notation, we have

$$T_f' = T_{f'} + \lambda\delta_a + \mu\delta_b. \tag{28.10}$$

$T_{f'}$ denotes the distribution associated with the usual derivative of f, which is defined everywhere except at a and b. In this case, we see that the usual derivative and the derivative in the sense of distributions give different results: The discontinuity of f at a point, say a, causes a point distribution $\lambda\delta_a$ to appear in the (distribution) derivative; the coefficient λ is equal to the size of the "jump" of the function at a. Formula (28.9) is a particular case.

28.4.5 An infinite number of discontinuities

If, for example, a function f has infinitely many points of discontinuity $(na)_{n\in\mathbb{Z}}$ equally spaced on $\mathbb{R}$, then formula (28.10) can be generalized. Since every test function φ has compact support, one is led to the finite case described above.

EXAMPLE: Consider the periodic function f with period a defined on $(0, a)$ by $f(x) = x/a$ (Figure 28.2).

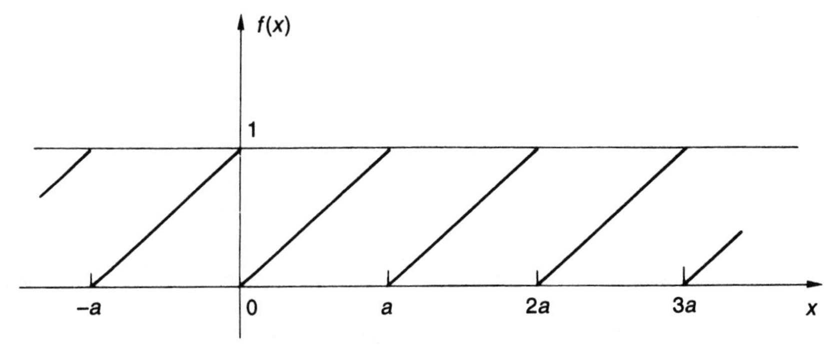

FIGURE 28.2.

Then we have (see Section 27.3.4(b))

$$f' = \frac{1}{a} - \sum_{n=-\infty}^{+\infty} \delta_{na}.$$

We are now going to see what can be done, from the point of view of distributions, with certain functions that are unbounded and not integrable in the neighborhood of a point.

28.5 Some new distributions

28.5.1 The distribution $\mathrm{pv}\left(\frac{1}{x}\right)$

The function $f(x) = 1/x$, defined for $x \neq 0$, is not integrable around the origin. Thus we cannot associate a distribution with f, and we will have the same problem with any rational function having a real pole. This is something of a wrench in our distribution machinery, for one of our main goals with distributions is to extend the notion of function. Furthermore, the functions in question often arise in practical problems. We will see here how this problem can be resolved. Although f is not locally integrable, its primitive $F(x) = \ln(|x|)$ is locally integrable. The distribution that comes to our aid is simply the derivative of F in the sense of distributions. This distribution is well-defined (Section 28.4.3) and is denoted by $\mathrm{pv}\left(\frac{1}{x}\right)$, where pv stands for "principal value." But what is the value of

$$\langle \mathrm{pv}\Big(\frac{1}{x}\Big), \varphi\rangle?$$

We have

$$\langle T_F', \varphi\rangle = -\langle T_F, \varphi'\rangle = -\int_{\mathbb{R}} \ln(|x|)\varphi'(x)\,dx.$$

This integral can be written as a limit,

$$\langle T_F', \varphi\rangle = \lim_{\varepsilon\to 0} J_\varepsilon,$$

with

$$J_\varepsilon = -\int_{-\infty}^{-\varepsilon} \ln(|x|)\varphi'(x)\,dx - \int_{\varepsilon}^{+\infty} \ln(|x|)\varphi'(x)\,dx.$$

Integrating by parts shows that

$$J_\varepsilon = [\varphi(\varepsilon) - \varphi(-\varepsilon)]\ln(\varepsilon) + \int_{|x|\geq\varepsilon} \frac{\varphi(x)}{x}\,dx.$$

The mean value theorem,

$$\varphi(\varepsilon) - \varphi(-\varepsilon) = 2\varepsilon\varphi'(c_\varepsilon), \quad |c_\varepsilon| < \varepsilon,$$

shows that the integrated term tends to 0 as $\varepsilon \to 0$. Thus we see that

$$\langle \mathrm{pv}\Big(\frac{1}{x}\Big), \varphi\rangle = \lim_{\varepsilon\to 0}\int_{|x|\geq\varepsilon} \frac{\varphi(x)}{x}\,dx = \lim_{\varepsilon\to 0}\int_{\varepsilon}^{+\infty} \frac{\varphi(x)-\varphi(-x)}{x}\,dx. \tag{28.11}$$

Note that the integral

$$\int_{\mathbb{R}} \frac{\varphi(x)}{x}\,dx$$

does not exist in general, but taking symmetric limits ($-\varepsilon$ and ε) around the origin guarantees the existence of the limit (28.11) as $\varepsilon \to 0$. This is a particular case of what is called the "principal value" of an integral.

One easily deduces from (28.11) that

$$x \cdot \mathrm{pv}\Big(\frac{1}{x}\Big) = 1. \tag{28.12}$$

Indeed, by using (28.11) again, we see that

$$\langle x \cdot \operatorname{pv}\Big(\frac{1}{x}\Big), \varphi\rangle = \langle \operatorname{pv}\Big(\frac{1}{x}\Big), x\varphi\rangle = \lim_{\varepsilon \to 0} \int_{|x| \geq \varepsilon} \varphi(x)\, dx = \int_{\mathbb{R}} \varphi = \langle 1, \varphi\rangle.$$

Finally, we repeat the defining relation:

$$\operatorname{pv}\Big(\frac{1}{x}\Big) = (\ln|x|)'. \tag{28.13}$$

28.5.2 The distribution $\operatorname{fp}\big(\frac{1}{x^2}\big)$

Here is a second example, which will give a better idea about the generality of the technique used in the last section. This time the function that is not locally integrable is $g(x) = 1/x^2$. The associated distribution, which remains to be defined, will be denoted by $\operatorname{fp}\big(\frac{1}{x^2}\big)$, where fp stands for "finite part." There is no sense in writing

$$\langle \operatorname{fp}\Big(\frac{1}{x^2}\Big), \varphi\rangle = \lim_{\varepsilon \to 0} \int_{|x| \geq \varepsilon} \frac{\varphi(x)}{x^2}\, dx,$$

since this time the limit generally does not exist. Again, the plan is to consider a primitive of g, but of a higher order. The function $F(x) = \ln(|x|)$ is, up to a sign, a second primitive of g, so we define

$$\operatorname{fp}\Big(\frac{1}{x^2}\Big) = -F'',$$

where the derivative is taken in the sense of distributions. The new distribution is thus the negative of the derivative of $\operatorname{pv}\big(\frac{1}{x}\big)$:

$$\operatorname{fp}\Big(\frac{1}{x^2}\Big) = -\operatorname{pv}\Big(\frac{1}{x}\Big)'. \tag{28.14}$$

This and (28.11) show that

$$\langle \operatorname{fp}\Big(\frac{1}{x^2}\Big), \varphi\rangle = \lim_{\varepsilon \to 0} \int_{\varepsilon}^{+\infty} \frac{\varphi'(x) - \varphi'(-x)}{x}\, dx.$$

Integrating by parts shows that

$$\langle \operatorname{fp}\Big(\frac{1}{x^2}\Big), \varphi\rangle = \lim_{\varepsilon \to 0} \Big\{ -\frac{1}{\varepsilon}[\varphi(\varepsilon) + \varphi(-\varepsilon) - 2\varphi(0)] + \int_{\varepsilon}^{+\infty} \frac{\varphi(x) + \varphi(-x) - 2\varphi(0)}{x^2}\, dx \Big\}.$$

It is not difficult to see that the integrated term tends to 0 as $\varepsilon \to 0$, and we are left with the formula

$$\langle \operatorname{fp}\Big(\frac{1}{x^2}\Big), \varphi \rangle = \lim_{\varepsilon \to 0} \int_\varepsilon^{+\infty} \frac{\varphi(x) + \varphi(-x) - 2\varphi(0)}{x^2}\, dx. \tag{28.15}$$

As before, one easily shows that

$$x^2 \cdot \operatorname{fp}\Big(\frac{1}{x^2}\Big) = 1. \tag{28.16}$$

In summary:

$$\langle \operatorname{pv}\Big(\frac{1}{x}\Big), \varphi \rangle = -\int_{\mathbb{R}} \ln|x| \varphi'(x)\, dx = \lim_{\varepsilon \to 0} \int_\varepsilon^{+\infty} \frac{\varphi(x) - \varphi(-x)}{x}\, dx,$$
$$\langle \operatorname{fp}\Big(\frac{1}{x^2}\Big), \varphi \rangle = -\int_{\mathbb{R}} \ln|x| \varphi''(x)\, dx = \lim_{\varepsilon \to 0} \int_\varepsilon^{+\infty} \frac{\varphi(x) + \varphi(-x) - 2\varphi(0)}{x^2}\, dx.$$

28.6 Exercises

Exercise 28.1

(a) Compute the successive derivatives in the sense of distributions of the following functions:

$$u(x); \quad f(x) = x^k u(x), \ \ k \in \mathbb{N}^*; \quad g(x) = |x|.$$

(b) Compute the first two derivatives in the sense of distributions of the functions

$$h(x) = |\sin x| \quad \text{and} \quad k(x) = u(x)\sin x.$$

Exercise 28.2 What is the parity of $\delta^{(n)}$?

Exercise 28.3 Suppose $T \in \mathscr{D}'(\mathbb{R})$ and $\alpha \in C^\infty(\mathbb{R})$. Prove the following results:

$$(\alpha T)' = \alpha' T + \alpha T',$$
$$(\alpha T)^{(n)} = \sum_{k=0}^{n} \binom{n}{k} \alpha^{(k)} T^{(n-k)}.$$

Exercise 28.4 Compare $(T')_\sigma$ and $(T_\sigma)'$. Compute $(\delta')_\sigma$ and $(\delta_\sigma)'$.

Exercise 28.5 Suppose $f \in L^1_{\text{loc}}(\mathbb{R})$. Show that $F(x) = \displaystyle\int_a^x f(t)\, dt$, $a \in \mathbb{R}$, has f as its derivative in the sense of distributions.

Exercise 28.6 Compute $x\delta'$, $x^2\delta'$, and $x\delta''$.

Exercise 28.7 For $f : \mathbb{R} \to \mathbb{C}$ measurable and $\lambda \in \mathbb{R}^*$, the operator h_λ is defined by

$$h_\lambda f(x) = f(\lambda x), \quad x \in \mathbb{R}.$$

Extend this definition to distributions.

Exercise 28.8

(a) Define the conjugate $\overline{T}$ of a distribution $T \in \mathscr{D}'(\mathbb{R})$.

(b) Define $\text{Re}(T)$ and $\text{Im}(T)$. Characterize a distribution as being real or imaginary.

(c) Verify that δ_a, $a \in \mathbb{R}$, and $\text{pv}\left(\frac{1}{x}\right)$ are real.

(d) Prove that the derivative of a real distribution is real.

Exercise 28.9 Prove that $x \cdot \text{fp}\left(\frac{1}{x^2}\right) = \text{pv}\left(\frac{1}{x}\right)$ two different ways.

***Exercise 28.10** Prove the equality

$$(e^x - 1)\,\text{pv}\left(\frac{1}{x}\right) = \frac{e^x - 1}{x}.$$

How can this be generalized? Deduce the decomposition

$$\text{pv}\left(\frac{1}{x}\right) = S + f,$$

where S is a distribution with bounded support and $f \in L^2(\mathbb{R})$.

***Exercise 28.11 (solution of $xT = 0$, $T \in \mathscr{D}'(\mathbb{R})$)**

(a) Show that a function $\chi \in \mathscr{D}(\mathbb{R})$ satisfies $\chi(0) = 0$ if and only if $\chi = x\psi$ with $\psi \in \mathscr{D}(\mathbb{R})$.

(b) Take $\theta \in \mathscr{D}(\mathbb{R})$ such that $\theta(0) = 1$. Show that for $\varphi \in \mathscr{D}(\mathbb{R})$ there exists a $\psi \in \mathscr{D}(\mathbb{R})$ such that $\varphi = \varphi(0)\theta + x\psi$.

(c) Deduce from this that $T \in \mathscr{D}'(\mathbb{R})$ is a solution of $xT = 0$ if and only if T is of the form $K\delta$ with $K \in \mathbb{C}$.

(d) Give an example of a distribution T such that $\text{supp}(T) \subset \{0\}$ and $xT \neq 0$.

Exercise 28.12 (solution of $xT = 1$, $T \in \mathscr{D}'(\mathbb{R})$) Deduce from Exercise 28.11 and equation (28.13) that $T \in \mathscr{D}'(\mathbb{R})$ is a solution of the equation $xT = 1$ if and only if T is of the form $\text{pv}\left(\frac{1}{x}\right) + K\delta$ with $K \in \mathbb{C}$.

Exercise 28.13 Solve the equations $x^2T = 0$ and $x^2T = 1$, $T \in \mathscr{D}'(\mathbb{R})$.
Hint: Use the methods of Exercises 28.11–28.12 and equation (28.17).

Exercise 28.14 (solution of $xT = U$; $T, U \in \mathscr{D}'(\mathbb{R})$)

(1) Suppose $U \in \mathscr{D}'(\mathbb{R})$ and $0 \notin \operatorname{supp}(U)$.

(a) Show that there exists an $\alpha > 0$ such that $\operatorname{supp}(U) \cap (-\alpha, \alpha) = \emptyset$.

(b) Show that the equation $xT = U$ has at least one solution $T_0 \in \mathscr{D}'(\mathbb{R})$.

(c) Give the general form of all solutions $T \in \mathscr{D}'(\mathbb{R})$.

(d) Find all distributions $T \in \mathscr{D}'(\mathbb{R})$ such that

$$xT = \sum_{k=-\infty,\, k\neq 0}^{+\infty} \delta_k.$$

(2) Show that one can find $T, U \in \mathscr{D}'(\mathbb{R})$ such that $0 \in \operatorname{supp}(U)$ and $xT = U$.

Exercise 28.15

Take $b \in L^1_{\text{loc}}(\mathbb{R})$ and define a by $a(x) = xb(x)$.

(a) Solve the equation $xT = a$, $T \in \mathscr{D}'(\mathbb{R})$.

(b) Apply this to $a(x) = \sin x$.

Hint: For (a), use Exercise 28.11(c) to show that $T = b + K\delta$, $K \in \mathbb{C}$.

Exercise 28.16

Assume that $a \in L^1_{\text{loc}}(\mathbb{R})$ is such that the function m defined by

$$m(x) = \frac{a(x) - a(0)}{x}$$

is also in $L^1_{\text{loc}}(\mathbb{R})$.

(a) Solve the equation $xT = a$, $T \in \mathscr{D}'(\mathbb{R})$.

(b) Apply this to $a(x) = e^x$.

Hint: Use the relation $x \cdot \operatorname{pv}\left(\frac{1}{x}\right) = 1$ to transform the equation and reduce it to the case in Exercise 28.15.

Exercise 28.17

(a) Assume $\varphi \in \mathscr{D}(\mathbb{R})$ with $\operatorname{supp}(\varphi) \subset R\backslash\{0\}$.

- Show that there exists an $\alpha > 0$ such that $\varphi(x) = 0$ for all $x \in [-\alpha, \alpha]$.
- Deduce from this that for each $k \in \mathbb{N}$ there exists $\psi \in \mathscr{D}(\mathbb{R})$ such that $\varphi(x) = x^k \psi(x)$ for all x.

(b) Show that

$$x^n T = 0 \text{ for some } n \in \mathbb{N} \quad \Longrightarrow \quad \operatorname{supp}(T) \subset \{0\}.$$

Hint: If $\varphi \in \mathcal{D}(\mathbb{R})$ is such that $\operatorname{supp}(\varphi) \subset \mathbb{R}\backslash\{0\}$, by (a) there exists $\psi \in \mathscr{D}(\mathbb{R})$ such that $\varphi(x) = x^n\psi(x)$ for all x. Then $\langle T, \varphi\rangle = \langle T, x^n\psi\rangle = \langle x^n T, \psi\rangle = 0$, and hence $\operatorname{supp}(T) \subset \{0\}$).

****Exercise 28.18** Let $g \in C^\infty(\mathbb{R})$ and let its derivative be odd and bounded.

(a) For $x < 0$, show that the integral

$$\int_A^x \frac{g'(t)}{t}\,dt, \quad A < 0,$$

has a finite limit as $A \to -\infty$. Let f be the even function defined for $x < 0$ by

$$f(x) = \int_{-\infty}^x \frac{g'(t)}{t}\,dt = \lim_{A\to-\infty} \int_A^x \frac{g'(t)}{t}\,dt.$$

(b) Show that

$$|f(x)| \le C(1 - \log|x|) \quad \text{for} \quad 0 < |x| \le 1.$$

Deduce that $f \in L^1_{\text{loc}}(\mathbb{R})$.

(c) Show that the derivative of f in the sense of distributions is

$$f'(x) = g'(x)\,\text{pv}\left(\frac{1}{x}\right).$$

Lesson 29

Convergence of a Sequence of Distributions

We mentioned in Section 26.4 that an essential property of derivation in the sense of distributions is its continuity:

$$T_n \to T \quad \Longrightarrow \quad T_n' \to T'.$$

We saw in Section 26.2 that from the point of view of physics, the impulse is a limit. For these and other reasons it is important to investigate the notion of limit in $\mathscr{D}'$.

29.1 The limit of a sequence of distributions

The limit of a sequence of distributions is simply the "pointwise limit" where the "points" are test functions.

29.1.1 Definition A sequence of distributions $(T_n)_{n\in\mathbb{N}}$ is said to converge to the distribution T if

$$\langle T_n, \varphi \rangle \to \langle T, \varphi \rangle$$

for all $\varphi \in \mathscr{D}$.

29.1.2 Examples

(a) If the real sequence $(a_n)_{n\in\mathbb{N}}$ converges to a, then δ_{a_n} converges to δ_a, since $\varphi(a_n) \to \varphi(a)$ for all φ in $\mathscr{D}$.

(b) Consider the sequence of functions $(f_n)_{n\in\mathbb{N}}$ defined by

$$f_n(x) = \sin 2\pi n x.$$

For fixed x (not an integer or half-integer), the sequence $f_n(x)$ does not converge as $n \to +\infty$. Thus the sequence of functions (f_n) does not tend pointwise to a limit. We know, however, from the Riemann–Lebesgue theorem that

$$\int_{\mathbb{R}} \varphi(x) \sin 2\pi n x \, dx \to 0$$

for all $\varphi \in \mathscr{D}$. We conclude that the sequence (f_n) tends to 0 in $\mathscr{D}'$, and we have the following negative result:

$$f_n \to f \text{ in } \mathscr{D}' \quad \not\Rightarrow f_n(x) \to f(x) \text{ for a.e. } x \in \mathbb{R}.$$

Finally, we wish to emphasize *the convergence of a sequence of functions in the sense of distributions:*

$$f_n \to f \text{ in } \mathscr{D}' \quad \Longleftrightarrow \quad \int f_n \varphi \to \int f \varphi \quad \text{for all } \varphi \in \mathscr{D}. \tag{29.1}$$

29.1.3 Theorem (continuity of derivation) *If the sequence of distributions $(T_n)_{n \in \mathbb{N}}$ converges to the distribution T, then the sequence of derivatives $(T_n')_{n \in \mathbb{N}}$ converges to T'.*

The proof is read directly from Definition 29.1.1 and the definition of the derivative (28.6). This theorem formalizes our statement in Section 26.4.

29.2 Revisiting Dirac's impulse

In Lesson 26, physical reasoning led us to consider an impulse as the limit as $n \to +\infty$ of a family of functions like this (see Figure 29.1):

$$f_n(x) = \begin{cases} \dfrac{n}{2} & \text{if } |x| \leq \dfrac{1}{n}, \\ 0 & \text{otherwise.} \end{cases}$$

We wish to see whether this sequence has a limit in the sense of distributions. For $\varphi \in \mathscr{D}$, the mean value theorem implies that

$$\langle f_n, \varphi \rangle = \frac{n}{2} \int_{-\frac{1}{n}}^{\frac{1}{n}} \varphi(x) \, dx = \varphi(c_n)$$

for some c_n, $-1/n < c_n < 1/n$. As $n \to +\infty$, $\varphi(c_n) \to \varphi(0)$; hence,

$$\langle f_n, \varphi \rangle \to \varphi(0) = \langle \delta, \varphi \rangle$$

for all $\varphi \in \mathscr{D}$. This means that

$$f_n \to \delta \text{ in } \mathscr{D}'.$$

Although we have taken a particular form for the functions f_n, this argument is easily generalized to more general sequences, and it shows that the

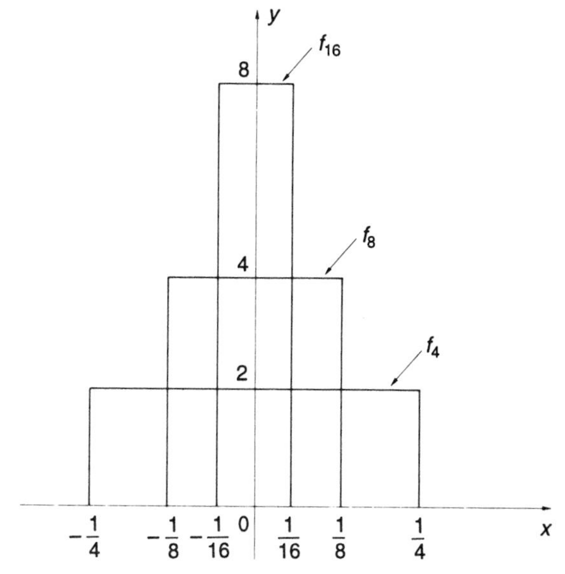

FIGURE 29.1. The impulse as the limit of functions f_n.

point distribution at the origin is the model of the impulse that we sought in Lesson 26.

The distribution δ models the unit impulse at the origin.

This argument justifies our calling δ the Dirac unit impulse (or mass) at the origin, and it agrees with the fact that δ is the derivative of Heaviside's unit step function (Sections 26.2 and 26.3).

29.3 Relations with the convergence of functions

The convergence of a sequence of functions in the sense of distributions is generally "weaker" than the notions of convergence we have previously encountered. Here are several results (not an exhaustive list) that allow us to conclude convergence in the sense of distributions.

29.3.1 Proposition *Let $(f_n)_{n\in\mathbb{N}}$ be a sequence of integrable functions that converges to f in $L^1(\mathbb{R})$. Then (f_n) converges to f in the sense of distributions.*

Proof. This is a consequence of the following inequalities:

$$\left|\int_{\mathbb{R}}(f_n - f)\varphi\right| \leq \int_{\mathbb{R}}|f_n - f||\varphi| \leq \|\varphi\|_\infty\|f_n - f\|_1. \qquad \square$$

29.3.2 Proposition *If a sequence $(f_n)_{n\in\mathbb{N}}$ of square-integrable functions converges in $L^2(\mathbb{R})$ to a function $f \in L^2(\mathbb{R})$, then (f_n) converges to f in the sense of distributions.*

Proof. This is a consequence of Schwarz's inequality:

$$\left|\int_{\mathbb{R}} (f_n - f)\varphi\right| \leq \int_{\mathbb{R}} |f_n - f||\varphi| \leq \|\varphi\|_2 \|f_n - f\|_2. \qquad \square$$

29.3.3 Proposition *Let $(f_n)_{n\in\mathbb{N}}$ be a sequence of measurable functions such that $f_n(x) \to f(x)$ almost everywhere. If there is a function $g \in L^1_{\text{loc}}(\mathbb{R})$ such that for all $n \in \mathbb{N}$*

$$|f_n(x)| \leq g(x) \quad \text{for a.e. } x,$$

then (f_n) tends to f in the sense of distributions.

This proposition is a direct consequence of the theorem on dominated convergence.

29.3.4 Proposition *Suppose $(f_n)_{n\in\mathbb{N}}$ is a sequence of functions in $L^1_{\text{loc}}(\mathbb{R})$ that converges to f uniformly on every bounded interval. Then (f_n) tends to f in the sense of distributions.*

The proof is left as an exercise.

29.4 Applications to the convergence of trigonometric series

Consider an arbitrary trigonometric series

$$\sum_{n=-\infty}^{+\infty} \alpha_n e^{2i\pi n \frac{x}{a}}, \tag{29.2}$$

where the α_n do not necessarily tend to 0. This series is not, a priori, the Fourier series of a function, and furthermore, it is not, in general, pointwise convergent. Nevertheless, this series converges in $\mathscr{D}'$ under rather general conditions and thus defines a distribution. Recall that the series (29.2) converges if the sums

$$f_N(x) = \sum_{n=-N}^{N} \alpha_n e^{2i\pi n \frac{x}{a}}$$

converge as $N \to +\infty$..

29.4.1 Definition A complex sequence $(\alpha_n)_{n\in\mathbb{Z}}$ is said to be slowly increasing if there exist a positive constant A and an integer k such that

$$|\alpha_n| \leq A|n|^k \tag{29.3}$$

for all sufficiently large $|n|$.

This is equivalent to saying there exist a constant C and an integer $k \in \mathbb{N}$ such that

$$|\alpha_n| \leq C(1+|n|^k) \tag{29.4}$$

for all $n \in \mathbb{Z}$.

29.4.2 Theorem *If the sequence (α_n) is slowly increasing, then the series (29.2) converges in $\mathscr{D}'$ to a periodic distribution with period a.*

Proof. Let k be the integer in (29.3) and consider the series

$$\sum_{n\neq 0} n^{-k-2}\alpha_n e^{2i\pi n\frac{x}{a}}. \tag{29.5}$$

By hypothesis, the modulus of the general term is dominated by A/n^2 for all sufficiently large n. Thus the series converges uniformly on $\mathbb{R}$ to a continuous function F:

$$F(x) = \sum_{n\neq 0} n^{-k-2}\alpha_n e^{2i\pi n\frac{x}{a}}.$$

From Proposition 29.3.4 we have convergence in $\mathscr{D}'$, and by continuity of derivation, the series differentiated $k+2$ times tends to $F^{(k+2)}$ in the sense of distributions:

$$\sum_{\substack{n=-N\\ n\neq 0}}^{N} \left(\frac{2i\pi}{a}\right)^{k+2} \alpha_n e^{2i\pi n\frac{x}{a}} \to F^{(k+2)}$$

as $N \to +\infty$. This proves that the series (29.2) converges in the sense of distributions to the distribution

$$T = \left(\frac{a}{2i\pi}\right)^{k+2} F^{(k+2)} + \alpha_0.$$ □

We will see in Proposition 36.1.3 that (29.2) converges in $\mathscr{D}'$ to f when it is the Fourier series of a function $f \in L^1_{\mathrm{loc}}(0,a)$.

REMARKS:

Note that the technique used here—going to a higher-order primitive and then descending by differentiation in the sense of distributions—is the same as we used to define $\mathrm{pv}\left(\frac{1}{x}\right)$ and $\mathrm{fp}\left(\frac{1}{x^2}\right)$.

The general term of the series (29.2) need not tend to 0 in the sense of functions, but it evidently tends to 0 in $\mathscr{D}'$.

29.4.3 Theorem (term-by-term differentiation)

Let $(u_n)_{n\in\mathbb{N}}$ be a sequence of functions that are absolutely continuous on bounded intervals. Assume that the series

$$\sum_{n=0}^{+\infty} u_n$$

converges in $\mathscr{D}'$ to a locally integrable function f. Then the differentiated series

$$\sum_{n=0}^{+\infty} u_n'$$

converges in $\mathscr{D}'$ to the derivative f' (in the sense of distributions) of f.

The proof is a direct application of the continuity of derivation.

29.5 The Fourier series of Dirac's comb

29.5.1 Given a real positive number a, Dirac's comb (Figure 29.2) with mesh a is the distribution defined by (Section 27.3.4)

$$\Delta_a = \sum_{n=-\infty}^{+\infty} \delta_{na}.$$

Thus

$$\langle \Delta_a, \varphi \rangle = \sum_{n=-\infty}^{+\infty} \varphi(na)$$

for all $\varphi \in \mathscr{D}$, and it is clear that Δ_a is periodic with period a.

29.5.2 The product $f\Delta_a$ is well-defined whenever $f \in C^\infty$ (Definition 28.3.1), and

$$\langle f\Delta_a, \varphi \rangle = \langle \Delta_a, f\varphi \rangle = \sum_{n=-\infty}^{+\infty} f(na)\varphi(na).$$

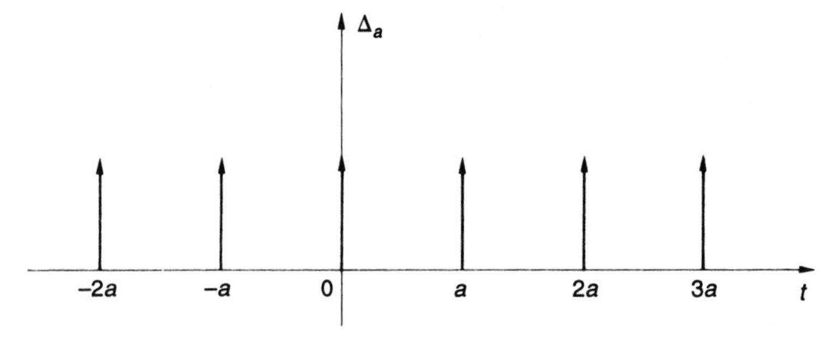

FIGURE 29.2. Dirac's comb.

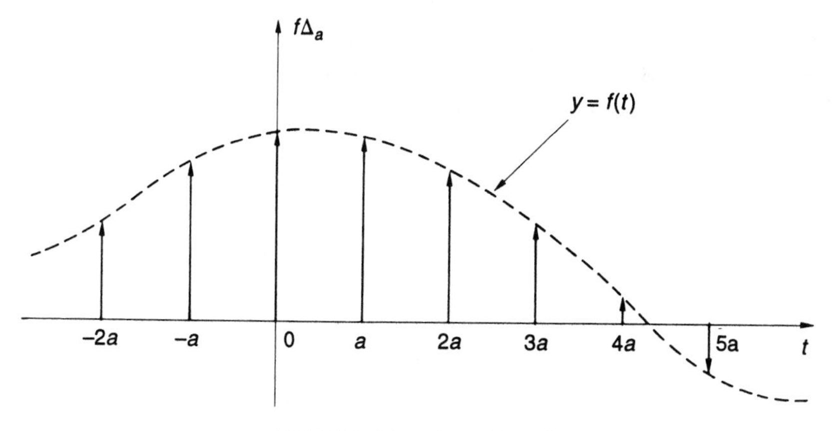

FIGURE 29.3. Sampling f.

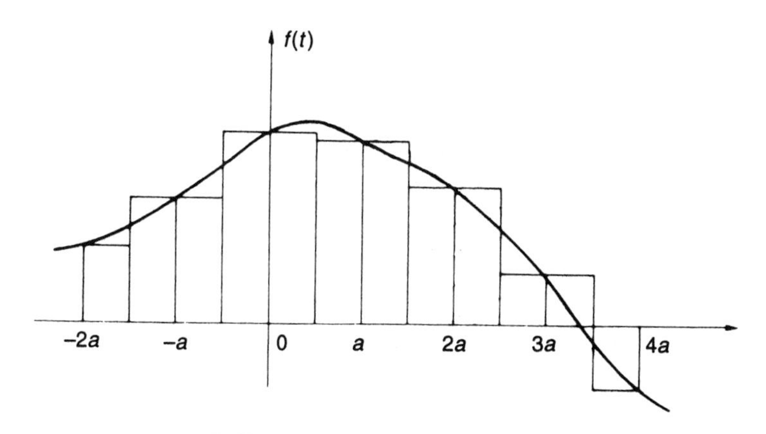

FIGURE 29.4. Approximating f.

Written as distributions,

$$f\Delta_a = \sum_{n=-\infty}^{+\infty} f(na)\delta_{na}.$$

Equation (29.6) is a sequence of impulses that represents the sampling of the signal f every a "seconds." (See Figure 29.3.) From the point of view of approximating f by a sequence of Dirac masses and letting a tend to zero, it is in fact the sequence of impulses $af\Delta_a$ that approximates f by step functions (Figure 29.4). In the sense of distributions,

$$f \approx \sum_{n=-\infty}^{+\infty} f(na)\ \tau_{na}\ \chi_{\left[-\frac{a}{2},\frac{a}{2}\right)} \approx \sum_{n=-\infty}^{+\infty} f(na)\delta_{na}, \quad \text{or} \quad f \approx af\Delta_a.$$

29.5.3 We will show that Δ_a can be expressed as a trigonometric series just as for a periodic function. We first develop the Fourier series of the function with period a defined on $(0, a)$ by $f(x) = x/a$ (Figure 29.5).

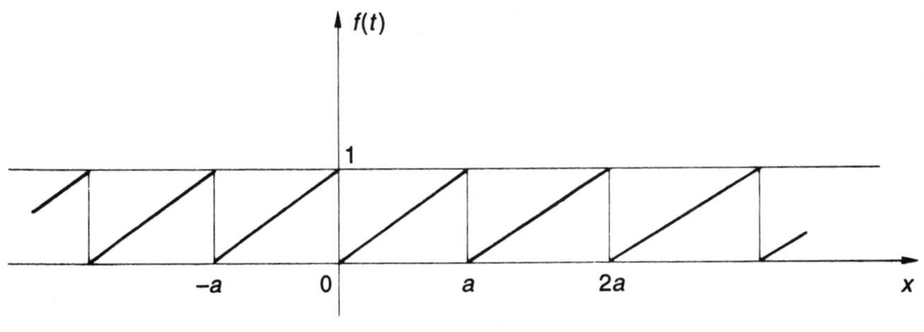

FIGURE 29.5.

An easy computation shows that

$$f(x) = \frac{1}{2} + \frac{i}{2\pi} \sum_{n \neq 0} \frac{1}{n} e^{2i\pi n \frac{x}{a}}.$$

This series converges in the norm of $L^2(0, a)$ (Proposition 16.3.4). By an argument similar to that used in Proposition 29.3.2, this series also converges in $\mathscr{D}'$; by continuity of derivation, we can differentiate term by term; hence

$$f'(x) = -\frac{1}{a} \sum_{n \neq 0} e^{2i\pi n \frac{x}{a}}.$$

On the other hand, from Section 28.4.5,

$$f' = \frac{1}{a} - \sum_{n=-\infty}^{+\infty} \delta_{na},$$

and we see that

$$\Delta_a = \sum_{n=-\infty}^{+\infty} \delta_{na} = \frac{1}{a} \sum_{n=-\infty}^{+\infty} e^{2i\pi n \frac{x}{a}}. \tag{29.6}$$

The series on the right diverges in the sense of functions, but it converges in the sense of distributions to Dirac's comb. This series is the Fourier series expansion of Dirac's comb, and it illustrates the general situation described in Section 29.4.

29.5.1 Remark The proof we used for the term-by-term differentiation of the Fourier series applies to any a-periodic function f that is square integrable over a period.

29.5.2 Proposition *The Fourier series of a function in $L^2_{\mathrm{p}}(0, a)$ can be differentiated term by term in the sense of distributions.*

29.6 Exercises

Exercise 29.1 Let T be a distribution and let (h_n) be a sequence of real numbers that tends to 0 ($h_n \neq 0$). Show that $\frac{1}{h_n}(T - \tau_{h_n} T) \to T'$ in $\mathscr{D}'(\mathbb{R})$ as $n \to +\infty$.

Exercise 29.2 Show that every sequence of functions (f_n) that satisfies the following three conditions converges to δ as $n \to +\infty$:

$$f_n \in L^1(\mathbb{R}) \text{ and } \int_{\mathbb{R}} f_n(x)\, dx = 1.$$

$$f_n \geq 0 \text{ a.e.}$$

$$f_n = 0 \text{ a.e. on } (-\infty, -1/n) \cup (1/n, +\infty).$$

***Exercise 29.3** Let $(f_n)_{n\in\mathbb{N}}$ be the sequence of functions defined by

$$f_n(x) = \frac{\sin nx}{x} \quad \text{for } x \neq 0.$$

(a) Show that $(f_n)_{n\in\mathbb{N}}$ does not converge pointwise.

(b) Using Exercise 27.2, show that f_n converges in $\mathscr{D}'(\mathbb{R})$ to $K\delta$, where K is a constant that can be computed from the property

$$\lim_{x\to+\infty} \int_0^x \frac{\sin x}{x}\, dx = \frac{\pi}{2}.$$

Exercise 29.4 Given $a \in \mathbb{R}$, consider the sequence of distributions

$$T_n = \sum_{k=-\infty}^{\infty} \lambda_k(n)\delta_{ka}.$$

Show that T_n tends to 0 in $\mathscr{D}'(\mathbb{R})$ if and only if for all $k \in \mathbb{Z}$,

$$\lim_{n\to\infty} \lambda_k(n) = 0.$$

Exercise 29.5 Prove Proposition 29.3.4.

Exercise 29.6 Assume that $f : \mathbb{R} \to \mathbb{C}$ is piecewise continuous on all closed, bounded intervals.

(a) Prove that the sequence of distributions

$$T_N = \frac{1}{N} \sum_{n=-\infty}^{+\infty} f\left(\frac{n}{N}\right)\delta_{\frac{n}{N}}$$

tends to f in the sense of distributions as $N \to +\infty$.

(b) Show that this is also true for the sequence

$$S_N = \frac{1}{N} \sum_{n=-N^2}^{N^2} f\left(\frac{n}{N}\right)\delta_{\frac{n}{N}}.$$

Exercise 29.7 Prove that

$$\sum_{n=-\infty}^{+\infty} (-1)^n e^{i\pi n x} = 2 \sum_{n=-\infty}^{+\infty} \delta_{2n+1}.$$

Hint: Use the Fourier series expansion of the function with period 2 defined on $(-1,1)$ by $f(x) = x$ (see Section 29.5.3).

Exercise 29.8 Show that the sequence of functions $f_n(x) = ne^{-\pi n^2 x^2}$ tends to δ as $n \to +\infty$.

Hint: First compute $\int_{\mathbb{R}} f_n(x)\, dx$ knowing that $\int_{\mathbb{R}} e^{-\pi x^2}\, dx = 1$; then show that $\lim_{n\to\infty} \int_{\mathbb{R}} f_n(x)(\varphi(x) - \varphi(0))\, dx = 0$ for all $\varphi \in \mathscr{D}(\mathbb{R})$.

Exercise 29.9 Let $(a_n)_{n\in\mathbb{Z}}$ be a real sequence and let $(\lambda_n)_{n\in\mathbb{Z}}$ be a complex sequence. Give a sufficient condition on (a_n) that implies that

$$T = \sum_{n=-\infty}^{+\infty} \lambda_n \delta_{a_n}$$

is a distribution.

Exercise 29.10 Prove directly that the sequence

$$f_n(x) = \alpha_n e^{2i\pi n x}, \quad n \in \mathbb{N},$$

where (α_n) is slowly increasing, tends to 0 in $\mathscr{D}'(\mathbb{R})$.

Lesson 30

Primitives of a Distribution

When studying physical systems governed by differential equations, physicists often consider the derivatives to be taken in the sense of distributions. This is necessary, for example, when the inputs are discontinuous. This leads one to define generalized solutions of differential equations in terms of distributions.

30.1 Distributions whose derivatives are zero

The question of whether a distribution has a primitive reduces to the following problem:

$$\text{Given } T \in \mathscr{D}', \text{ find } U \in \mathscr{D}' \text{ such that } U' = T.$$

We will see that the results for distributions are the same as those for functions. We first consider the case $T = 0$.

30.1.1 Theorem *The derivative of a distribution U is the zero element of $\mathscr{D}'$ if and only if U is a constant; that is, if and only if $U(\varphi) = K \int_{\mathbb{R}} \varphi$.*

Proof. Assume that $U' = 0$. Then

$$\langle U, \varphi' \rangle = 0$$

for all $\varphi \in \mathscr{D}$.

If $\varphi \in \mathscr{D}$, then $\varphi' \in \mathscr{D}$ and $\int_{\mathbb{R}} \varphi' = 0$. Conversely, if $\psi \in \mathscr{D}$ and $\int_{\mathbb{R}} \psi = 0$, then the function

$$\varphi(x) = \int_{-\infty}^{x} \psi(t)\, dt$$

is in $\mathscr{D}$ and is a primitive of ψ. This prompts us to define

$$\mathscr{D}_0 = \{\varphi' \mid \varphi \in \mathscr{D}\} = \{\psi \mid \psi \in \mathscr{D},\ \textstyle\int_{\mathbb{R}} \psi = 0\}.$$

Then for all $\psi \in \mathscr{D}_0$,

$$\langle U, \psi \rangle = 0.$$

Let θ be an arbitrary but fixed function in $\mathscr{D}$ such that

$$\int_{\mathbb{R}} \theta = 1 \quad \text{and} \quad \theta(x) = 0 \ \text{ if } \ |x| \geq 1. \tag{30.1}$$

For $\varphi \in \mathscr{D}$, define $\psi_\varphi = \varphi - I(\varphi)\theta$, where $I(\varphi) = \int_{\mathbb{R}} \varphi$. Then

$$\varphi = \psi_\varphi + I(\varphi)\theta, \tag{30.2}$$

and $\psi_\varphi \in \mathscr{D}_0$. Applying U to (30.2), we have

$$\langle U, \varphi \rangle = I(\varphi)\langle U, \theta \rangle.$$

This shows that U is the constant distribution

$$\varphi \mapsto \int_{\mathbb{R}} K\varphi$$

with $K = \langle U, \theta \rangle$. Conversely, if U is a constant, it is clear that $U' = 0$. □

30.1.2 Remark Let f be a continuous function on $\mathbb{R}$ whose derivative in the sense of distributions is also a continuous function. It is not obvious that this implies that f is continuously differentiable in the usual sense. This is, however, a consequence of Theorem 30.1.1:

Let g be the continuous function for which $T_f' = T_g$ and let

$$G(x) = \int_0^x g(t)\, dt.$$

G is continuously differentiable in the usual sense and $G' = g$. By Theorem 30.1.1,

$$T_f = G + \text{a constant},$$

and so

$$f(x) = f(0) + \int_0^x g(t)\, dt.$$

30.2 Primitives of a distribution

30.2.1 Theorem *Every distribution T has a primitive U in $\mathscr{D}'$, and all the primitives of T are of the form $U + C$, where C is some constant.*

Proof. If U and V are primitives of T, then $V' - U' = 0$, so by Theorem 30.1.1,

$$V = U + C.$$

We need to show that T has at least one primitive U. If U exists, then necessarily

$$\langle U', \varphi \rangle = -\langle U, \varphi' \rangle = \langle T, \varphi \rangle \tag{30.3}$$

for all $\varphi \in \mathscr{D}$. To determine U, we must know the value of $\langle U, \varphi \rangle$ for each φ in $\mathscr{D}$. We know how U acts on ψ in $\mathscr{D}_0$ from (30.3). Now suppose that φ is in $\mathscr{D}$ and write

$$\psi = \psi_\varphi = \varphi - I(\varphi)\theta$$

using the notation of Theorem 30.1.1. One primitive of ψ is

$$F_\varphi(x) = \int_{-\infty}^{x} [\varphi(t) - I(\varphi)\theta(t)]\, dt.$$

Clearly,

$$F_{\varphi'} = \varphi \quad \text{and} \quad (F_\varphi)' = \varphi - I(\varphi)\theta = \psi.$$

Since the derivative of F_φ is in $\mathscr{D}_0$, we know that F_φ is in $\mathscr{D}$.

Now define $U : \mathscr{D} \to \mathbb{C}$ by

$$\langle U, \varphi \rangle = -\langle T, F_\varphi \rangle.$$

(At this point we do not know that U is a distribution, so writing $\langle U, \varphi \rangle$ is a slight abuse of notation.) If U is a distribution, then it is a primitive of T, since

$$\langle U', \varphi \rangle = -\langle U, \varphi' \rangle = \langle T, F_{\varphi'} \rangle = \langle T, \varphi \rangle.$$

To prove that U is a distribution, it is sufficient to show that the mapping $\varphi \mapsto F_\varphi$ is linear and continuous from $\mathscr{D}$ to $\mathscr{D}$.

The mapping is clearly linear, so we focus on continuity. Let (φ_n) be a sequence in $\mathscr{D}$ that tends to 0. The sequence of integrals $(I(\varphi_n))$ tends to 0; hence

$$\psi_n = \varphi_n - I(\varphi_n)\theta \tag{30.4}$$

tends to 0 in $\mathscr{D}$. If $[A, B]$ is an interval containing the supports of all the ψ_n, then the supports of the functions

$$F_{\varphi_n}(x) = \int_{-\infty}^{x} \psi_n(t)\, dt$$

are also in $[A, B]$. Since

$$\|F_{\varphi_n}\|_\infty \le \|\psi_n\|_1 \le (B - A)\|\psi_n\|_\infty,$$

it follows that $F_{\varphi_n}(x) \to 0$ uniformly on $\mathbb{R}$. For any integer $p \ge 1$,

$$(F_{\varphi_n})^{(p)} = \psi_n^{(p-1)},$$

which tend to 0 uniformly on $\mathbb{R}$. Thus the sequence (F_{φ_n}) tends to 0 in $\mathscr{D}$, and this proves continuity. □

30.3 Exercises

*Exercise 30.1

(a) Solve the differential equation

$$T' + aT = 0 \quad (a \neq 0)$$

in $\mathscr{D}'$. (One can make the change of unknown distribution $T = e^{-ax}S$.)

(b) Deduce from this the general solutions of

$$T' + aT = \delta \quad \text{and} \quad T' + aT = u \quad (a \neq 0).$$

***Exercise 30.2** Use Exercise 28.1 to derive the general solutions of the following differential equations:

$$T'' + T = \delta; \quad T'' + T = \Delta_\pi; \quad T'' + \omega^2 T = \omega \Delta_{\frac{\pi}{\omega}} \quad (\omega > 0).$$

***Exercise 30.3** For $-\infty < a < b < +\infty$ we consider the space

$$W^{1,1}(a,b) = \left\{ f \in L^1(a,b) \mid f' \in L^1(a,b) \right\},$$

where the derivative is taken in the sense of distributions. Show that $W^{1,1}(a,b)$ is the space of absolutely continuous (AC) functions on $[a,b]$.

Remark: If $f \in W^{1,1}(a,b)$, one shows that f has a unique representative in the space of absolutely continuous functions by using $g(x) = \int_a^x f'(t)\,dt$, $x \in [a,b]$.

****Exercise 30.4 (the Sobolev space $H^1(a,b)$)** Let (a,b) an interval, bounded or not. The vector space $H^1(a,b)$ is defined by

$$H^1(a,b) = \left\{ f \in L^2(a,b) \mid f' \in L^2(a,b) \right\},$$

where the derivative f' is taken in the sense of distributions.

(a) Show that $u \notin H^1(-1,1)$, that $g(t) = \sqrt{t}$ is not in $H^1(0,1)$, and that the polynomials are in $H^1(a,b)$ if $b - a < +\infty$ but not in $H^1(\mathbb{R})$ (except 0). Give some examples of elements of $H^1(\mathbb{R})$.

(b) Assume that (a,b) is bounded.

- Use Schwarz's inequality to show that $L^2(a,b) \subset L^1(a,b)$.
- If $g \in L^2(a,b)$, show that f defined by $f(x) = \int_a^x g(t)\,dt$ is in $H^1(a,b)$.

(c) Let (a,b) be arbitrary, $f \in H^1(a,b)$, and $c \in (a,b)$. Define $h(x) = \int_c^x f'(t)\,dt$ for $x \in (c,b)$.

- Show that $h - f$ is constant on (c,b) (use Theorem 30.1.1). Deduce from this that f is absolutely continuous on all intervals $[c,d] \subset (a,b)$.
- Is h in $H^1(a,b)$?

(d) A scalar product can be defined on $H^1(a,b)$ by

$$((f,g))_1 = \int_a^b f(t)\overline{g}(t)\,dt + \int_a^b f'(t)\overline{g'}(t)\,dt.$$

We wish to show that $H^1(a,b)$ is a Hilbert space. If (f_n) is a Cauchy sequence in $H^1(a,b)$, show that there are f and g in $L^2(a,b)$ such that

$$f_n \to f \quad \text{and} \quad f_n' \to g \quad \text{in} \quad L^2(a,b).$$

Show that $g = f'$ using convergence in the sense of distributions as in Proposition 29.3.2 and finish the proof.

***Exercise 30.5** Let f be a function in $H^1(0,a)$ and let $c_n = c_n(f)$, $n \in \mathbb{Z}$, be its Fourier coefficients (see Lesson 6).

(a) Show that $\sum_{n=-\infty}^{+\infty} |c_n|^2 < +\infty$.

(b) Show that $\sum_{n=-\infty}^{+\infty} n^2 |c_n|^2 < +\infty$ if and only if $f(0^+) = f(a^-)$.

***Exercise 30.6** Show that if $f \in H^1(\mathbb{R})$, then $\lim_{|x|\to\infty} f(x) = 0$ (apply the argument used in Proposition 17.2.1(ii) to the function f^2). Verify that $f \in H^1(\mathbb{R})$ does not imply $f \in L^1(\mathbb{R})$.

Chapter IX

Convolution and the Fourier Transform of Distributions

Lesson 31

The Fourier Transform of Distributions

We are going to extend the Fourier transform to distributions by interpreting the transpose formula we obtained for functions. Consider the following formal computation. Let f be a function in $L^1(\mathbb{R})$ and let φ be in $\mathscr{D}$. Then $\widehat{f}$ is a continuous function, and in the sense of distributions we have

$$\begin{aligned}\langle \widehat{f}, \varphi \rangle &= \int_{\mathbb{R}} \Big(\int_{\mathbb{R}} e^{-2i\pi\xi x} f(x)\, dx \Big) \varphi(\xi)\, d\xi = \int_{\mathbb{R}} f(x) \int_{\mathbb{R}} e^{-2i\pi x\xi} \varphi(\xi)\, d\xi\, dx \\ &= \int_{\mathbb{R}} f(x) \widehat{\varphi}(x)\, dx = \langle f, \widehat{\varphi} \rangle.\end{aligned}$$

This suggests that the Fourier transform for distributions should be defined by transposition:

$$\langle \widehat{T}, \varphi \rangle = \langle T, \widehat{\varphi} \rangle. \tag{31.1}$$

We know that the expression $\langle T, \widehat{\varphi} \rangle$ make sense when $\widehat{\varphi} \in \mathscr{D}(\mathbb{R})$, and we have seen that $\widehat{\varphi} \in C^\infty$ (Proposition 17.2.1). But does $\widehat{\varphi}$ have compact support? We will see in Proposition 31.5.4 that this is never the case. We have, however, shown that the space of rapidly decreasing functions $\mathscr{S}(\mathbb{R})$ is invariant under the Fourier transform (Theorem 19.2.3). This leads to the introduction of the subspace of tempered distributions.

31.1 The space $\mathscr{S}'(\mathbb{R})$ of tempered distributions

The function $\widehat{\varphi}$ in equation (31.1) is in $\mathscr{S}(\mathbb{R})$. Although $\widehat{\varphi}$ does not have compact support, it is very small at infinity.

31.1.1 Definition $\mathscr{S}'(\mathbb{R})$ denotes the vector space of continuous linear functionals T defined on $\mathscr{S}(\mathbb{R})$. Thus

$$\varphi_n \to 0 \text{ in } \mathscr{S} \quad \Longrightarrow \quad \langle T, \varphi_n \rangle \to 0 \text{ in } \mathbb{C}.$$

If we take φ in $\mathscr{D}(\mathbb{R})$, then $T(\varphi)$ is well-defined, since $\mathscr{D}(\mathbb{R}) \subset \mathscr{S}(\mathbb{R})$ (Figure 31.1). Furthermore, convergence in $\mathscr{D}(\mathbb{R})$ implies convergence in $\mathscr{S}(\mathbb{R})$ (see Definitions 19.2.4 and 27.3.2). This means that the elements of $\mathscr{S}'(\mathbb{R})$ restricted to $\mathscr{D}(\mathbb{R})$ are distributions. Since $\mathscr{D}(\mathbb{R})$ is dense in $\mathscr{S}(\mathbb{R})$ (Exercise 19.7), we can identify $\mathscr{S}'(\mathbb{R})$ with a subspace of $\mathscr{D}'(\mathbb{R})$ (see Exercise 31.1).

31.1.2 Definition The elements of $\mathscr{S}'(\mathbb{R})$ are called tempered distributions.

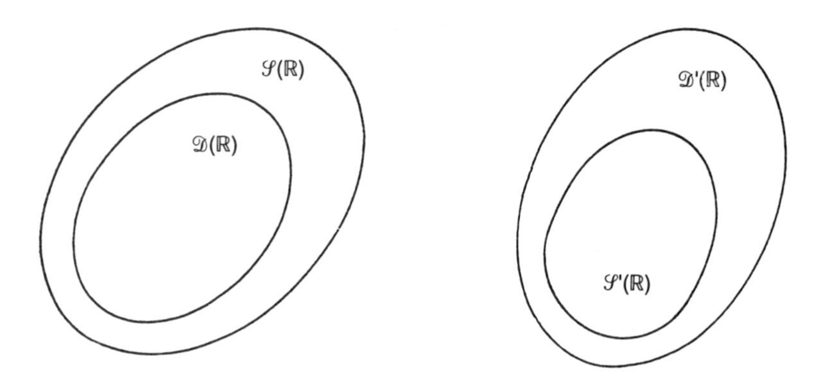

FIGURE 31.1.

The following result provides a practical way to characterize the tempered distributions.

31.1.3 Proposition *Suppose that T is a distribution, i.e., $T \in \mathscr{D}'(\mathbb{R})$. Then T is a tempered distribution, $T \in \mathscr{S}'(\mathbb{R})$, if and only if T is continuous on $\mathscr{D}(\mathbb{R})$ in the topology of $\mathscr{S}(\mathbb{R})$.*

Proof. Clearly the condition is necessary: If T is a tempered distribution, then $\varphi_n \in \mathscr{D} \subset \mathscr{S}$ and $\varphi_n \to 0$ in $\mathscr{S}$ imply that $T(\varphi_n) \to 0$.

The proof in the other direction depends on the fact that $\mathscr{D}(\mathbb{R})$ is dense in $\mathscr{S}(\mathbb{R})$ in the topology of $\mathscr{S}(\mathbb{R})$. We have already seen in Proposition 22.1.3 how a linear operator that is defined and continuous on a dense subspace can be uniquely extended to a continuous linear operator on the whole space. Proposition 22.1.3 was about normed spaces, which is not the case for $\mathscr{D}$ and $\mathscr{S}$. The topologies of $\mathscr{D}$ and $\mathscr{S}$ (which have not been discussed, since we are mainly interested in convergence of sequences) are more complicated than a topology given by a norm. With a full understanding of the topology of $\mathscr{S}$, the result follows directly. However, it is also possible to give a proof using sequences. This is left as Exercise 31.12, where there are plenty of hints. We also refer to [Kho72]. □

We will use the following notion of convergence for sequences in $\mathscr{S}'(\mathbb{R})$.

31.1.4 Definition Suppose T_n is a sequence in $\mathscr{S}'(\mathbb{R})$. We say that T_n tends to 0 in $\mathscr{S}'(\mathbb{R})$ if

$$\lim_{n\to\infty} \langle T_n, \varphi\rangle = 0 \quad \text{for all } \varphi \in \mathscr{S}(\mathbb{R}).$$

Note that convergence in $\mathscr{S}'(\mathbb{R})$ implies convergence in $\mathscr{D}'(\mathbb{R})$, since $\mathscr{D}(\mathbb{R}) \subset \mathscr{S}(\mathbb{R})$.

Here are some important properties that follow directly from the definitions.

31.1.5 Proposition *It T is a tempered distribution, then we have the following results:*

(i) *For each $k \in \mathbb{N}$, $x^k T$ is in $\mathscr{S}'(\mathbb{R})$.*

(ii) *For each $k \in \mathbb{N}$, the derivative $T^{(k)}$ is in $\mathscr{S}'(\mathbb{R})$.*

(iii) *The mappings $T \to x^k T$ and $T \to T^{(k)}$ are continuous from $\mathscr{S}'(\mathbb{R})$ to $\mathscr{S}'(\mathbb{R})$.*

Proof.

(i) The mapping $\varphi \to x^k\varphi$ is continuous in $\mathscr{S}(\mathbb{R})$ (Proposition 19.2.5). We have $\langle x^kT, \varphi\rangle = \langle T, x^k\varphi\rangle$ for all $\varphi \in \mathscr{D}(\mathbb{R})$. If a sequence $\varphi_n \in \mathscr{D}(\mathbb{R})$ tends to 0 in $\mathscr{S}$, then $x^k\varphi_n \to 0$ in $\mathscr{S}$. Thus $\langle x^kT, \varphi_n\rangle \to 0$, which shows that $x^kT \in \mathscr{S}'(\mathbb{R})$ (Proposition 31.1.3).

(ii) Take $\varphi_n \in \mathscr{D}(\mathbb{R})$ such that $\varphi_n \to 0$ in $\mathscr{S}$. The mapping $\varphi \to \varphi^{(k)}$ is continuous in $\mathscr{S}(\mathbb{R})$ (Proposition 19.2.5). Since $\langle T^{(k)}, \varphi_n\rangle = (-1)^k\langle T, \varphi_n^{(k)}\rangle$, we can pass to the limit and conclude, as in (i), that $T^{(k)} \in \mathscr{S}'(\mathbb{R})$.

(iii) Suppose that T_n is in $\mathscr{S}'(\mathbb{R})$ and that $T_n \to T$ in $\mathscr{S}'(\mathbb{R})$. Then for all φ in $\mathscr{S}(\mathbb{R})$, $\langle x^kT_n, \varphi\rangle = \langle T_n, x^k\varphi\rangle \to \langle T, x^k\varphi\rangle = \langle x^kT, \varphi\rangle$. This proves the continuity of $T \mapsto x^kT$. The same technique is used to show that $T \mapsto T^{(k)}$ is continuous in $\mathscr{S}'(\mathbb{R})$. □

We will present several useful examples of tempered distributions.

31.1.6 Definition Suppose $f : \mathbb{R} \to \mathbb{C}$ is a measurable function. Then f is said to be slowly increasing if there exist $C > 0$ and $N \in \mathbb{N}$ such that

$$|f(x)| \leq C(1 + x^2)^N \quad \text{for all } x \in \mathbb{R}.$$

If f is slowly increasing, then it is clearly in $L^1_{\text{loc}}(\mathbb{R})$.

31.1.7 Proposition *Every slowly increasing function f is a tempered distribution.*

Proof. Since f is in $L^1_{\text{loc}}(\mathbb{R})$, we know that T_f is a distribution (Proposition 27.4.1). We use Proposition 31.1.3 to show that T_f is tempered. Thus let

φ_n be a sequence in $\mathscr{D}(\mathbb{R})$ that tends to 0 in $\mathscr{S}(\mathbb{R})$. We have

$$\begin{aligned}
|\langle T, \varphi_n \rangle| &\leq \int_{\mathbb{R}} |f(x)||\varphi_n(x)|\, dx \\
&= \int_{\mathbb{R}} \frac{|f(x)|}{(1+x^2)^N} (1+x^2)^N |\varphi_n(x)|\, dx \\
&\leq C \sup_{x\in\mathbb{R}} |(1+x^2)^{2N} \varphi_n(x)| \int_{\mathbb{R}} \frac{dx}{(1+x^2)^N}.
\end{aligned}$$

From Proposition 19.2.5 we know that $(1+x^2)^{2N}\varphi_n(x)$ tends to 0 in $\mathscr{S}(\mathbb{R})$. Hence $\lim_{n\to\infty} |\langle T, \varphi_n\rangle| = 0$, and T is tempered by Proposition 31.1.3. □

Being a slowly increasing function is not a necessary condition for f to be in $\mathscr{S}'(\mathbb{R})$. Here is another criterion.

31.1.8 Proposition *The functions in $L^p(\mathbb{R})$, $p \geq 1$, are tempered distributions.*

Proof. Since $L^p(\mathbb{R}) \subset L^1_{\mathrm{loc}}(\mathbb{R})$, every element f of $L^p(\mathbb{R})$ is a distribution. We proceed as in the proof of Proposition 31.1.7. Let φ_n be a sequence in $\mathscr{D}(\mathbb{R})$ that tends to 0 in $\mathscr{S}(\mathbb{R})$. Then

$$|\langle T, \varphi_n\rangle| \leq \left(\int_{\mathbb{R}} |f(x)|^p\, dx\right)^{1/p} \left(\int_{\mathbb{R}} |\varphi_n(x)|^q\, dx\right)^{1/q} = \|f\|_p \|\varphi_n\|_q.$$

If $\varphi_n \to 0$ in $\mathscr{S}(\mathbb{R})$, then $\lim_{n\to\infty} \|\varphi_n\|_q = 0$, and this proves the result. □

So far, the examples of tempered distributions have been functions. Here is a different example.

31.1.9 Proposition *If the complex sequence $(y_n)_{n\in\mathbb{Z}}$ is slowly increasing and $a > 0$, then*

$$T = \sum_{n=-\infty}^{+\infty} y_n\, \delta_{na}$$

is a tempered distribution.

Proof. We have already seen that T is a distribution (Section 27.3.4(b)). Let φ_p be a sequence in $\mathscr{D}(\mathbb{R})$ that tends to 0 in $\mathscr{S}(\mathbb{R})$. We must show that $\lim_{p\to\infty} |\langle T, \varphi_p\rangle| = 0$.

Since $(y_n)_{n\in\mathbb{Z}}$ is slowly increasing (see Definition 29.4.1), there is an $N \in \mathbb{N}$ such that $|y_n| \leq C(1+n^2)^N$ for all $n \in \mathbb{Z}$ (take any $N \geq k/2$ for the k in (29.4)). With this we have

$$\begin{aligned}
|y_n \langle \delta_{na}, \varphi_p\rangle| &\leq C|1+n^2|^N |\varphi_p(na)| \\
&= \frac{C|1+n^2|^N}{|1+(an)^2|^N} |1+(an)^2|^{N+1} \frac{|\varphi_p(na)|}{1+(an)^2}.
\end{aligned}$$

Since φ_p is in $\mathscr{S}(\mathbb{R})$,

$$|1+(an)^2|^{N+1}|\varphi_p(na)| \leq \sup_{x\in\mathbb{R}} |(1+x^2)^{N+1}\varphi_p(x)| = M_p$$

for all $p \in \mathbb{N}$ and all $n \in \mathbb{Z}$. Also,

$$\frac{C(1+n^2)^N}{(1+(an)^2)^N} \leq C \sup_{x\in\mathbb{R}} \frac{(1+x^2)^N}{1+(ax)^2)^N} = C_1.$$

Combining these inequalities, we have $|y_n\langle\delta_{na},\varphi_p\rangle| \leq C_1 \dfrac{M_p}{1+(an)^2}$; hence

$$|\langle T,\varphi_p\rangle| \leq C_2 M_p,$$

where $C_2 = C_1 \sum_{n=-\infty}^{+\infty} \dfrac{1}{1+(an)^2}$. By hypothesis, $\lim_{p\to\infty} M_p = 0$, and this proves that $T \in \mathscr{S}'(\mathbb{R})$. □

As an example, Dirac's comb $\Delta_a = \sum_{n=-\infty}^{+\infty} \delta_{na}$ is a tempered distribution.

To finish this section, we state without proof the structure theorem for tempered distributions [Sch65b, Kho72].

31.1.10 Theorem *If $T \in \mathscr{S}'(\mathbb{R})$, then there are integers $n_1, n_2, \ldots, n_p$ and slowly increasing continuous functions $f_1, f_2, \ldots, f_p$ such that*

$$T = \sum_{k=1}^{p} f_k^{(n_k)}.$$

31.2 The Fourier transform on $\mathscr{S}'(\mathbb{R})$

We are now in position to validate equation (31.1).

31.2.1 Definition Suppose that $T \in \mathscr{S}'(\mathbb{R})$. The Fourier transform of T, which is denoted by $\widehat{T}$ or by $\mathscr{F}(T)$, is defined by

$$\langle\widehat{T},\varphi\rangle = \langle T,\widehat{\varphi}\rangle \quad \text{for all } \varphi \in \mathscr{S}(\mathbb{R}). \tag{31.2}$$

$\widehat{T}$ is a tempered distribution because the Fourier transform is a continuous operator on $\mathscr{S}(\mathbb{R})$ (Proposition 19.2.5). Formula (31.2) extends the Fourier transform from $L^1(\mathbb{R})$ or $L^2(\mathbb{R})$ to tempered distributions.

31.2.2 Proposition *If f is in $L^1(\mathbb{R})$ or $L^2(\mathbb{R})$, then $\widehat{T_f} = T_{\widehat{f}}$.*

Proof. Take $f \in L^1(\mathbb{R})$ and $\varphi \in \mathscr{S}(\mathbb{R})$. Then $\widehat{f} \in L^\infty(\mathbb{R})$ is a tempered distribution and $\widehat{\varphi} \in \mathscr{S}(\mathbb{R})$. By Proposition 17.1.4,

$$\langle T_{\widehat{f}}, \varphi\rangle = \int_{\mathbb{R}} \widehat{f}(\xi)\varphi(\xi)\,d\xi = \int_{\mathbb{R}} f(x)\widehat{\varphi}(x)\,dx = \langle T_f, \widehat{\varphi}\rangle = \langle \widehat{T_f}, \varphi\rangle.$$

If $f \in L^2(\mathbb{R})$, then $\widehat{f} \in L^2(\mathbb{R})$ is a tempered distribution. For all $\varphi \in \mathscr{S}(\mathbb{R})$, $\varphi, \widehat{\varphi} \in L^2(\mathbb{R})$. Thus by Proposition 22.1.5,

$$\begin{aligned}\langle T_{\widehat{f}}, \varphi\rangle &= \int_{\mathbb{R}} \widehat{f}(\xi)\varphi(\xi)\,d\xi = \int_{\mathbb{R}} f(x)\widehat{\varphi}(x)\,dx \\ &= \langle T_f, \widehat{\varphi}\rangle = \langle \widehat{T_f}, \varphi\rangle.\end{aligned}$$

□

The Fourier transform is a linear 1-to-1 mapping on $\mathscr{S}(\mathbb{R})$ that is continuous in both directions (bicontinuous). The next result follows from transposition.

31.2.3 Theorem *The Fourier transform is a linear, 1-to-1, bicontinuous mapping from $\mathscr{S}'(\mathbb{R})$ to $\mathscr{S}'(\mathbb{R})$. The inverse mapping, $\mathscr{F}^{-1} = \overline{\mathscr{F}}$, is defined for all $\varphi \in \mathscr{S}(\mathbb{R})$ by*

$$\langle \overline{\mathscr{F}}\, T, \varphi\rangle = \langle T, \overline{\mathscr{F}}\,\varphi\rangle.$$

For all $T \in \mathscr{S}'(\mathbb{R})$,

$$\mathscr{F}\,\overline{\mathscr{F}}\,T = \overline{\mathscr{F}}\,\mathscr{F}\,T = T. \tag{31.3}$$

Proof. The mapping $T \mapsto \mathscr{F}\,T$ from $\mathscr{S}'(\mathbb{R})$ to $\mathscr{S}'(\mathbb{R})$ is clearly linear. It is also continuous: If $T_n \to 0$ in $\mathscr{S}'(\mathbb{R})$, then

$$\langle \mathscr{F}\,T_n, \varphi\rangle = \langle T_n, \mathscr{F}\,\varphi\rangle \to 0 \qquad \text{as } n \to \infty.$$

The same proof works for $\overline{\mathscr{F}}$. For all $\varphi \in \mathscr{S}(\mathbb{R})$,

$$\langle \mathscr{F}\,\overline{\mathscr{F}}\,T, \varphi\rangle = \langle T, \overline{\mathscr{F}}\,\mathscr{F}\,\varphi\rangle = \langle T, \mathscr{F}\,\overline{\mathscr{F}}\,\varphi\rangle = \langle \overline{\mathscr{F}}\,\mathscr{F}\,T, \varphi\rangle = \langle T, \varphi\rangle,$$

which proves (31.3). Thus $\mathscr{F}$ is 1-to-1 and $\mathscr{F}^{-1} = \overline{\mathscr{F}}$. □

Here are several important properties of the Fourier transform on $\mathscr{S}'(\mathbb{R})$; they are direct consequences of the definition.

31.2.4 Proposition *Let T be a tempered distribution.*

(i) *For all $k \in \mathbb{N}$,*

$$\begin{aligned}\widehat{T}^{(k)} &= [(-2i\pi x)^k T]\widehat{\ }, \\ \widehat{T^{(k)}} &= (2i\pi\xi)^k \widehat{T}.\end{aligned}$$

(ii) *For $a \in \mathbb{R}$,*

$$\begin{aligned}\tau_a\widehat{T} &= [e^{2i\pi a x}T]\widehat{\ }, \\ \widehat{\tau_a T} &= e^{-2i\pi a\xi}\widehat{T}.\end{aligned}$$

Proof.

(i) Since $T \in \mathscr{S}'(\mathbb{R})$, $x^k T \in \mathscr{S}'(\mathbb{R})$ (Proposition 31.1.5(i)). Thus $\widehat{x^k T}$ exists, and we have $\langle \widehat{x^k T}, \varphi \rangle = \langle T, x^k \widehat{\varphi} \rangle$ for all $\varphi \in \mathscr{S}(\mathbb{R})$. From Proposition 17.2.1(ii) we have $x^k \widehat{\varphi} = \dfrac{1}{(2i\pi)^k} \widehat{\varphi^{(k)}}$; hence

$$\langle \widehat{x^k T}, \varphi \rangle = \frac{1}{(2i\pi)^k} \langle T, \widehat{\varphi^{(k)}} \rangle = \frac{(-1)^k}{(2i\pi)^k} \langle \widehat{T}^{(k)}, \varphi \rangle,$$

and $\widehat{T}^{(k)} = [(-2i\pi x)^k T]\widehat{\ }$. The relation $\widehat{T^{(k)}} = (2i\pi\xi)^k \widehat{T}$ is obtained similarly from Proposition 17.2.1(i).

(ii) The function $x \mapsto e^{+2i\pi a x}$ is in $C^\infty(\mathbb{R})$ and is bounded. Thus $e^{+2i\pi a \xi} T$ is in $\mathscr{S}'(\mathbb{R})$, and

$$\langle [e^{2i\pi a x} T]\widehat{\ }, \varphi \rangle = \langle T, e^{2i\pi a \xi} \widehat{\varphi} \rangle = \langle T, \widehat{\tau_{-a}\varphi} \rangle$$

for all $\varphi \in \mathscr{S}(\mathbb{R})$ (Proposition 17.2.4(i)). But we know that

$$\langle T, \widehat{\tau_{-a}\varphi} \rangle = \langle \widehat{T}, \tau_{-a}\varphi \rangle = \langle \tau_a \widehat{T}, \varphi \rangle$$

(Definition 28.1.3), so $\tau_a \widehat{T} = [e^{2i\pi a x} T]\widehat{\ }$. The proof that $\widehat{\tau_a T} = e^{-2i\pi a \xi} \widehat{T}$ is similar. □

31.2.5 Proposition *For $T \in \mathscr{S}'(\mathbb{R})$ we have the following relations:*

(i) $\mathscr{F}(T_\sigma) = (\mathscr{F} T)_\sigma = \overline{\mathscr{F}}\, T$.

(ii) $\mathscr{F}(\mathscr{F} T) = T_\sigma$.

Proof.

(i) Take $\varphi \in \mathscr{S}(\mathbb{R})$. By Proposition 17.2.3(ii),

$$(\mathscr{F} \varphi)_\sigma = \mathscr{F}(\varphi_\sigma) = \overline{\mathscr{F}}\, \varphi,$$

so for $T \in \mathscr{S}'(\mathbb{R})$ we have

$$\langle \mathscr{F}(T_\sigma), \varphi \rangle = \langle T_\sigma, \mathscr{F}(\varphi) \rangle = \langle T, (\mathscr{F} \varphi)_\sigma \rangle = \langle T, \overline{\mathscr{F}}\, \varphi \rangle = \langle \overline{\mathscr{F}}\, T, \varphi \rangle.$$

Similarly, $\langle (\mathscr{F} T)_\sigma, \varphi \rangle = \langle T, \mathscr{F}(\varphi_\sigma) \rangle = \langle T, \overline{\mathscr{F}}\, \varphi \rangle = \langle \overline{\mathscr{F}}\, T, \varphi \rangle$, which completes the proof of (i).

(ii) For $\varphi \in \mathscr{S}(\mathbb{R})$, we know from Proposition 18.2.1 that $\mathscr{F}(\mathscr{F} \varphi) = \varphi_\sigma$. Consequently,

$$\langle \mathscr{F}(\mathscr{F} T), \varphi \rangle = \langle T, \mathscr{F}\mathscr{F} \varphi \rangle = \langle T, \varphi_\sigma \rangle = \langle T_\sigma, \varphi \rangle,$$

and $\mathscr{F}(\mathscr{F} T) = T_\sigma$. □

31.3 Examples of Fourier transforms in $\mathscr{S}'(\mathbb{R})$

31.3.1 Dirac's impulse

It is easy verify that δ_a is a tempered distribution for all $a \in \mathbb{R}$. We wish to compute the Fourier transform $\widehat{\delta}$. For $\varphi \in \mathscr{S}(\mathbb{R})$,

$$\langle \widehat{\delta}, \varphi \rangle = \langle \delta, \widehat{\varphi} \rangle = \widehat{\varphi}(0) = \int_{\mathbb{R}} \varphi(x)\, dx = \langle \mathbf{1}, \varphi \rangle;$$

hence

$$\widehat{\delta} = \mathbf{1}. \tag{31.4}$$

In the same way,

$$\langle \widehat{\delta_a}, \varphi \rangle = \langle \delta_a, \widehat{\varphi} \rangle = \int_{\mathbb{R}} e^{-2i\pi a\xi} \varphi(\xi)\, d\xi,$$

so

$$\widehat{\delta_a}(\xi) = e^{-2i\pi a\xi}. \tag{31.5}$$

We note that the Fourier transform of δ_a is a C^∞ function. For the derivatives of the Dirac impulse we have (Proposition 31.2.4(i))

$$\widehat{\delta^{(k)}}(\xi) = (2i\pi\xi)^k, \tag{31.6}$$

and by taking the Fourier transform of both sides of (31.6) we see that

$$\widehat{x^k} = \frac{1}{(-2i\pi)^k} \delta^{(k)}. \tag{31.7}$$

The Fourier transform of a polynomial (which cannot be computed by integration because $x^k \notin L^1(\mathbb{R})$) is a linear combination of the derivatives of the Dirac distribution at the origin.

31.3.2 Sinusoidal signals

Let $f(x) = e^{2i\pi ax}$ with a real and fixed. f is in $L^\infty(\mathbb{R})$ and is a tempered distribution. For $\varphi \in \mathscr{S}(\mathbb{R})$,

$$\langle \widehat{T_f}, \varphi \rangle = \int_{\mathbb{R}} e^{2i\pi ax} \widehat{\varphi}(x)\, dx = \overline{\mathscr{F}}\, \widehat{\varphi}(a) = \varphi(a)$$

by Theorem 19.3.1. Thus $\langle \widehat{T_f}, \varphi \rangle = \langle \delta_a, \varphi \rangle$ and $\widehat{T_f} = \delta_a$. In particular, the Fourier transform of the constant function $\mathbf{1}$ (which is in neither $L^1(\mathbb{R})$ nor $L^2(\mathbb{R})$) is δ:

$$\widehat{e^{2i\pi ax}} = \delta_a, \tag{31.8}$$

$$\widehat{\mathbf{1}} = \delta. \tag{31.9}$$

31.3.3 Dirac's comb

We have seen that $\Delta_a = \sum_{n=-\infty}^{+\infty} \delta_{na}$ is in $\mathscr{S}'(\mathbb{R})$. Since the Fourier transform is continuous,

$$\widehat{\Delta_a}(\xi) = \sum_{n=-\infty}^{+\infty} \widehat{\delta_{na}}(\xi) = \sum_{n=-\infty}^{+\infty} e^{2i\pi na\xi}.$$

But from (29.7) we know that

$$\sum_{n=-\infty}^{+\infty} e^{2i\pi na\xi} = \frac{1}{a}\Delta_{\frac{1}{a}};$$

thus

$$\widehat{\Delta_a} = \frac{1}{a}\Delta_{\frac{1}{a}}. \tag{31.10}$$

The Fourier transform of the Dirac comb with grid a is a^{-1} times the Dirac comb with grid a^{-1}. For $a = 1$, it is a distribution equal to its Fourier transform, which is a property shared by the Gaussian $g(t) = e^{-\pi t^2}$ (Proposition 1(.3.2).

31.4 The space $\mathscr{E}'(\mathbb{R})$ of distributions with compact support

We have seen that the Fourier transform of a function f in $L^1(\mathbb{R})$ is continuous and zero at infinity (Theorem 17.1.3). If in addition f has compact support, then $\widehat{f} \in C^\infty(\mathbb{R})$ (Proposition 17.2.1(iii)). We will show that $\widehat{T} \in C^\infty(\mathbb{R})$ for distributions with compact support.

31.4.1 Definition $\mathscr{E}'(\mathbb{R})$ denotes the subspace of $\mathscr{D}'(\mathbb{R})$ of those distributions that have compact support.

It can be shown that $\mathscr{E}'(\mathbb{R})$ is the dual of $\mathscr{E}(\mathbb{R}) = C^\infty(\mathbb{R})$ when $C^\infty(\mathbb{R})$ is endowed with the following topology: A sequence φ_n tends to 0 if and only if for each $p \in \mathbb{N}$, $\varphi_n^{(p)}$ tends to 0 uniformly on every compact set $K \subset \mathbb{R}$.

T is a continuous linear functional on $\mathscr{E}(\mathbb{R})$ if and only if there exists a compact set K, a constant $C > 0$, and an integer $m \in \mathbb{N}$ such that

$$|\langle T, \varphi\rangle| \le C \sup_{k\le m} \sup_{x\in K} |\varphi^{(k)}(x)| \tag{31.11}$$

for all $\varphi \in C^\infty(\mathbb{R})$. $\mathscr{E}'(\mathbb{R})$ is a linear subspace of $\mathscr{S}'(\mathbb{R})$ [Kho72].

31.4.2 Examples

(a) A function $f \in L^1_{\text{loc}}(\mathbb{R})$ with compact support is in $\mathscr{E}'(\mathbb{R})$.

(b) Let δ_a be the Dirac distribution at a. For all $p \in \mathbb{N}$, $\delta_a^{(p)}$ is in $\mathscr{E}'(\mathbb{R})$.

Working with the elements of $\mathscr{E}'(\mathbb{R})$ is facilitated by knowing their structure; thus we are going to assume the following theorem [Sch65b, Kho72].

31.4.3 Theorem (representation of $\mathscr{E}'(\mathbb{R})$) *If $T \in \mathscr{E}'(\mathbb{R})$ and the support of T is in the interior of some compact set K, there exist integers $n_1, n_2, \ldots, n_p$ and continuous functions $f_1, f_2, \ldots, f_p$ whose supports are in K, and*

$$T = \sum_{j=1}^{p} f_j^{(n_j)}.$$

31.5 The Fourier transform on $\mathscr{E}'(\mathbb{R})$

Since $\mathscr{E}'(\mathbb{R})$ is a subspace of $\mathscr{S}'(\mathbb{R})$, the Fourier transform of a distribution with compact support is well-defined. In fact, it is a well-behaved function.

31.5.1 Theorem *If T is in $\mathscr{E}'(\mathbb{R})$, then $\widehat{T}$ and all of its derivatives are slowly increasing functions in $C^\infty(\mathbb{R})$. Furthermore, $\widehat{T}^{(k)}(\xi) = \langle T, \varphi_\xi^{(k)} \rangle$ for $k = 0, 1, 2, \ldots$, where $\varphi_\xi(x) = e^{-2i\pi x\xi}$; that is,*

$$\widehat{T}^{(k)}(\xi) = \langle T_x, (-2i\pi x)^k \, e^{-2i\pi x\xi} \rangle,$$

where T_x indicates that the "integration" is with respect to x.

Proof. We use Theorem 31.4.3 and write $T = \sum_{j=1}^{p} f_j^{(n_j)}$. Since $f_j^{(n_j)}$ is in $\mathscr{E}'(\mathbb{R})$ for $j = 1, 2, \ldots, p$, we have

$$\begin{aligned}
\langle T_x, e^{-2i\pi x\xi} \rangle &= \sum_{j=1}^{p} \langle f_j^{(n_j)}, e^{-2i\pi x\xi} \rangle \\
&= \sum_{j=1}^{p} (-1)^{n_j} \int_{\mathbb{R}} (-2i\pi\xi)^{n_j} e^{-2i\pi x\xi} f_j(x)\, dx \\
&= \sum_{j=1}^{p} (2i\pi\xi)^{n_j} \widehat{f_j}(\xi).
\end{aligned}$$

Since f_j is continuous and has compact support, we know from Proposition 17.2.1(iii) that the function $\xi \mapsto \langle T_x, e^{-2i\pi x\xi} \rangle$ is infinitely differentiable. On the other hand, $\widehat{T} = \sum_{j=1}^{p} \widehat{f_j^{(n_j)}} = \sum_{j=1}^{p} (2i\pi\xi)^{n_j} \widehat{f_j}(\xi)$. We conclude that $\widehat{T}(\xi) = \langle T_x, e^{-2i\pi x\xi} \rangle$, which proves the case for $k = 0$.

For $k \geq 1$ we compute $\widehat{T}^{(k)}$ in the sense of distributions. If $\varphi \in \mathscr{S}(\mathbb{R})$, then

$$\begin{aligned}
\langle \widehat{T}^{(k)}, \varphi \rangle &= (-1)^k \langle \widehat{T}, \varphi^{(k)} \rangle = (-1)^k \langle T, \widehat{\varphi^{(k)}} \rangle = \langle T, (-2i\pi x)^k \widehat{\varphi}(x) \rangle \\
&= \sum_{j=1}^{p} \int_{\mathbb{R}} f_j^{(n_j)}(x)\, (-2i\pi x)^k \left(\int_{\mathbb{R}} e^{-2i\pi x\xi} \varphi(\xi)\, d\xi \right) dx \\
&= \int_{\mathbb{R}} \varphi(\xi) \left(\sum_{j=1}^{p} \int_{\mathbb{R}} f_j^{(n_j)}(x)(-2i\pi x)^k\, e^{-2i\pi x\xi}\, dx \right) d\xi,
\end{aligned}$$

which shows that $\widehat{T}^{(k)}(\xi) = \langle T_x, (-2i\pi x)^k e^{-2i\pi x\xi} \rangle$.

Finally, to show that $\widehat{T}^{(k)}$ is slowly increasing, we use the continuity of T (31.11): There exist $C > 0$, K compact, and $m \in \mathbb{N}$ such that

$$|\langle T, \varphi \rangle| \leq C \sup_{j \leq m} \sup_{x \in K} |\varphi^{(j)}(x)|$$

for all $\varphi \in C^\infty(\mathbb{R})$. By taking $\varphi(x) = e^{-2i\pi x\xi}(-2i\pi x)^k$ we see that

$$|T^{(k)}(\xi)| \leq C \sup_{j \leq m} \sup_{x \in K} \left| \frac{d^j}{dx^j} \left((-2i\pi x)^k e^{-2i\pi x\xi} \right) \right|,$$

from which it follows that $\widehat{T}^{(k)}(\xi)$ is slowly increasing. □

31.5.2 Theorem (the Paley–Wiener theorem)

Suppose that $T \in \mathscr{E}'(\mathbb{R})$ and that $\operatorname{supp} T \subset [-M, +M]$ for some $M > 0$. Then $f(\xi) = \widehat{T}(\xi)$ can be extended to a holomorphic function $\widetilde{f} : \mathbb{C} \to \mathbb{C}$ that satisfies the following estimate: There exist $C > 0$ and $M \in \mathbb{N}$ such that for all $z \in \mathbb{C}$,

$$|\widetilde{f}(z)| \leq C(1 + |z|^2)^{\frac{m}{2}} e^{2\pi M |\mathrm{Im}(z)|}.$$

Proof. We know that $f(\xi) = \widehat{T}(\xi)$ is the C^∞ function $f(\xi) = \langle T_x, e^{-2i\pi x\xi} \rangle$. Define $\widetilde{f}(z) = \langle T_x, e^{-2i\pi xz} \rangle$ for $z \in \mathbb{C}$. One shows by direct computation (as above) that $\widetilde{f}$ is holomorphic on $\mathbb{C}$, i.e., it is infinitely differentiable on $\mathbb{C}$. The inequality follows from the continuity of T. □

31.5.3 Remark The converse of the Paley–Wiener theorem is also true [Kho72]. The Paley–Wiener theorem has the following important consequence.

31.5.4 Proposition *The Fourier transform of a distribution T with compact support ($T \neq 0$) cannot have compact support.*

Proof. If $T \in \mathscr{E}'(\mathbb{R})$ has compact support, the function $f = \widehat{T}$ is the restriction to $\mathbb{R}$ of a holomorphic function on $\mathbb{C}$. Thus f is analytic on $\mathbb{R}$. If f had compact support, it would vanish on some nonempty open interval. But being analytic, this implies that f vanishes everywhere. Thus f cannot have compact support. □

This result will be used in Section 38.5.1.

31.6 Formulary

(i)

$$T^{(k)} \overset{\mathscr{F}}{\longmapsto} (2i\pi\xi)^k \widehat{T}$$

$$(-2i\pi x)^k T \overset{\mathscr{F}}{\longmapsto} \widehat{T}^{(k)}$$

(ii)

$$\tau_a T \overset{\mathscr{F}}{\longmapsto} e^{-2i\pi\xi a}\widehat{T}$$

$$e^{2i\pi x a} T \overset{\mathscr{F}}{\longmapsto} \tau_a \widehat{T}$$

(iii)

$$\delta_a \overset{\mathscr{F}}{\longmapsto} e^{-2i\pi a\xi}$$

$$e^{2i\pi x a} \overset{\mathscr{F}}{\longmapsto} \delta_a$$

$$\delta^{(k)} \overset{\mathscr{F}}{\longmapsto} (2i\pi\xi)^k$$

$$x^k \overset{\mathscr{F}}{\longmapsto} (-2i\pi)^{-k}\delta^{(k)}$$

(iv)

$$u(x) \overset{\mathscr{F}}{\longmapsto} \frac{1}{2}\delta + \frac{1}{2i\pi}\,\mathrm{pv}\left(\frac{1}{\xi}\right)$$

$$\mathrm{sign}(x) \overset{\mathscr{F}}{\longmapsto} \frac{1}{i\pi}\,\mathrm{pv}\left(\frac{1}{\xi}\right)$$

$$\mathrm{pv}\left(\frac{1}{x}\right) \overset{\mathscr{F}}{\longmapsto} -i\pi\,\mathrm{sign}(\xi)$$

(v)

$$\mathscr{F}(\mathscr{F}\,T) = T_\sigma$$

$$(\mathscr{F}\,T)_\sigma = \mathscr{F}\,T_\sigma = \overline{\mathscr{F}}\,T$$

$$\mathscr{F}\,\overline{\mathscr{F}}\,T = \overline{\mathscr{F}}\,\mathscr{F}\,T = T$$

31.7 Exercises

Exercise 31.1 ($\mathscr{S}'(\mathbb{R})$ as a subspace of $\mathscr{D}'(\mathbb{R})$)

For $T \in \mathscr{S}'\mathbb{R})$ we write $j(T) = T_{|\mathscr{S}(\mathbb{R})}$ to denote the restriction of T to $\mathscr{D}(\mathbb{R})$.

(a) Show that $j(T) \in \mathscr{D}'(\mathbb{R})$.

(b) Show that $j : \mathscr{S}'(\mathbb{R}) \mapsto \mathscr{D}'(\mathbb{R})$ is injective ($j(T) = 0$ implies $T = 0$).

(c) Show that j is continuous.

*Exercise 31.2

(1) Suppose that $f \in \mathscr{S}(\mathbb{R})$ and that $T \in \mathscr{S}'(\mathbb{R})$. For $\varphi \in \mathscr{S}(\mathbb{R})$ we define $\langle fT, \varphi \rangle = \langle T, f\varphi \rangle$.

   (a) Show that $fT \in \mathscr{S}'(\mathbb{R})$.

   (b) Show that the mapping $(f, T) \mapsto fT$ from $\mathscr{S}(\mathbb{R}) \times \mathscr{S}'(\mathbb{R})$ to $\mathscr{S}'(\mathbb{R})$ is continuous with respect to each variable.

(2) Show that $fT \in \mathscr{S}'(\mathbb{R})$ if T is in $\mathscr{S}'(\mathbb{R})$ and f is in $C^\infty(\mathbb{R})$ with $f^{(j)}$ slowly increasing for all j.

Exercise 31.3 Show that a primitive of a tempered distribution is a tempered distribution.

Exercise 31.4 Show that $\log|x|$ is a tempered distribution. Deduce from this that $\operatorname{pv}\left(\frac{1}{x}\right)$ and $\operatorname{fp}\left(\frac{1}{x^2}\right)$ are also tempered distributions.

***Exercise 31.5 (the Fourier transform of u)** Take the Fourier transform of the equality $u' = \delta$ and use Exercise 28.12 to show that $\widehat{u}$ is of the form $\widehat{u}(\xi) = \frac{1}{2i\pi}\operatorname{pv}\left(\frac{1}{\xi}\right) + \lambda\delta$. Show that $\lambda = \frac{1}{2}$ by establishing that $\operatorname{pv}\left(\frac{1}{\xi}\right)$ is an odd distribution.

Use this to show that $\operatorname{pv}\left(\frac{1}{x}\right) \in \mathscr{S}'(\mathbb{R})$.

Exercise 31.6 Compute the Fourier transform of $\operatorname{pv}\left(\frac{1}{x}\right)$. Use this to compute the Fourier transform of $\operatorname{sign}(x)$.

Exercise 31.7 Compute the Fourier transform of $\operatorname{fp}\left(\frac{1}{x^2}\right)$. Use this to compute the Fourier transform of $|x|$.

Exercise 31.8 Show that

$$[\cos 2\pi\lambda t]\widehat{\ } = \frac{1}{2}(\delta_\lambda + \delta_{-\lambda}) \quad \text{and} \quad [\sin 2\pi\lambda t]\widehat{\ } = \frac{1}{2i}(\delta_\lambda - \delta_{-\lambda}).$$

Hint: Use Section 31.3.2.

Exercise 31.9 Let $f(x) = \arctan x$.

(a) To which of the spaces $L^1(\mathbb{R})$, $L^2(\mathbb{R})$, $\mathscr{S}(\mathbb{R})$, $\mathscr{E}'(\mathbb{R})$, $\mathscr{S}'(\mathbb{R})$ does f belong?

(b) Compute $\widehat{f'}(\xi)$.

(c) Deduce from this and Exercise 28.16 that

$$\widehat{f}(\xi) = \frac{1}{2i}\operatorname{pv}\left(\frac{1}{\xi}\right) - \frac{1 - e^{2\pi|\xi|}}{2i\xi}.$$

Exercise 31.10

(a) Compute the Fourier transform of $\arctan \frac{1}{x}$ from Exercises 31.6 and 31.9 and the formula

$$\arctan x + \arctan \frac{1}{x} = \frac{\pi}{2} \operatorname{sign}(x), \quad x \neq 0.$$

(b) Prove this result starting with the derivative of $\arctan \frac{1}{x}$ and proceeding as in Exercise 31.9.

Exercise 31.11 We wish to study the limit, in the sense of distributions, of the sequence of functions $f_n(x) = \left(x + \frac{i}{n}\right)^{-1}$.

(a) Verify that $f_n \in \mathscr{S}'(\mathbb{R})$ and compute $\widehat{f_n}$.

(b) Prove that $\widehat{f_n}$ converges in $\mathscr{S}'(\mathbb{R})$ to $-2i\pi u$ and deduce that f_n converges in $\mathscr{S}'(\mathbb{R})$ to $\operatorname{pv}\left(\frac{1}{x}\right) - i\pi\delta$.

***Exercise 31.12** We wish to prove Proposition 31.1.3. Thus assume that $T \in \mathscr{D}'(\mathbb{R})$ is continuous on $\mathscr{D}(\mathbb{R})$ in the topology of $\mathscr{S}(\mathbb{R})$. The plan is to extend T to a continuous linear functional on $\mathscr{S}(\mathbb{R})$ using the fact that $\mathscr{D}(\mathbb{R})$ is dense in $\mathscr{S}(\mathbb{R})$.

(a) Extending T to $\mathscr{S}(\mathbb{R})$: Let ψ be an arbitrary element of $\mathscr{S}(\mathbb{R})$ and let φ_n be a sequence in $\mathscr{D}(\mathbb{R})$ that converges to ψ in the topology of $\mathscr{S}(\mathbb{R})$. Show that $T(\varphi_n)$ converges and define $T(\psi) = \lim_{n\to\infty} T(\varphi_n)$. (Show that $T(\varphi_n)$ is a Cauchy sequence by arguing indirectly.)

(b) T is well-defined: Show that if ϕ_n is another sequence in $\mathscr{D}(\mathbb{R})$ that converges to ψ in the topology of $\mathscr{S}(\mathbb{R})$, then $\lim_{n\to\infty} T(\phi_n) = T(\psi)$.

(c) T is continuous on $\mathscr{S}(\mathbb{R})$: We must show that $T(\psi_n) \to 0$ whenever $\psi_n \to 0$ in $\mathscr{S}(\mathbb{R})$. Define α_m as in Exercise 19.7. Then $\alpha_m \psi_n \in \mathscr{D}(\mathbb{R})$, and for each n, $\alpha_m \psi_n \to \psi_n$ in $\mathscr{S}(\mathbb{R})$ as $m \to \infty$. Hence, given $\varepsilon > 0$, for each n there is an $m(n)$ such that

$$|T(\alpha_{m(n)}\psi_n - \psi_n)| \leq \varepsilon.$$

Use this and estimates developed from the equation

$$x^p (\alpha_{m(n)}\psi_n)^{(q)}(x) = x^p \sum_{j=0}^{q} \binom{q}{j} m(n)^{-j} \alpha^{(j)}\left(\frac{x}{m(n)}\right) \psi_n^{(q-j)}(x)$$

to show that $T(\psi_n) \to 0$.

Lesson 32

Convolution of Distributions

We discussed the convolution of functions in Lesson 20. There we saw that it is not always possible to take the convolution of two functions; it is the same for distributions. We will study the convolution of distributions and its basic properties for the more important cases.

32.1 The convolution of a distribution and a C^∞ function

When f and g are in $L^1(\mathbb{R})$, the convolution $f * g(x) = \int_{\mathbb{R}} f(x-t)g(t)\,dt$ is well-defined and $f * g \in L^1(\mathbb{R})$. Now consider $f * g$ as a distribution. For $\varphi \in \mathscr{D}(\mathbb{R})$ we have

$$\begin{aligned}\langle f * g, \varphi\rangle &= \int_{\mathbb{R}}\left(\int_{\mathbb{R}} f(x-t)g(t)\,dt\right)\varphi(x)\,dx \\ &= \int_{\mathbb{R}}\left(\int_{\mathbb{R}} f(x-t)\varphi(x)\,dx\right)g(t)\,dt \qquad (32.1)\\ &= \int_{\mathbb{R}}\left(\int_{\mathbb{R}} g(x-u)\varphi(x)\,dx\right)f(u)\,du. \qquad (32.2)\end{aligned}$$

From this it appears that one should study the quantities $f_\sigma * \varphi$ and $g_\sigma * \varphi$ when f and g are distributions.

32.1.1 Proposition *Suppose that $\varphi \in C^\infty(\mathbb{R})$ and $T \in \mathscr{D}'(\mathbb{R})$ satisfy one of the following three conditions:*

(i) $\varphi \in \mathscr{D}(\mathbb{R})$ *and* $T \in \mathscr{D}'(\mathbb{R})$.

(ii) $\varphi \in \mathscr{S}(\mathbb{R})$ *and* $T \in \mathscr{S}'(\mathbb{R})$.

(iii) $\varphi \in C^\infty(\mathbb{R})$ *and* $T \in \mathscr{E}'(\mathbb{R})$.

Then the function ψ defined by

$$\psi(x) = \langle \tau_x T, \varphi\rangle \qquad (32.3)$$

is infinitely differentiable, and

$$\psi^{(k)}(x) = \langle \tau_x T, \varphi^{(k)} \rangle \quad \textit{for } k = 1, 2, \ldots. \tag{32.4}$$

Proof.

(i) When $\varphi \in \mathscr{D}(\mathbb{R})$, the expression $\langle \tau_x T, \varphi \rangle$ makes sense for all $x \in \mathbb{R}$. We wish to show that the function $\psi(x) = \langle \tau_x T, \varphi \rangle$ is differentiable. Thus let h_n be a sequence of nonzero reals that tends to 0 as $n \to \infty$. Define

$$\alpha_n(y) = \frac{1}{h_n}[\varphi(y + x + h_n) - \varphi(y + x)];$$

then

$$\frac{1}{h_n}[\psi(x + h_n) - \psi(x)] = \langle T, \alpha_n \rangle.$$

Now, $\lim_{n\to\infty} \alpha_n(y) = \varphi'(x + y) = \tau_{-x}\varphi'(y)$. To prove that ψ is differentiable it is sufficient to show that α_n converges to $\tau_{-x}\varphi'$ in $\mathscr{D}(\mathbb{R})$.

We consider the support of the α_n. If $\mathrm{supp}(\varphi) \subset [-M, M]$ and $|h_n| \leq 1$, then $\mathrm{supp}(\alpha_n) \subset [-x - M - 1, -x + M + 1]$, which is a fixed compact interval K. The following inequality, which is based on the mean value theorem, shows that α_n and of all its derivatives converge uniformly on K: For each $q \in \mathbb{N}$,

$$\begin{aligned} |\alpha_n^{(q)}(y) - \varphi^{(q+1)}(x + y)| &= |\varphi^{(q+1)}(x + y + \theta_n h_n) - \varphi^{(q+1)}(x + y)| \\ &\leq |h_n| \, \|\varphi^{(q+2)}\|_\infty, \quad 0 < \theta_n < 1. \end{aligned}$$

This proves that ψ is differentiable and that

$$\psi'(x) = \langle T, \tau_{-x}\varphi' \rangle = \langle \tau_x T, \varphi' \rangle.$$

Similarly, one proves that $\psi \in C^\infty(\mathbb{R})$ and that equation (32.4) holds for $k > 1$.

(ii) If $\varphi \in \mathscr{S}(\mathbb{R})$ and $T \in \mathscr{S}'(\mathbb{R})$, then $\langle \tau_x T, \varphi \rangle$ makes sense. To show that ψ is differentiable it is sufficient to verify that α_n converges to $\tau_{-x}\varphi'$ in $\mathscr{S}(\mathbb{R})$. As in (i), the mean value theorem leads to the inequality

$$\begin{aligned} &|y^p(\alpha_n - \tau_{-x}\varphi')^{(q)}(y)| \leq |h_n||y^p \varphi^{(q+2)}(x + y + \rho_n h_n)| \\ &\quad = \frac{|h_n||y|^p}{1 + |x + y + \rho_n h_n|^p}\left[(1 + |x + y + \rho_n h_n|^p)\varphi^{(q+2)}(x + y + \rho_n h_n)\right] \end{aligned}$$

where $0 < \rho_n < 1$. Consequently,

$$\sup_{y\in\mathbb{R}} |y^p(\alpha_n - \tau_{-x}\varphi')^{(q)}(y)| \leq C|h_n|\Big(\|\varphi^{(q+2)}\|_\infty + \sup_{t\in\mathbb{R}} |t^p \varphi^{(q+2)}(t)|\Big)$$

for some constant C. Since φ is in $\mathscr{S}(\mathbb{R})$, we see that α_n converges in $\mathscr{S}(\mathbb{R})$ to $\tau_{-x}\varphi'$ as $n \to \infty$. That $\psi \in C^\infty(\mathbb{R})$ and (32.4) are proved similarly.

(iii) Again, $\langle \tau_x T, \varphi\rangle$ makes sense because $\varphi \in C^\infty(\mathbb{R})$ and $T \in \mathscr{E}'(\mathbb{R})$. To prove that ψ is differentiable it is sufficient to show that $\alpha_n^{(q)}$ converges to $(\tau_{-x}\varphi')^{(q)}$ uniformly on all compact subsets of $\mathbb{R}$. This is done using inequalities similar to those used in the proofs of (i) and (ii). □

32.1.2 Definition Assume that $\varphi \in C^\infty(\mathbb{R})$ and $T \in \mathscr{D}'(\mathbb{R})$ satisfy one of the conditions in Proposition 32.1.1. The convolution of φ and T is the function $\varphi * T$ defined by

$$\varphi * T(x) = \langle T_y, \varphi(x-y)\rangle. \tag{32.5}$$

The "y" in this definition indicates the variable of "integration." Written differently,

$$\langle T_y, \varphi(x-y)\rangle = \langle \tau_{-x}T, \varphi_\sigma\rangle,$$

which is the function we studied in Proposition 32.1.1. Thus we know the meaning of the convolutions

$$\mathscr{D} * \mathscr{D}', \qquad \mathscr{S} * \mathscr{S}', \qquad C^\infty * \mathscr{E}'.$$

32.1.3 Proposition (convolution $\mathscr{S} * \mathscr{S}'$) *Assume $\varphi \in \mathscr{S}(\mathbb{R})$ and $T \in \mathscr{S}'(\mathbb{R})$. The convolution $\varphi * T$ and all of its derivatives are slowly increasing C^∞ functions.*

Proof. By definition $\varphi * T(x) = \langle T_y, \varphi(x-y)\rangle$. We use Theorem 31.1.10 and write $T = \sum_{k=1}^p f_k^{(n_k)}$, where the continuous slowly increasing functions f_k satisfy $|f_k(x)| \le C_k(1+x^2)^{N_k}$. Then

$$\begin{aligned}\varphi * T(x) &= \sum_{k=1}^p \langle f_k^{(n_k)}(y), \varphi(x-y)\rangle = \sum_{k=1}^p \int_{\mathbb{R}} f_k(y)\varphi^{(n_k)}(x-y)\,dy\\ &= \sum_{k=1}^p \int_{\mathbb{R}} f_k(x-y)\varphi^{(n_k)}(y)\,dy,\end{aligned}$$

and

$$\begin{aligned}|\varphi * T(x)| &\le \sum_{k=1}^p C_k \int_{\mathbb{R}} (1+(x-y)^2)^{N_k} |\varphi^{(n_k)}(y)|\,dy\\ &\le \sum_{k=1}^p C_k \int_{\mathbb{R}} \sum_{j=1}^{2N_k} |x|^j |\beta_{kj}(y)\varphi^{(n_k)}(y)|\,dy,\end{aligned}$$

where $\beta_{kj}(y)$ is a polynomial in y.

Since $\varphi \in \mathscr{S}(\mathbb{R})$, $\beta_{kj}\varphi^{(n_k)}$ is in $\mathscr{S}(\mathbb{R})$ and hence in $L^1(\mathbb{R})$. Thus $|\varphi * T(x)|$ is bounded by a polynomial in x, which proves that it is slowly increasing.

One obtains similar estimates for the derivatives of $\varphi * T(x)$ by repeating the computation for $(\varphi * T)^{(k)}(x) = \langle T_y, \varphi^{(k)}(x-y)\rangle$. □

We are now going state a few of the convolution's essential properties. The convolution defined in Definition 32.1.2 is an operator that is continuous in each variable. The proof of this result [Sch65b], which we assume, depends on the topologies of the three spaces involved.

32.1.4 Proposition (continuity) *The mapping $(\varphi, T) \to \varphi * T$ defined under one of the hypotheses of Proposition 32.1.1 is continuous with respect to each variable.*

32.1.5 Proposition (derivation) *If $\varphi \in C^\infty(\mathbb{R})$ and $T \in \mathscr{D}'(\mathbb{R})$ satisfy one of the hypotheses of Proposition 32.1.1, then $\varphi * T \in C^\infty(\mathbb{R})$ and*

$$(\varphi * T)^{(k)} = \varphi^{(k)} * T = \varphi * T^{(k)}, \quad k = 1, 2, \ldots . \tag{32.6}$$

Proof. From (32.4), the kth derivative of the function

$$H(x) = \langle T_y, \varphi(x-y)\rangle = \langle \tau_{-x}T, \varphi_\sigma\rangle$$

is

$$\begin{aligned} H^{(k)}(x) &= (-1)^k \langle \tau_{-x}T, (\varphi_\sigma)^{(k)}\rangle = \langle \tau_{-x}T, (\varphi^{(k)})_\sigma\rangle \\ &= \langle T_y, \varphi^{(k)}(x-y)\rangle = \varphi^{(k)} * T(x). \end{aligned}$$

On the other hand,

$$\begin{aligned} \varphi * T^{(k)}(x) &= \langle \tau_{-x}T^{(k)}, \varphi_\sigma\rangle = (-1)^k \langle T, (\varphi_\sigma)^{(k)}(y-x)\rangle \\ &= \langle T, \varphi^{(k)}(x-y)\rangle = \varphi^{(k)} * T(x). \end{aligned}$$

□

32.1.6 Proposition (support) *If $\varphi \in C^\infty(\mathbb{R})$ and $T \in \mathscr{E}'(\mathbb{R})$, then $\operatorname{supp}(\varphi * T) \subset \operatorname{supp}(\varphi) + \operatorname{supp}(T)$, where "+" denotes the algebraic sum of the two sets.*

Proof. Since $\operatorname{supp}(T)$ is compact, $\operatorname{supp}(\varphi) + \operatorname{supp}(T)$ is closed. Define $\Omega = \mathbb{R}\setminus\big(\operatorname{supp}(\varphi) + \operatorname{supp}(T)\big)$. For $x \in \Omega$ and $y \in \operatorname{supp}(\varphi)$, $(x-y) \notin \operatorname{supp}(T)$, and hence $\langle T_y, \varphi(x-y)\rangle = 0$. This proves the required inclusion. □

32.1.7 Corollary *If $\varphi \in \mathscr{D}(\mathbb{R})$ and $T \in \mathscr{E}'(\mathbb{R})$ then the convolution $\varphi * T$ has compact support.*

The results of this section show that the convolution of a distribution and a C^∞ function is a smoothing operation. We will see in the next section, where we introduce the convolution of two distributions, that $\mathscr{D}(\mathbb{R})$ is even dense in $\mathscr{D}'(\mathbb{R})$.

32.2 The convolution $\mathscr{E}' * \mathscr{D}'$

The expressions (32.1) and (32.2) suggest a way to generalize the convolution to distributions. Given two distributions S and T we write

$$\begin{aligned}\langle S * T, \varphi\rangle &= \big\langle S_t, \langle T_x, \varphi(x+t)\rangle\big\rangle \\ &= \big\langle T_u, \langle S_x, \varphi(x+u)\rangle\big\rangle.\end{aligned}$$

We have seen that the function $\psi(t) = \langle T_x, \varphi(x+t)\rangle$ is C^∞ when $T \in \mathscr{D}'(\mathbb{R})$. Thus the expression $\langle S_t, \psi(t)\rangle$ makes sense when $S \in \mathscr{E}'(\mathbb{R})$. Similarly, the function $\alpha(u) = \langle S_x, \varphi(x+u)\rangle$ is in $\mathscr{D}(\mathbb{R})$ by Proposition 32.1.6, and the expression $\langle T_u, \alpha(u)\rangle$ makes sense for all $T \in \mathscr{D}'(\mathbb{R})$. It is not clear, however, that $\langle S_t, \psi(t)\rangle = \langle T_u, \alpha(u)\rangle$, as is the case for functions. This is, in fact, true [Sch65b]; we state without proof the next result.

32.2.1 Theorem ($\mathscr{E}' * \mathscr{D}'$) *Assume $S \in \mathscr{E}'(\mathbb{R})$ and $T \in \mathscr{D}'(\mathbb{R})$.*

(i) *There exists a distribution called the convolution of S and T and denoted by $S * T$ such that for all $\varphi \in \mathscr{D}(\mathbb{R})$,*

$$\langle S * T, \varphi\rangle = \big\langle S_t, \langle T_x, \varphi(x+t)\rangle\big\rangle = \big\langle T_u, \langle S_x, \varphi(x+u)\rangle\big\rangle. \quad (32.7)$$

(ii) *The mapping $(S, T) \mapsto S * T$ from $\mathscr{E}'(\mathbb{R}) \times \mathscr{D}'(\mathbb{R})$ to $\mathscr{D}'(\mathbb{R})$ is continuous with respect to each variable.*

Formula (32.7) is important because it allows us to develop a calculus for the convolution. The convolution is a commutative operation; we next consider its other important properties.

32.2.2 Proposition (Dirac distributions) *Take $T \in \mathscr{D}'(\mathbb{R})$.*

(i) *Then*

$$\delta_a * T = T * \delta_a = \tau_a T, \quad (32.8)$$

and in particular, δ acts like a unit element for convolution.

(ii)

$$\delta^{(k)} * T = T * \delta^{(k)} = T^{(k)}, \quad k = 1, 2, \dots . \quad (32.9)$$

Proof. One needs to be careful not to confuse the index a in δ_a with the "dummy variables" u and x in (32.7). Both results follow from simple computations. To prove (i) we have

$$\begin{aligned}\langle \delta_a * T, \varphi\rangle &= \big\langle T_u, \langle \delta_a, \varphi(x+u)\rangle\big\rangle = \langle T_u, \varphi(a+u)\rangle \\ &= \langle T, \tau_{-a}\varphi\rangle = \langle \tau_a T, \varphi\rangle.\end{aligned}$$

For the proof of (ii) we have

$$\begin{aligned}\langle \delta^{(k)} * T, \varphi\rangle &= \big\langle T_u, \langle \delta^{(k)}, \varphi(x+u)\rangle\big\rangle = (-1)^k \langle T_u, \varphi^{(k)}(u)\rangle \\ &= \langle T^{(k)}, \varphi\rangle.\end{aligned}$$

□

32.2.3 Proposition (derivatives) *If $S \in \mathscr{E}'(\mathbb{R})$ and $T \in \mathscr{D}'(\mathbb{R})$, then*

$$(S * T)^{(k)} = S^{(k)} * T = S * T^{(k)}, \quad k = 1, 2, \ldots . \tag{32.10}$$

Proof. For $\varphi \in \mathscr{D}(\mathbb{R})$,

$$\begin{aligned}\langle (S * T)^{(k)}, \varphi\rangle &= (-1)^k \langle S * T, \varphi^{(k)}\rangle = (-1)^k \left\langle S_t, \langle T_x, \varphi^{(k)}(x+t)\rangle\right\rangle \\ &= \left\langle S_t, \left\langle T_x^{(k)}, \varphi(x+t)\right\rangle\right\rangle = \langle S * T^{(k)}, \varphi\rangle.\end{aligned}$$

Similarly, $(S * T)^{(k)} = S^{(k)} * T$. □

32.2.4 Remark If $T \in \mathscr{E}'(\mathbb{R})$, one can find a primitive of T by writing $T * U$ where U is the Heaviside distribution. Indeed, from (32.10) we see that $(T * U)' = T * U' = T * \delta = T$.

32.2.5 Proposition (support of a convolution)

(i) *Assume $S \in \mathscr{E}'(\mathbb{R})$ and $T \in \mathscr{D}'(\mathbb{R})$. If $\operatorname{supp}(S) = A$ and $\operatorname{supp}(T) = B$, then $\operatorname{supp}(S * T) \subset A + B$.*

(ii) *If S and T are in $\mathscr{E}'(\mathbb{R})$, then $S * T \in \mathscr{E}'(\mathbb{R})$.*

Proof.

(i) Since A is compact, $A + B$ is closed. Let $\Omega = \mathbb{R} \setminus (A + B)$ and take $\varphi \in \mathscr{D}(\mathbb{R})$ with $\operatorname{supp}(\varphi) \subset \Omega$. We will show that $\langle S * T, \varphi\rangle = 0$. We have $\langle S * T, \varphi\rangle = \left\langle T_u, \langle S_x, \varphi(x+u)\rangle\right\rangle$ and $\psi(u) = \langle S_x, \varphi(x+u)\rangle = S * \varphi_\sigma(-u)$. Thus we wish to show that $\operatorname{supp}(\psi) \cap B = \emptyset$.

If $u \in \operatorname{supp}(\psi) \cap B$, then $-u \in \operatorname{supp}(\varphi_\sigma) + \operatorname{supp}(S)$ (recall that by Proposition 32.1.6 $\operatorname{supp}(S * \varphi_\sigma) \subset \operatorname{supp}\varphi_\sigma + \operatorname{supp}S$). This means that $-u = y + x$ with $-y \in \operatorname{supp}(\varphi)$ and $x \in \operatorname{supp}(S)$. But then $-y = u + x$ with $u \in B$, $x \in A$, and hence

$$\operatorname{supp}(\varphi) \cap (A + B) \neq \emptyset,$$

which is a contradiction, since $\operatorname{supp}(\varphi) \subset \Omega = \mathbb{R} \setminus (A + B)$.

(ii) If S and T are in $\mathscr{E}'(\mathbb{R})$, then $A + B$ is compact. □

32.2.6 Proposition (density of $\mathscr{D}(\mathbb{R})$ in $\mathscr{D}'(\mathbb{R})$) *If $T \in \mathscr{D}'(\mathbb{R})$, then there exists a regularizing sequence $\theta_n \in \mathscr{D}(\mathbb{R})$ such that θ_n converges to T in $\mathscr{D}'(\mathbb{R})$.*

Proof. Choose the usual regularizing sequence ρ_n (Definition 21.3.1). We know that ρ_n tends to δ in $\mathscr{D}'(\mathbb{R})$. Let $\alpha_n = \rho_n * T$. By Theorem 32.2.1(ii), α_n converges to T in $\mathscr{D}'(\mathbb{R})$. This proves that $C^\infty(\mathbb{R})$ is dense in $\mathscr{D}'(\mathbb{R})$. To prove that $\mathscr{D}(\mathbb{R})$ is dense in $\mathscr{D}'(\mathbb{R})$, we must show that the functions α_n can be chosen with compact support. We fix this by multiplying α_n by a function $\beta_n \in \mathscr{D}(\mathbb{R})$ such that $\beta_n(x) = 1$ for $|x| \leq n$ and $\beta_n(x) = 0$ otherwise. Then $\theta_n = \alpha_n\beta_n \in \mathscr{D}(\mathbb{R})$, and it converges to T in $\mathscr{D}'(\mathbb{R})$. □

32.3 The convolution $\mathscr{E}' * \mathscr{S}'$

The convolution $\mathscr{E}' * \mathscr{S}'$ is a particular case of the convolution $\mathscr{E}' * \mathscr{D}'$. But in this case the distribution that one obtains is tempered.

Referring to (32.7) we see that the functions $\psi(t) = \langle T_x, \varphi(x+t)\rangle$ and $\alpha(u) = \langle S_x, \varphi(x+u)\rangle$ are well-defined when $\varphi \in \mathscr{S}(\mathbb{R})$, $T \in \mathscr{S}'(\mathbb{R})$, and $S \in \mathscr{E}'(\mathbb{R})$ (Proposition 32.1.1). We establish a preliminary result.

32.3.1 Proposition *If $S \in \mathscr{E}'(\mathbb{R})$, the mapping $\varphi \mapsto \alpha$ defined by $\alpha(u) = \langle S_x, \varphi(x+u)\rangle$ is continuous from $\mathscr{S}(\mathbb{R})$ to $\mathscr{S}(\mathbb{R})$.*

Proof. We use the theorem about the structure of elements in $\mathscr{E}'(\mathbb{R})$ (Theorem 31.4.3). Thus we write $S = \sum_{j=1}^{p} f_j^{(n_j)}$, where the f_j are continuous with support in some compact set K. Then

$$\alpha(u) = \sum_{j=1}^{p} \langle f_j^{(n_j)}(x), \varphi(x+u)\rangle = \sum_{j=1}^{p} (-1)^{n_j} \int_K f_j(x) \varphi^{(n_j)}(x+u)\, dx,$$

and it is not difficult to verify that $\alpha \in \mathscr{S}(\mathbb{R})$.

To prove continuity, we assume that $\varphi_n \to 0$ in $\mathscr{S}(\mathbb{R})$ and show that the corresponding sequence α_n converges to 0 in $\mathscr{S}(\mathbb{R})$. It is clear that we can differentiate under the integral sign, and consequently

$$\begin{aligned}
|u^m \alpha_n^{(q)}(u)| &\le \sum_{j=1}^{p} \int_K |f_j(x)| |u^m \varphi_n^{(n_j+q)}(x+u)|\, dx \\
&\le \sum_{j=1}^{p} \int_K |f_j(x)| \frac{|u|^m}{1+|x+u|^m} (1+|x+u|^m) |\varphi_n^{(n_j+q)}(x+u)|\, dx \\
&\le C \sum_{j=1}^{p} \left(\|\varphi_n^{(n_j+q)}\|_\infty + \sup_{t\in\mathbb{R}} |t^m \varphi_n^{(n_j+q)}(t)| \right)
\end{aligned}$$

for some constant C. This shows that $\varphi_n \to 0$ in $\mathscr{S}(\mathbb{R})$ implies that α_n converges to 0 in $\mathscr{S}(\mathbb{R})$. □

We use this result to prove the next one.

32.3.2 Proposition *If $S \in \mathscr{E}'(\mathbb{R})$ and $T \in \mathscr{S}'(\mathbb{R})$, then the convolution $S * T$ is a tempered distribution.*

Proof. We know that $S*T$ is a distribution. Let φ_n be a sequence in $\mathscr{D}(\mathbb{R})$ that tends to 0 in $\mathscr{S}(\mathbb{R})$. Then $\langle S * T, \varphi_n\rangle = \big\langle T_u, \langle S_x, \varphi_n(x+u)\rangle\big\rangle$, and by Proposition 32.3.1 the sequence $\alpha_n(u) = \langle S_x, \varphi_n(x+u)\rangle$ is in $\mathscr{S}(\mathbb{R})$ and converges to 0. Hence $\lim_{n\to\infty} \langle T, \alpha_n\rangle = 0$. The result follows from Proposition 31.1.3. □

The next step is to examine continuity.

32.3.3 Proposition

(i) *Let S_n be a sequence in $\mathscr{E}'(\mathbb{R})$ that converges to 0 in $\mathscr{E}'(\mathbb{R})$; that is, $\langle S_n, \varphi\rangle \to 0$ for all $\varphi \in C^\infty(\mathbb{R})$. Then $S_n * T \to 0$ in $\mathscr{S}'(\mathbb{R})$, and hence in $\mathscr{D}'(\mathbb{R})$, for all $T \in \mathscr{S}'(\mathbb{R})$.*

(ii) *Let T_n be a sequence in $\mathscr{S}'(\mathbb{R})$ that converges to 0 in $\mathscr{S}'$. Then for all $S \in \mathscr{E}'(\mathbb{R})$, $S * T_n \to 0$ in $\mathscr{S}'(\mathbb{R})$, and hence in $\mathscr{D}'(\mathbb{R})$.*

Proof.

(i) By Proposition 32.1.1, the function $\psi(t) = \langle T_x, \varphi(x+t)\rangle$ is in $C^\infty(\mathbb{R})$ for all φ in $\mathscr{S}(\mathbb{R})$. Thus

$$\langle S_n * T, \varphi\rangle = \langle S_n, \psi\rangle \to 0 \quad \text{as } n \to \infty.$$

(ii) Similarly, $\alpha(u) = \langle S_x, \varphi(x+u)\rangle$ is in $\mathscr{S}(\mathbb{R})$ for all φ is $\mathscr{S}(\mathbb{R})$ (Proposition 32.3.1). Hence $\lim_{n\to\infty}\langle S * T_n, \varphi\rangle = \lim_{n\to\infty}\langle T_n, \varphi\rangle = 0$. □

It is necessary to pay attention to the various notions of convergence. Here is a simple example:

$$\delta_n * \mathbf{1} = \mathbf{1}$$

for all n, and δ_n converges to 0 in $\mathscr{D}'(\mathbb{R})$. In this case the result of Proposition 32.3.3 is not true. This is because δ_n does not converge in $\mathscr{E}'(\mathbb{R})$.

32.3.4 Proposition

(i) *Let S_n be a sequence in $\mathscr{E}'(\mathbb{R})$ that converges to 0 in $\mathscr{D}'(\mathbb{R})$. Assume that there exists a compact set K such that $\operatorname{supp}(S_n) \subset K$ for all n. Then $S_n * T \to 0$ in $\mathscr{D}'(\mathbb{R})$ for all $T \in \mathscr{S}'(\mathbb{R})$.*

(ii) *Let T_n a sequence in $\mathscr{S}'(\mathbb{R})$ that converges to 0 in $\mathscr{D}'(\mathbb{R})$. Then for all $S \in \mathscr{E}'(\mathbb{R})$, $S * T_n \to 0$ in $\mathscr{D}'(\mathbb{R})$.*

Proof.

(i) Take $\varphi \in \mathscr{D}(\mathbb{R})$. The function $\psi(t) = \langle T_x, \varphi(x+t)\rangle$ is in $C^\infty(\mathbb{R})$. Since $\operatorname{supp}(S_n) \subset K$, $\langle S_n, \psi\rangle = \langle S_n, \theta\psi\rangle$, where θ is a function in $\mathscr{D}(\mathbb{R})$ such that $\theta(x) = 1$ for $x \in K$. Then $\theta\psi$ is in $\mathscr{D}(\mathbb{R})$ and $\langle S_n, \theta\psi\rangle \to 0$.

(ii) If $\varphi \in \mathscr{D}(\mathbb{R})$, then the function $\alpha(u) = \langle S_x, \varphi(x+u)\rangle$ is in $\mathscr{D}(\mathbb{R})$, and hence $\langle T_n, \alpha\rangle \to 0$. □

32.4 The convolution $\mathscr{D}'_+ * \mathscr{D}'_+$

We have studied the convolution of two distributions where at least one of them has compact support. Without this condition on the support, the convolution is not generally defined. However, as is the case for functions, the convolution is defined when both distributions are in $\mathscr{D}'_+$ (or $\mathscr{D}'_-$).

We begin with a preliminary result about $\mathscr{D}'_+$.

32.4.1 Proposition *Suppose that $T \in \mathscr{D}'_+$ and $\varphi \in C^\infty(\mathbb{R})$ and that* $\operatorname{supp}(T) \subset [a, +\infty)$ *and* $\operatorname{supp}(\varphi) \subset (-\infty, b]$. *Then $\langle T, \varphi\rangle$, defined by*

$$\langle T, \varphi\rangle = \langle T, \theta\varphi\rangle, \tag{32.11}$$

where θ is a function in $\mathscr{D}(\mathbb{R})$ equal to 1 on an interval $[-M, M]$ containing a and b in its interior, is well-defined.

Proof. $\theta\varphi \in \mathscr{D}(\mathbb{R})$, so $\langle T, \theta\varphi\rangle$ makes sense. We must show that the definition of $\langle T, \varphi\rangle$ does not depend on the choice of θ. Let θ_1 be another function in $\mathscr{D}(\mathbb{R})$ equal to 1 on $[-M_1, M_1]$ containing a and b. Then $(\theta - \theta_1)\varphi$ vanishes on $[-m, +\infty)$, where $m = \min\{M, M_1\}$. Since $\operatorname{supp}(T) \subset [a, +\infty)$, we have $\operatorname{supp}(T) \cap \operatorname{supp}((\theta - \theta_1)\varphi) = \emptyset$ and $\langle T, (\theta - \theta_1)\varphi\rangle = 0$. □

To define the convolution, it is necessary to give meaning to the expressions $\big\langle S_t, \langle T_x, \varphi(x + t)\rangle\big\rangle$ and $\big\langle T_u, \langle S_x, \varphi(x + u)\rangle\big\rangle$ for S and T in $\mathscr{D}'_+$ and φ in $\mathscr{D}(\mathbb{R})$.

32.4.2 Proposition *Suppose that $T \in \mathscr{D}'_+$ and $\varphi \in C^\infty(\mathbb{R})$ and that* $\operatorname{supp}(T) \subset [a, +\infty)$ *and* $\operatorname{supp}(\varphi) \subset (-\infty, b]$. *Then $\psi(t) = \langle T_x, \varphi(x + t)\rangle$ is defined of all $t \in \mathbb{R}$,* $\operatorname{supp}(\psi) \subset (-\infty, b - a]$, *and $\psi \in C^\infty(\mathbb{R})$.*

Proof. The function $\tau_{-t}\varphi$ is in $C^\infty(\mathbb{R})$ with $\operatorname{supp}(\tau_{-t}\varphi) \subset (-\infty, b - t]$ for all $t \in \mathbb{R}$. Thus by Proposition 32.4.1, $\langle T_x, \varphi(x + t)\rangle$ is well-defined.

Now, $\psi(t) = 0$ if $\operatorname{supp}(\tau_{-t}\varphi) \cap \operatorname{supp}(T) = \emptyset$, which is the case when $b - t < a$. Hence, $\operatorname{supp}(\psi) \subset (-\infty, b - a]$. That $\psi \in C^\infty$ is a consequence of (32.11) and Proposition 32.1.1(i). □

These preliminary results lead to the next theorem, which we state without proof (see, for example, [Sch65b]).

32.4.3 Theorem (the convolution $\mathscr{D}'_+ * \mathscr{D}'_+$) *Suppose that S and T are in $\mathscr{D}'_+$.*

(i) *There exists a distribution called the convolution of S and T and denoted by $S * T$ such that for all $\varphi \in \mathscr{D}(\mathbb{R})$,*

$$\langle S * T, \varphi\rangle = \big\langle S_t, \langle T_x, \varphi(x + t)\rangle\big\rangle = \big\langle T_u, \langle S_x, \varphi(x + u)\rangle\big\rangle. \tag{32.12}$$

(ii) $(S * T)^{(k)} = S^{(k)} * T = S * T^{(k)}, \quad k = 1, 2, 3 \ldots .$

(iii) *The mapping $(S, T) \mapsto S * T$ of $\mathscr{D}'_+ \times \mathscr{D}'_+$ into $\mathscr{D}'$ is continuous with respect to each variable. (The convergence of S_n to 0 in $\mathscr{D}'_+$ means that $S_n \to 0$ in $\mathscr{D}'(\mathbb{R})$ and that there exists a c such that for all n,* $\operatorname{supp}(S_n) \subset [c, +\infty)$.*)*

32.4.4 Proposition (support of $S * T$) *If S and $T \in \mathscr{D}'_+$ with* $\operatorname{supp}(S) \subset [a_1, +\infty)$ *and* $\operatorname{supp}(T) \subset [a_2, +\infty)$, *then*

$$\operatorname{supp}(S * T) \subset [a_1 + a_2, +\infty),$$

and hence $S * T$ is in $\mathscr{D}'_+$.

Proof. Take $\varphi \in \mathscr{D}(\mathbb{R})$ with $\operatorname{supp}(\varphi) \subset (-\infty, a_1 + a_2)$. The support of $\psi(t) = \langle T_x, \varphi(x+t)\rangle$ is in $(-\infty, a_1)$ by Proposition 32.4.2. Thus

$$\operatorname{supp}(\psi) \cap \operatorname{supp}(S) = \emptyset$$

and $\langle S * T, \varphi\rangle = 0$, which proves that $\operatorname{supp}(S * T) \subset [a_1 + a_2, \infty)$. □

32.4.5 Remark As in Section 32.2.4, we can obtain a primitive of T in $\mathscr{D}'_+$ by taking the convolution of T with the Heaviside distribution U. Since $U \in \mathscr{D}'_+$, the primitive of T is in $\mathscr{D}'_+$.

32.5 The associativity of convolution

We have defined the convolution of two distributions and seen that it is a commutative operation. If we wish to convolve three or more distributions, we run into two problems: existence and associativity. Here is a classic example: We wish to compute $\mathbf{1} * \delta' * u$. From Proposition 32.2.3,

$$\mathbf{1} * \delta' = \mathbf{1}' * \delta = 0.$$

Thus $(\mathbf{1} * \delta') * u$ makes sense and is equal to 0. On the other hand, by Proposition 32.2.2,

$$\delta' * u = \delta * u' = \delta * \delta = \delta;$$

hence $\mathbf{1} * (\delta' * u)$ makes sense and is equal to δ. This shows that the convolution product is not associative in general. Nevertheless, convolution is associative in several cases.

32.5.1 Proposition *The convolution of n distributions of which at least $n-1$ have compact support is associative and commutative.*

Proof. The proof is based directly on the definitions. We take S and T in $\mathscr{E}'(\mathbb{R})$ and U in $\mathscr{D}'(\mathbb{R})$; we show, for example, that $(S*T)*U = S*(T*U)$. First, these convolutions make sense, since $S * T \in \mathscr{E}'(\mathbb{R})$ (Proposition 32.2.5(ii)) and $(S * T) * U$ is an $\mathscr{E}' * \mathscr{D}'$ convolution. Similarly, $T * U$ is an $\mathscr{E}' * \mathscr{D}'$ convolution, so $T * U \in \mathscr{D}'$; and $S * (T * U)$ is another $\mathscr{E}' * \mathscr{D}'$ convolution. Take $\varphi \in \mathscr{D}(\mathbb{R})$. From equation (32.7),

$$\langle (S * T) * U, \varphi\rangle = \big\langle (S * T)_t, \langle U_x, \varphi(x+t)\rangle\big\rangle.$$

Now, $\psi(t) = \langle U_x, \varphi(x+t)\rangle$ is in C^∞, and $S * T \in \mathscr{E}'(\mathbb{R})$. Then

$$\begin{aligned}
\langle S * T, \psi\rangle &= \big\langle S_z, \langle T_u, \psi(u+z)\rangle\big\rangle \\
&= \Big\langle S_z, \big\langle T_u, \langle U_x, \varphi(x+u+z)\rangle\big\rangle\Big\rangle \\
&= \big\langle S_z, \langle (T * U)_t, \varphi(t+z)\rangle\big\rangle \\
&= \langle S * (T * U), \varphi\rangle,
\end{aligned}$$

from which we see that $(S * T) * U = S * (T * U)$. □

The space $\mathscr{E}'(\mathbb{R})$ endowed with the convolution operation is a *convolution algebra*, since $T, S \in \mathscr{E}'(\mathbb{R})$ implies that $T * S \in \mathscr{E}'(\mathbb{R})$. This algebra is commutative and associative, and it contains a unit element, δ. $\mathscr{D}'_+(\mathbb{R})$ is another important convolution algebra.

32.5.2 Proposition *The convolution in $\mathscr{D}'_+(\mathbb{R})$ is associative.*

The proof is similar to that of Proposition 32.5.1, and δ is also a unit element for this algebra.

Since these two distribution algebras have unit elements, it is natural to ask whether a distribution S in $\mathscr{E}'(\mathbb{R})$ or $\mathscr{D}'_+(\mathbb{R})$ has an inverse; that is, is there a T such that

$$S * T = \delta.$$

This question is important for solving differential equations with constant coefficients. With this in mind, we introduce the differential operator

$$P = \sum_{m=0}^{p} a_m D^{(m)},$$

where $a_m \in \mathbb{C}$ and $D^{(m)}$ denotes the mth derivative. Given $U \in \mathscr{D}'(\mathbb{R})$, we wish to find a $T \in \mathscr{D}'(\mathbb{R})$ such that

$$P(T) = U. \tag{32.13}$$

We saw in Proposition 32.2.2(ii) that $T^{(k)} = \delta^{(k)} * T$, and hence we can write (32.13) as

$$\sum_{m=0}^{p} a_m \delta^{(m)} * T = U. \tag{32.14}$$

A distribution E is said to be an *elementary solution* of (32.14) if

$$\left(\sum_{m=0}^{p} a_m \delta^{(m)} \right) * E = \delta.$$

This means that E is the inverse of $S = \sum_{m=0}^{p} a_m \delta^{(m)}$. (We will develop methods for computing elementary solutions in Lesson 35.) If we have associativity, then knowing E yields all the solutions of (32.13), since

$$S * E * U = (S * E) * U = \delta * U = U,$$

and hence $T = E * U$. Furthermore, this solution is unique: If T_1 and T_2 satisfy (32.13), then we have

$$S * T_1 = S * T_2 = U.$$

By taking the convolution with E this becomes, thanks to associativity,

$$(E * S) * T_1 = (E * S) * T_2,$$

and since $E * S = \delta$, we are left with

$$T_1 = T_2.$$

In the language of filters, we will see that E is the impulse response (Lessons 34 and 35). These techniques also apply to finding solutions of partial differential equations [Sch65a].

32.6 Exercises

Exercise 32.1 Let S and T be two even (or two odd) distributions. Show that $S * T$ is even.

Exercise 32.2 If T is a distribution in $\mathscr{D}'_+(\mathbb{R})$, show that it has a unique primitive in $\mathscr{D}'_+(\mathbb{R})$, which is $P = u * T$.

Exercise 32.3 Let P be a polynomial of degree less than or equal to m. For which distributions S does $S * P$ make sense? Show that $S * P$ is a polynomial of degree less than or equal to m.

Exercise 32.4 Suppose $f(x) = (1 - x)u(x)$ and $g(x) = e^x u(x)$. Show that $f * g$ makes sense and compute this convolution.

Exercise 32.5 Consider the differential operator defined by

$$P(f) = f'' + a^2 f, \quad a \in \mathbb{R}.$$

Show that $u(x)\dfrac{\sin ax}{a}$ is an elementary solution of P in $\mathscr{D}'_+(\mathbb{R})$ if $a \neq 0$ and that $u(x)x$ is an elementary solution if $a = 0$.

Exercise 32.6 Keeping in mind the last exercise, show that it is possible to find an elementary solution for the operator

$$P = \frac{d^m}{dx^m} + a_{m-1}\frac{d^{m-1}}{dx^{m-1}} + \cdots + a_1\frac{d}{dx} + a_0$$

of the form $E(x) = u(x)g(x)$ where g is a C^∞ function that is the solution of a differential equation in the classical sense. What is this equation and what are the boundary conditions?

Exercise 32.7

(a) If A_1 and A_2 have inverses in $\mathscr{D}'_+(\mathbb{R})$, show that $A_1 * A_2$ is invertible. What is the inverse of $A_1 * A_2$?

(b) Use this result to find the elementary solution of the operator

$$P = \frac{d^2}{dx^2} - 2\lambda\frac{d}{dx} + \lambda^2.$$

Exercise 32.8 What are the inverses in $\mathscr{D}'_+(\mathbb{R})$ of u, δ', and $\delta' - \alpha\delta$?

Lesson 33

Convolution and the Fourier Transform of Distributions

As is the case for functions (Lesson 23), the Fourier transform interchanges convolution and multiplication of distributions. We wish to determine under what conditions the relations $\widehat{T * U} = \widehat{T} \cdot \widehat{U}$ and $\widehat{T \cdot U} = \widehat{T} * \widehat{U}$ are true. The first thing to notice is that one must be careful manipulating these relations, since the product of two distributions is not generally defined. We faced a similar problem with the convolution in the last lesson. There we were able to establish several conditions under which the convolution is well-defined and consistent with the convolution for functions.

33.1 The Fourier transform and convolution $\mathscr{S} * \mathscr{S}'$

The interchange properties established for $\mathscr{S}(\mathbb{R})$ and the definition of the Fourier transform on $\mathscr{S}'(\mathbb{R})$ lead to our first result.

33.1.1 Proposition *If* $\psi \in \mathscr{S}(\mathbb{R})$ *and* $T \in \mathscr{S}'(\mathbb{R})$, *then*

(i)
$$\widehat{\psi * T} = \widehat{\psi} \cdot \widehat{T}; \tag{33.1}$$

(ii)
$$\widehat{\psi \cdot T} = \widehat{\psi} * \widehat{T}. \tag{33.2}$$

Proof.

(i) $\psi * T$ is in $\mathscr{S}'(\mathbb{R})$ by Proposition 32.1.3. For all $\varphi \in \mathscr{S}(\mathbb{R})$,

$$\langle \widehat{\psi * T}, \varphi \rangle = \langle \psi * T, \widehat{\varphi} \rangle = \big\langle T_u, \langle \psi_x, \widehat{\varphi}(x+u) \rangle \big\rangle.$$

On the other hand, since $\widehat{\psi} \in \mathscr{S}(\mathbb{R})$ and $\widehat{T} \in \mathscr{S}'(\mathbb{R})$, the product $\widehat{\psi} \cdot \widehat{T}$ makes sense. Thus

$$\langle \widehat{\psi} \cdot \widehat{T}, \varphi \rangle = \langle \widehat{T}, \widehat{\psi} \cdot \varphi \rangle = \langle T, \widehat{\widehat{\psi} \cdot \varphi} \rangle.$$

Applying Proposition 23.1.2(ii), we see that

$$\widehat{\widehat{\psi} \cdot \varphi} = \widehat{\widehat{\psi}} * \widehat{\varphi} = \psi_\sigma * \widehat{\varphi} = \langle \psi_x, \widehat{\varphi}(x+u) \rangle;$$

hence $\widehat{\psi * T} = \widehat{\psi} \cdot \widehat{T}$.

(ii) $\widehat{\psi} \in \mathscr{S}(\mathbb{R})$ and $\widehat{T} \in \mathscr{S}'(\mathbb{R})$ imply that $\widehat{\psi} * \widehat{T}$ is defined and is in $\mathscr{S}'(\mathbb{R})$. Applying the operator $\overline{\mathscr{F}}$ to (i), we have

$$\overline{\mathscr{F}}(\widehat{\psi} * \widehat{T}) = \overline{\mathscr{F}}\,\widehat{\psi} \cdot \overline{\mathscr{F}}\,\widehat{T} = \psi \cdot T.$$

Since $\mathscr{F}$ is an isomorphism on $\mathscr{S}'(\mathbb{R})$ (Theorem 31.2.3), $\widehat{\psi} * \widehat{T} = \widehat{\psi \cdot T}$. □

33.2 The Fourier transform and convolution $\mathscr{E}' * \mathscr{S}'$

If we take $S \in \mathscr{E}'(\mathbb{R})$ and $T \in \mathscr{S}'(\mathbb{R})$, then $\widehat{S}$ is in $C^\infty(\mathbb{R})$ (Theorem 31.5.1) and $\widehat{T}$ is in $\mathscr{S}'(\mathbb{R})$. The product $\widehat{S} \cdot \widehat{T}$ makes sense, $S * T \in \mathscr{S}'(\mathbb{R})$ (Proposition 32.3.2), and we can compute $\widehat{S * T}$.

33.2.1 Proposition *If $S \in \mathscr{E}'(\mathbb{R})$ and $T \in \mathscr{S}'(\mathbb{R})$, then*

$$\widehat{S * T} = \widehat{S} \cdot \widehat{T}. \tag{33.3}$$

Proof. For all $\varphi \in \mathscr{S}(\mathbb{R})$,

$$\langle \widehat{S} \cdot \widehat{T}, \varphi \rangle = \langle \widehat{T}, \widehat{S} \cdot \varphi \rangle.$$

Since $S \in \mathscr{E}'(\mathbb{R})$, the Fourier transform $\widehat{S}$ and all of its derivatives are slowly increasing C^∞ functions (Theorem 31.5.1). Thus the product $\widehat{S} \cdot \varphi$ is in $\mathscr{S}(\mathbb{R})$ and

$$\langle \widehat{T}, \widehat{S} \cdot \varphi \rangle = \langle T, \widehat{\widehat{S} \cdot \varphi} \rangle.$$

By applying (33.2) we see that

$$\widehat{\widehat{S} \cdot \varphi} = \widehat{\widehat{S}} * \widehat{\varphi} = S_\sigma * \widehat{\varphi}.$$

Thus

$$\begin{aligned}
\langle T, \widehat{\widehat{S} \cdot \varphi} \rangle &= \big\langle T_u, \langle S_\sigma(x), \widehat{\varphi}(u-x) \rangle \big\rangle \\
&= \big\langle T_u, \langle S_x, \widehat{\varphi}(u+x) \rangle \big\rangle \\
&= \langle T * S, \widehat{\varphi} \rangle \\
&= \langle \widehat{T * S}, \varphi \rangle,
\end{aligned}$$

and $\widehat{S * T} = \widehat{S} \cdot \widehat{T}$. □

33.3 The Fourier transform and convolution $L^2 * L^2$

We studied the convolution $L^2 * L^2$ in Section 23.2, but it was not possible then to prove the formula $\widehat{f * g} = \widehat{f} \cdot \widehat{g}$ (Section 23.2.2). In fact, when f and g are in $L^2(\mathbb{R})$, the best one can say in general is that $f * g$ is in $L^\infty \cap C^0$ (Section 20.5), and it is not possible to take the Fourier transform of such a function. Now, however, we know that $f * g$ is in $\mathscr{S}'(\mathbb{R})$ (Proposition 31.1.8), and it is possible to take the Fourier transform of such a distribution. Since $f * g(t) = \overline{\mathscr{F}}(\widehat{f} \cdot \widehat{g})(t)$ for all $t \in \mathbb{R}$ (Proposition 23.2.1(i)), we have $\widehat{f * g} = \widehat{f} \cdot \widehat{g}$ simply by taking the Fourier transform.

33.3.1 Proposition *If f and g are in $L^2(\mathbb{R})$, then*

(i)
$$\widehat{f * g} = \widehat{f} \cdot \widehat{g};$$

(ii)
$$\widehat{f \cdot g} = \widehat{f} * \widehat{g}.$$

Equation (i) is just a restatement of Proposition 23.2.1(ii).

33.4 The Hilbert transform

Let f be a function in $L^2(\mathbb{R})$ with $\operatorname{supp}(f) \subset [0, +\infty)$, which means that f is a causal signal with finite energy. Suppose in addition that f is real-valued. We intend to study the real and imaginary parts of its Fourier transform

$$\widehat{f}(\xi) = A(\xi) + iB(\xi).$$

In particular, we wish to find expressions for the functions $A(\xi)$ and $B(\xi)$.

Since $\operatorname{supp}(f) \subset [0, +\infty)$, we can write $f(x) = \operatorname{sign}(x) \cdot f(x)$. Operating formally, we have

$$\widehat{f}(\xi) = \widehat{\operatorname{sign}} * \widehat{f}(\xi) = \frac{1}{i\pi} \operatorname{pv}\Big(\frac{1}{\xi}\Big) * \widehat{f}(\xi). \tag{33.4}$$

This is a $\mathscr{S}' * L^2$ convolution, which is not one of the cases we have discussed. We can, however, decompose $\frac{1}{i\pi} \operatorname{pv}\big(\frac{1}{x}\big)$ and write it as

$$\frac{1}{i\pi} \operatorname{pv}\Big(\frac{1}{x}\Big) = S + g$$

with $S \in \mathscr{E}'(\mathbb{R})$ and $g \in L^2(\mathbb{R})$ (see Exercise 28.10). Equation (33.4) is valid because $\widehat{\operatorname{sign}} * \widehat{f}$ makes sense, and

$$\begin{aligned} \overline{\mathscr{F}}(\widehat{\operatorname{sign}} * \widehat{f}) &= \overline{\mathscr{F}}\big((S + g) * \widehat{f}\big) = \overline{\mathscr{F}}(S) \cdot f + \overline{\mathscr{F}}(g) \cdot f \\ &= \overline{\mathscr{F}}(S + g) \cdot f = \operatorname{sign} \cdot f = f. \end{aligned}$$

Thus we see that

$$A(\xi) + iB(\xi) = \frac{1}{i\pi}\,\mathrm{pv}\Big(\frac{1}{\xi}\Big) * \big(A(\xi) + iB(\xi)\big),$$

and thus

$$A = \frac{1}{\pi}\,\mathrm{pv}\Big(\frac{1}{\xi}\Big) * B, \tag{33.5}$$

$$B = -\frac{1}{\pi}\,\mathrm{pv}\Big(\frac{1}{\xi}\Big) * A. \tag{33.6}$$

33.4.1 Proposition (the Hilbert transform)

(i) *The operator* $H : L^2(\mathbb{R}) \to L^2(\mathbb{R})$ *defined by* $Hf = \frac{1}{\pi}\,\mathrm{pv}\big(\frac{1}{x}\big) * f$ *is linear, 1-to-1, and bicontinuous. It is an isometry on* $L^2(\mathbb{R})$, *and* $H^{-1} = -H$.

(ii) *If* $f \in L^2(\mathbb{R})$ *is real and if* $\mathrm{supp}(f) \subset [0, +\infty)$, *then the real part of* $\widehat{f}$ *is the Hilbert transform of the imaginary part of* $\widehat{f}$.

Proof.

(i) We have just seen that H is well-defined on $L^2(\mathbb{R})$. $H(f)$ is in $L^2(\mathbb{R})$, since

$$\mathscr{F}\left[\frac{1}{\pi}\,\mathrm{pv}\left(\frac{1}{x}\right) * f\right] = \mathscr{F}\left[\frac{1}{\pi}\,\mathrm{pv}\left(\frac{1}{x}\right)\right] \cdot \widehat{f} = -i\,\mathrm{sign}(\xi)\widehat{f}(\xi) \tag{33.7}$$

and $\widehat{f} \in L^2(\mathbb{R})$. H is clearly linear; from (33.7) and Theorem 22.1.4(iii) we have

$$\|Hf\|_2 = \|\widehat{Hf}\|_2 = \|\widehat{f}\,\|_2 = \|f\|_2,$$

which proves that H is an isometry. To find the inverse, apply (33.7) to $H(f)$:

$$\mathscr{F}\,[H(H(f))(\xi)] = -i\ \mathrm{sign}(\xi)\widehat{H(f)}(\xi) = -\widehat{f}(\xi).$$

Thus $H(H(f)) = -f$ for all $f \in L^2(\mathbb{R})$, and $H^{-1} = -H$.

(ii) This is a restatement of (33.5). □

33.5 The analytic signal associated with a real signal

If $f \in L^2(\mathbb{R})$ is a real signal (real-valued), then the real part of its Fourier transform is even and the imaginary part is odd:

$$\widehat{f}(\xi) = A(\xi) + iB(\xi),$$
$$A \text{ even },\ B \text{ odd}.$$

It follows that $\widehat{f}$, and hence f, is completely determined by the restriction of $\widehat{f}$ to $[0,+\infty)$. Let G denote this restriction:

$$G = \widehat{f} \cdot u.$$

33.5.1 Definition The signal whose Fourier transform is $2G$, which is the signal $g = 2\overline{\mathscr{F}}\, G$, is called the analytic signal associated with the real signal f.

From the definition we have

$$g = 2\overline{\mathscr{F}}\,(\widehat{f}\cdot u) = 2f * \overline{\mathscr{F}}\, u = f * \left(\delta - \frac{1}{i\pi}\operatorname{pv}\Big(\frac{1}{x}\Big)\right);$$

hence

$$g = f + iHf.$$

Thus the analytic signal associated with f is obtained by adding an imaginary part to f equal to its Hilbert transform. In summary,

$$\begin{array}{c} g \text{ is the analytic signal} \\ \text{associated with } f \end{array} \iff \widehat{g}(\xi) = \begin{cases} 0 & \text{if } \xi < 0 \\ 2\widehat{f}(\xi) & \text{if } \xi > 0 \end{cases} \iff g = f + iHf.$$

33.6 Exercises

Exercise 33.1 Compute the Fourier transform of $f(x) = \cos\frac{\pi}{2}x \cdot \chi_{[-1,1]}(x)$ and justify the equation $\widehat{f} = \widehat{\cos\frac{\pi}{2}x} * \widehat{\chi}_{[-1,1]}$.

Exercise 33.2 Show that the convolution $\operatorname{pv}\big(\frac{1}{x}\big) * \operatorname{pv}\big(\frac{1}{x}\big)$ makes sense and compute it.

Exercise 33.3 Compute the Fourier transform of $\operatorname{pv}\big(\frac{1}{x}\big)$ from the relation

$$x\operatorname{pv}\Big(\frac{1}{x}\Big) = 1.$$

Exercise 33.4 Consider the sequence of distributions defined by

$$T_1 = \frac{1}{2}(\delta_{-1} + \delta_1),$$
$$T_n = T_{n-1} * T_1, \quad n \geq 2.$$

(a) Express T_n in terms of Dirac distributions.

(b) Compute $\widehat{T}_n$.

(c) Define $f_n(\xi) = \widehat{T}_n\left(\frac{\xi}{2\pi\sqrt{n}}\right)$. Show that f_n converges in $\mathscr{S}'(\mathbb{R})$ and find its limit.

(d) Study the convergence of $\overline{\mathscr{F}} f_n$ in $\mathscr{S}'(\mathbb{R})$.

Remark: Exercise 33.4 shows that any function of the form e^{-ax^2}, $a > 0$, is the limit in the sense of $\mathscr{D}'(\mathbb{R})$ of a sequence of finite linear combinations of Dirac distributions. In fact, this is an immediate result of Exercise 28.7 and the fact that for $T \in \mathscr{S}'(\mathbb{R})$ and $\lambda \in \mathbb{R}\backslash 0$, $\mathscr{F}(h_\lambda T) = \frac{1}{|\lambda|} h_{\frac{1}{\lambda}} \mathscr{F} T$.

Exercise 33.5 Suppose $f \in L^2(\mathbb{R})$, $\operatorname{supp} f \subset [0, +\infty)$, and $\mathscr{F} f = A + iB$ with A and B real-valued. Also assume that A is even. Show that $f = 2\mathscr{F} A \cdot u$.

Chapter X

Filters and Distributions

Lesson 34

Filters, Differential Equations, and Distributions

Filters for functions have been studied in Lessons 1, 2, 24, and 25. We are going to recast and complete this analysis in the light of what we now know about distributions. We will see that the basic tools developed so far, namely, convolution and the Fourier transform, play an essential role in the study of generalized filters, in the same way they did in the study of filters for functions.

34.1 Filters revisited

34.1.1 Definition (analog filter) Let X be a translation-invariant linear subspace of $\mathscr{D}'(\mathbb{R})$. Assume that X has a topology that is at least as fine as the topology induced by $\mathscr{D}'(\mathbb{R})$. An analog filter is a mapping $A : X \to \mathscr{D}'$ that is linear, invariant, and continuous (recall the definitions in Sections 2.1 and 2.2).

We want X to be endowed with a topology that gives the best chance for a mapping A to be continuous: We want the topology on X to be a fine as possible and that on $\mathscr{D}'$ to be as coarse as is reasonable. Note that Definition 34.1.1 is more general than the one given in Section 2.2. Thus X can be a function space ($X = L^p(\mathbb{R})$, $1 \le p \le +\infty, \dots$) or a distribution space ($X = \mathscr{E}'$, $\mathscr{S}'$, $\mathscr{D}'_+$, $\mathscr{D}', \dots$), and in particular, X can contain discrete signals.

34.1.2 Proposition (examples: convolution filters) *The convolution system*

$$A : X \to \mathscr{D}',$$

where $Af = h * f$, is an analog filter in the following cases:

(i) *h is a distribution with bounded support, that is, $h \in \mathscr{E}'$.*

(ii) *$h \in \mathscr{D}'_+$ and $X \subset \mathscr{D}'_+$.*

Proof. In case (i), A is well-defined for any subspace $X \subset \mathscr{D}'$, and it is linear. The invariance comes from Propositions 32.5.1 and 32.2.2: For all $a \in \mathbb{R}$,

$$\tau_a(h * f) = \delta_a * (h * f) = h * (\delta_a * f) = h * (\tau_a f),$$

and in terms of A,

$$\tau_a(Af) = A(\tau_a f)$$

for all $f \in X$. A is continuous by Theorem 32.2.1. For case (ii), Af is in $\mathscr{D}'_+$. A is linear and invariant by Proposition 32.5.2; for continuity, we invoke Theorem 32.4.3. □

34.1.3 Definition The impulse response of a filter A is its response $h = A\delta$ to the Dirac impulse (when $\delta \in X$). Similarly, its step response is $h_1 = Au$ (when $u \in X$), and its transfer function is $H = \widehat{h}$ (when $h \in \mathscr{S}'$).

When δ is not in X, for example when $X = L^2(\mathbb{R})$, one can still define the impulse response because in practice all of the filters encountered will be convolution systems.

34.1.4 Consistency of the two definitions of transfer function

In Section 2.3 we gave another definition for the transfer function. It was the function H such that for all $\lambda \in \mathbb{R}$,

$$A(e_\lambda) = H(\lambda)e_\lambda,$$

where $e_\lambda(t) = e^{2i\pi\lambda t}$. These two definitions are equivalent for convolution systems. One assumes, of course, that the sinusoidal signals e_λ are in the set of input signals. We look at several cases.

Case 1: $h \in \mathscr{E}'$.

From Proposition 33.2.1, Section 31.6, and (28.5),

$$\widehat{A(e_\lambda)} = \widehat{h * e_\lambda} = \widehat{h} \cdot \widehat{e}_\lambda = \widehat{h} \cdot \delta_\lambda = \widehat{h}(\lambda)\delta_\lambda \tag{34.1}$$

for all $\lambda \in \mathbb{R}$. By taking the inverse Fourier transform, we have

$$A(e_\lambda) = \widehat{h}(\lambda)e_\lambda,$$

so

$$H = \widehat{h}. \tag{34.2}$$

Case 2: $h \in \mathscr{S}$.

The equalities in (34.1) are valid in this case by Proposition 33.1.1; hence (34.2) is also true.

The consistency of the two definitions thus extends to all convolution systems where $h \in \mathscr{E}' + \mathscr{S}$. This is the case for the *RC* and *RLC* filters, and more generally for all of the filters studied in Lessons 24 and 25.

Case 3: $h \in L^1(\mathbb{R})$.

We have

$$(h * e_\lambda)(t) = \int_{\mathbb{R}} h(s) e^{2i\pi\lambda(t-s)}\, ds = \widehat{h}(\lambda) e_\lambda(t).$$

(See Section 20.3 regarding the convolution $L^1 * L^\infty$.)

34.2 Realizable, or causal, filters

The next definition generalizes the definition in Section 2.1.2 and Definition 24.5.1.

34.2.1 Definition An analog filter is said to be realizable if

$$\operatorname{supp}(f) \subset [t_0, +\infty) \quad \Longrightarrow \quad \operatorname{supp}(Af) \subset [t_0, +\infty)$$

for all $t_0 \in \mathbb{R}$.

34.2.2 Proposition *For the convolution system*

$$A : X \to \mathscr{D}', \qquad Af = h * f$$

to be realizable, it is necessary and sufficient that $\operatorname{supp}(h) \subset [0, +\infty)$.

Proof. We prove the result in the simple case where $\delta \in X$. If A is realizable, the relations $h = A\delta$ and $\operatorname{supp}(\delta) \subset [0, +\infty)$ imply that the support of h is in $[0, +\infty)$. Conversely, if $\operatorname{supp}(h) \subset [0, +\infty)$ and $\operatorname{supp}(f) \subset [t_0, +\infty)$, then we have $\operatorname{supp}(h * f) \subset [t_0, +\infty)$ by Proposition 32.4.4. □

34.3 Tempered solutions of linear differential equations

We will generalize, in the larger context of tempered distributions, what was done in Lesson 24 under the assumption that $p < q$. This limitation was imposed because the rational function $P(x)/Q(x)$ is not in $L^2(\mathbb{R})$ when $p \geq q$. This condition is no longer necessary, since we can now work in the space $\mathscr{S}'$ of tempered distributions.

Let A be a system whose input f and output $g = Af$ are related by the differential equation

$$\sum_{k=0}^{q} b_k g^{(k)} = \sum_{j=0}^{p} a_j f^{(j)}, \quad a_p \cdot b_q \neq 0, \tag{34.3}$$

taken in the sense of distributions. The coefficients a_j and b_k are fixed complex numbers, and f is a given tempered distribution. We will see that in general (34.3), has a unique solution g in $\mathscr{S}'$, which defines the output of the system. As before, we write

$$P(x) = \sum_{j=0}^{p} a_j x^j \quad \text{and} \quad Q(x) = \sum_{k=0}^{q} b_k x^k.$$

34.3.1 Proposition *If $f \in \mathscr{S}'$ and if $P(x)/Q(x)$ has no poles on the imaginary axis, then (34.3) has a unique solution $g \in \mathscr{S}'$.*

Proof. The proof copies that of Proposition 24.1.1. If there exists a solution $g \in \mathscr{S}'$, then by taking the Fourier transform, we obtain

$$\widehat{g}(\lambda) = H(\lambda)\widehat{f}(\lambda), \tag{34.4}$$

where

$$H(\lambda) = \frac{P(2i\pi\lambda)}{Q(2i\pi\lambda)}. \tag{34.5}$$

This shows that $\widehat{g}$ is uniquely determined, and Theorem 31.1.7 implies that g is uniquely determined. Thus (34.3) has at most one solution in $\mathscr{S}'$. On the other hand, H is slowly increasing, and it is easy to see that $H\widehat{f} \in \mathscr{S}'$ and that (34.4) defines a solution g in $\mathscr{S}'$. □

It is easy to show that $A : f \mapsto g$ is a filter. However, as in Lesson 24, our main interest is in computing the impulse response.

34.3.2 The solution of (34.3) is a convolution

Since H is slowly increasing, it is a tempered distribution (Proposition 31.1.7) and has an inverse Fourier transform $h = \mathscr{F}^{-1}H$, which we can compute by decomposing H into partial fractions.

Case 1 : H has only simple poles.

Then

$$H(\lambda) = \sum_{j=0}^{p-q} \alpha_j (2i\pi\lambda)^j + \sum_{k=0}^{q} \frac{\beta_k}{2i\pi\lambda - z_k},$$

where we define $\alpha_j = 0$ if $j < 0$ (the polynomial part is zero if $p < q$) and where $z_1, \dots, z_q$ are the simple poles of $P(x)/Q(x)$ in $\mathbb{C}$. Then

$$h(t) = \sum_{j=0}^{p-q} \alpha_j \delta^{(j)} + \sum_{k \in K_-} \beta_k e^{z_k t} u(t) - \sum_{k \in K_+} \beta_k e^{z_k t} u(-t), \qquad (34.6)$$

where K_- are K_+ are defined in Section 24.3.1.

Case 2: H has multiple poles.

The polynomial part contributes a sum $\sum_{j=0}^{p-q} \alpha_j \delta^{(j)}$ to h as in Case 1. Thus we can limit ourselves to the case $p < q$. Using the same notation we used in Section 24.3, we obtain the same result:

$$h(t) = \Big(\sum_{k \in K_-} P_k(t) e^{z_k t} \Big) u(t) - \Big(\sum_{k \in K_+} P_k(t) e^{z_k t} \Big) u(-t) \qquad (34.7)$$

with

$$P_k(t) = \sum_{m=1}^{m_k} \beta_{k,m} \frac{t^{m-1}}{(m-1)!}.$$

At this point we know that $g = Af$ satisfies the relation

$$\widehat{g}(\lambda) = \widehat{h}(\lambda) \widehat{f}(\lambda).$$

We would like to take the inverse Fourier transform and apply the results of Sections 33.1 and 33.2, but first we need to represent h as en element of $\mathscr{E}' + \mathscr{S}$. For this we write

$$h = h\theta + h(1-\theta) = h_1 + h_2,$$

where $\theta \in \mathscr{D}$ satisfies

$$\theta(t) = \begin{cases} 1 & \text{if} \quad |t| \le 1, \\ 0 & \text{if} \quad |t| \ge 2. \end{cases}$$

Now apply Propositions 33.1.1 and 33.2.1:

$$\widehat{h * f} = \widehat{h_1 * f} + \widehat{h_2 * f} = (\widehat{h}_1 + \widehat{h}_2) \widehat{f},$$

and

$$\widehat{h * f} = \widehat{h} \cdot \widehat{f} = \widehat{g}.$$

Taking the inverse Fourier transform, we obtain

$$g = h * f,$$

which shows that A is a convolution system.

34.3.3 Remarks

(a) The generalized solution in terms of functions that was given for this equation in Section 24.3.5 is the tempered solution that we have just obtained. This does not lessen our interest in Lesson 24 where we found that the solution is a function when the input f is a function ($p \leq q$). However, in Lesson 24 we assumed that $f \in L^1 \cap L^2 \cap L^\infty$; here we have a much wider range of inputs f, even if we restrict f to be a function.

(b) If $P(x)/Q(x)$ has a pole on the imaginary axis, then the solution g in $\mathscr{S}\,'$ is no longer unique. For example,

$$g'' + \omega^2 g = \delta, \quad \omega > 0,$$

has as solutions in $\mathscr{S}\,'$ all of the functions

$$g(t) = \begin{cases} A\cos\omega t + B\sin\omega t, & t < 0, \\ A\cos\omega t + \left(B + \dfrac{1}{\omega}\right)\sin\omega t, & t > 0. \end{cases}$$

This equation will be studied in Lesson 35. Here we merely note that the moderated growth at infinity imposed by the space $\mathscr{S}\,'$ is not enough to guarantee uniqueness. It is causality that will be determinant.

34.3.4 Causality

Proposition 34.2.2 and equations (34.6) and (43.7) give us a necessary and sufficient condition for causality.

34.3.5 Proposition *The filter $A : \mathscr{S}\,' \to \mathscr{S}\,'$ defined by equation (34.3) is realizable (or causal) if and only if the real parts of the poles of $P(x)/Q(x)$ are strictly negative.*

34.4 Exercises

Exercise 34.1 Suppose the filter $A : \mathscr{S}\,'(\mathbb{R}) \mapsto \mathscr{S}\,'(\mathbb{R})$ is governed by the equation

$$g'' - \omega^2 g = f'', \quad \omega > 0.$$

(a) Compute its impulse response.

(b) Compute and represent graphically the step response.

Exercise 34.2 Suppose the filter $A : \mathscr{S}\,'(\mathbb{R}) \to \mathscr{S}\,'(\mathbb{R})$ is governed by the equation

$$g'' + \sqrt{2}g' + g = f.$$

(a) Compute the impulse response of A.

(b) Is the filter A realizable?

Lesson 35

Realizable Filters and Differential Equations

This lesson is a direct continuation of the last one. We are going to look for the causal solutions of a linear differential equation with constant coefficients; thus by assumption, the filter will be realizable (Section 34.2). For convenience we write the equation with $b_q = 1$:

$$\sum_{k=0}^{q-1} b_k g^{(k)} + g^{(q)} = \sum_{j=0}^{p} a_j f^{(j)}. \tag{35.1}$$

We assume $f \in \mathscr{D}'_+$, and we wish to find a solution $g \in \mathscr{D}'_+$.

35.1 Representation of the causal solution

We cannot use the Fourier transform here as we did in Section 34.3 because the signals f and g are not assumed to be tempered. This lack of restriction is essential, since we will find solutions that grow exponentially.

35.1.1 Existence and uniqueness of a causal solution

We are going to transform equation (35.1) into a first-order linear system. By introducing the auxiliary functions $g_1, \dots, g_{q-1}$ and calling the right-hand side φ, (35.1) becomes

$$\begin{aligned} g' &= g_1, \\ g_1' &= g_2, \\ &\vdots \\ g_{q-2}' &= g_{q-1}, \\ g_{q-1}' &= -(b_{q-1} g_{q-1} + \cdots + b_0 g) + \varphi. \end{aligned} \tag{35.2}$$

If we define

$$M = \begin{bmatrix} 0 & 1 & 0 & \dots & 0 \\ 0 & 0 & 1 & \dots & 0 \\ \vdots & & & & \\ 0 & 0 & 0 & \dots & 1 \\ -b_0 & -b_1 & -b_2 & \dots & -b_{q-1} \end{bmatrix}, \quad G = \begin{bmatrix} g \\ g_1 \\ \vdots \\ g_{q-1} \end{bmatrix}, \quad \Phi = \begin{bmatrix} 0 \\ 0 \\ \vdots \\ \varphi \end{bmatrix},$$

then (35.2) can be written as the matrix equation

$$G' = MG + \Phi. \tag{35.3}$$

The procedure for transforming a linear differential equation of order q into a first-order system of linear equations is well known. The advantage here is that (35.3) can be solved as a first-order scalar equation. The only difference is that the exponents are matrices. We review this technique.

One proves that the series of matrices

$$e^{tM} = I + tM + \frac{t^2}{2!}M^2 + \cdots$$

converges for all real t in the (normed) space of $q \times q$ matrices. One also shows that e^{tM} is invertible, that the inverse is e^{-tM}, and that the derivative of the vector-valued function $t \mapsto e^{tM}$ is the function $t \mapsto M \cdot e^{tM}$. Finally, one has $e^{tM}e^{sM} = e^{(t+s)M}$.

Changing the unknown function to $G(t) = e^{tM}X(t)$ reduces (35.3) to

$$e^{tM} \cdot X'(t) = \Phi(t),$$

and we have

$$X'(t) = e^{-tM}\Phi(t) \quad \text{and} \quad X(t) = X_0 + \int_{-\infty}^{t} e^{-sM}\Phi(s)\,ds. \tag{35.4}$$

All solutions of (35.3) are of the form

$$G(t) = e^{tM}X_0 + \int_{-\infty}^{t} e^{(t-s)M}\Phi(s)\,ds, \tag{35.5}$$

where X_0 is an arbitrary fixed vector. Assume $\operatorname{supp}(\varphi) \subset [t_0, +\infty)$. Then $\Phi(t) = 0$ and $G(t) = e^{tM}X_0$ for all $t < t_0$. If $G(t)$ is causal, then necessarily $X_0 = 0$. Equation (35.3) thus has the unique causal solution

$$G(t) = \int_{-\infty}^{t} e^{(t-s)M}\Phi(s)\,ds, \tag{35.6}$$

and (35.1) has a unique causal solution g. Since $g(t) = 0$ for $t \in (-\infty, t_0)$, the system is realizable.

We have acted here as if φ were a function. If it is a distribution, then (35.4) is solved the same way using Theorem 30.2.1. One obtains a solution G whose support is limited on the left (Section 32.4). We can now state the main result.

35.1.2 Proposition *If $f \in \mathscr{D}'_+$, then equation (35.1) has a unique solution g in $\mathscr{D}'_+$. The system*

$$B : \mathscr{D}'_+ \to \mathscr{D}'_+,$$
$$f \mapsto g$$

is a convolution system and hence a filter.

Proof. Formula (35.6) shows that B is a convolution system. However, one can obtain the response g by computing the impulse response h of (35.1) directly. If $f = \delta$, then

$$\sum_{k=0}^{q-1} b_k h^{(k)} + h^{(q)} = \sum_{j=0}^{p} a_j \delta^{(j)}.$$

Taking the convolution with the input f gives (Theorem 32.4.3)

$$\sum_{k=0}^{q-1} b_k (h * f)^{(k)} + (h * f)^{(q)} = \sum_{j=0}^{p} a_j f^{(j)}.$$

Thus $h * f$ is a causal solution of (35.1), and indeed it is the unique solution g that was sought:

$$g = h * f, \quad h \in \mathscr{D}'_+. \tag{35.7}$$

□

35.1.3 Remark Notice once again, the fact that the differential equation has a unique solution is a consequence of a constraint on g; in this case, it is that g have support limited on the left. This restriction takes the place of initial conditions.

We will look at some examples of how to find actual solutions.

35.2 Examples

35.2.1 The RC filter $RCg' + g = f$

We are going to find the same result as we did in Section 25.1 but by a different method. The impulse response h is the solution in $\mathscr{D}'_+$ of

$$RCh' + h = \delta. \tag{35.8}$$

Since $h = 0$ on $(-\infty, 0)$ and since $\mathrm{supp}(\delta) = \{0\}$, it follows that

$$RCh' + h = 0$$

on the interval $(0, +\infty)$. Thus for all $t > 0$,

$$h(t) = ke^{-\frac{t}{RC}}$$

for some constant k. These relations imply that

$$h(t) = ke^{-\frac{t}{RC}}u(t)$$

for all $t \in \mathbb{R}$. The derivative of h is (Section 28.4.4)

$$h'(t) = -\frac{k}{RC}e^{-\frac{t}{RC}}u(t) + k\delta,$$

and substitution in (35.8) shows that

$$kRC\delta = \delta,$$

so

$$k = \frac{1}{RC}.$$

The impulse response (Figure 25.1) is

$$h(t) = \frac{1}{RC}e^{-\frac{t}{RC}}u(t). \tag{35.9}$$

Thus by (35.7), the causal solution of the equation is

$$g(t) = \frac{1}{RC}\int_{-\infty}^{t} e^{-\frac{t-s}{RC}} f(s)\, ds,$$

which is indeed what we found in Section 25.1. Taking $f = u$ gives the step response (Figure 25.2)

$$h_1(t) = \left(1 - e^{-\frac{t}{RC}}\right)u(t).$$

35.2.2 The filter $-\frac{1}{\omega^2}g'' + g = f$ $(\omega > 0)$

This second-order filter was studied in Section 25.3 where the conditions imposed by the use of the Fourier transform led to noncausal, slowly increasing solutions. Here, in contrast, the causality assumption will lead to a solution that grows exponentially.

The impulse response h is the solution in $\mathscr{D}'_+$ of

$$h'' - \omega^2\, h = -\omega^2 \delta. \tag{35.10}$$

To find h, we use the same method that we used above. This time, however, we will look directly for a solution h of the form $h = yu$ where y is a function in $C^\infty\mathbb{R}$. Thus

$$h' = y'u + y(0)\delta, \quad h'' = y''u + y'(0)\delta + y(0)\delta',$$

and (35.10) becomes

$$(y'' - \omega^2 y)u + y'(0)\delta + y(0)\delta' = -\omega^2\delta.$$

This prompts us to look for a function y that satisfies the equation

$$y'' - \omega^2 y = 0$$

with the boundary conditions

$$y(0) = 0,$$
$$y'(0) = -\omega^2,$$

which is a problem completely in terms of functions. The general solution is

$$y(t) = \lambda \cosh \omega t + \mu \sinh \omega t;$$

with the initial conditions this becomes

$$y(t) = -\omega \sinh \omega t.$$

The impulse response is

$$h(t) = -\omega \sinh \omega t \cdot u(t),$$

and the solution g is given by

$$g(t) = (h * f)(t) = -\omega \int_{-\infty}^{t} \sinh \omega(t-s) f(s)\, ds$$

when f is locally integrable with support limited on the left. If f is a distribution, then g is given by a convolution in $\mathscr{D}'_+$.

Note that h is not in $\mathscr{S}'$ and that the solution found here is completely different from that found in Section 25.3. The step response (Figure 35.1) is obtained by taking $f = u$:

$$h_1(t) = (1 - \cosh \omega t)u(t).$$

This filter is realizable but unstable: The input is bounded, but the output is not. We note in this regard that the implications of the positions of the poles given in Sections 24.4 and 24.5 do not apply here.

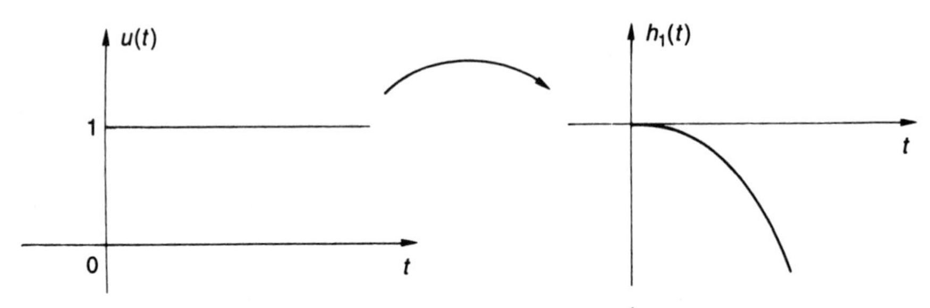

FIGURE 35.1. Causal step response of $-\frac{1}{\omega^2}g'' + g = f$.

35.2.3 The resonator $\frac{1}{\omega^2}g'' + g = f$

We encountered this filter in Lesson 24 (Section 24.3.3), but we were not able to analyze it using the Fourier transform because the poles were on the imaginary axis. This equation describes the mechanical example Section 1.3.6 where the friction is negligible (zero in the equation). The equation also represents a weight suspended on a spring where there is no "air" friction, which means that the coefficient of the first derivative is zero.

We wish to find the impulse response h using the method we used in the last section. Thus

$$h'' + \omega^2 h = \omega^2 \delta,$$

and by assuming $h = yu$, we have the system

$$\begin{aligned} y'' + \omega^2 y &= 0, \\ y(0) &= 0, \\ y'(0) &= \omega^2. \end{aligned}$$

The solution is

$$y(t) = \lambda \cos \omega t + \mu \sin \omega t$$

with $\lambda = 0$ and $\mu = \omega$. Thus

$$h(t) = \omega \sin \omega t \cdot u(t),$$

and the output is

$$g(t) = \omega \int_{-\infty}^{t} \sin \omega(t-s) f(s)\, ds$$

for a locally integrable, causal input f. The step response (Figure 35.2) is

$$h_1(t) = (h * u)(t) = (1 - \cos \omega t) u(t).$$

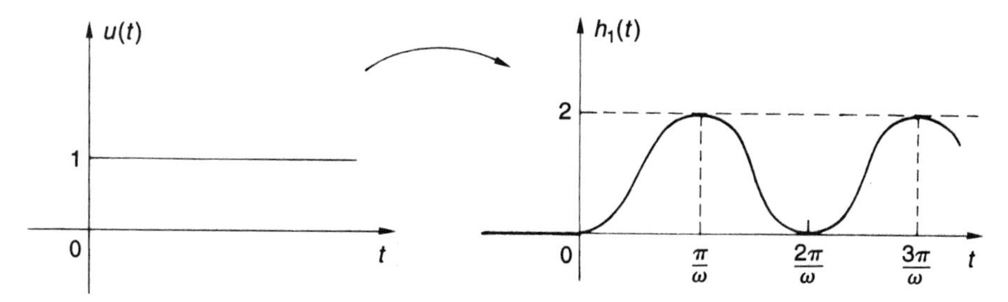

FIGURE 35.2. Resonator.

It is interesting to study the response to a sinusoidal input. For the input $f(t) = \sin \omega_0 t \cdot u(t)$, the output g is

$$g(t) = \begin{cases} \dfrac{\omega}{\omega^2 - \omega_0^2}(\omega \sin \omega_0 t - \omega_0 \sin \omega t)u(t) & \text{if} \quad \omega_0 \neq \omega, \\ \dfrac{1}{2}(\sin \omega t - \omega t \cos \omega t)u(t) & \text{if} \quad \omega_0 = \omega. \end{cases}$$

If $\omega_0 = \omega$, the amplitude of the oscillations tends to infinity. This is an instability, and it is an example of the phenomenon called *resonance*. It can cause the system to "explode."

35.2.4 The integrator $g' = f$

We saw in Section 25.4.1 that causality imposed the solution

$$g(t) = \int_{-\infty}^{t} f(s)\, ds.$$

We now see directly that the impulse response must satisfy $h' = \delta$, so $h = u$ and $g = u * f$.

35.2.5 The differentiator $g = f'$

The impulse response is $h = \delta'$, and we have $g = f * \delta' = f'$. No surprise! The step response is $h_1 = u$. Thus we have a realizable filter, but it is not stable.

35.3 Exercises

Exercise 35.1 We wish to solve the equation

$$-\frac{1}{\omega^2} g'' + g = f$$

by the matrix method described in Section 35.1.

(a) Write the equation as the first-order matrix system

$$G'(t) = MG(t) + \Phi(t) \text{ with } G = \begin{bmatrix} g \\ g' \end{bmatrix} \text{ and } \Phi = \begin{bmatrix} 0 \\ -\omega^2 f \end{bmatrix}.$$

(b) To compute e^{tM}, write $M = P\Delta P^{-1}$, where Δ is the matrix of eigenvalues $\lambda_1 = -\omega$, $\lambda_2 = \omega$ of M. Show that

$$e^{t\Delta} = \begin{bmatrix} e^{-\omega t} & 0 \\ 0 & e^{\omega t} \end{bmatrix}$$

and that $e^{tM} = Pe^{t\Delta}P^{-1}$.

(c) Compute e^{tM}.

(d) Deduce the integral expression for $g(t)$ found in Section 35.2.2.

Exercise 35.2 Consider the differential equation

$$g'' + \omega^2 g = f'', \quad \omega > 0. \tag{1}$$

(1) Solve (1) by the matrix method of Section 35.1 (see Exercise 35.1).

(2) Show that changing the unknown to $g_0 = g - f$ leads to the differential equation

$$g_0'' + \omega^2 g_0 = -\omega^2 f. \tag{2}$$

(a) Use Exercise 32.6 to compute the impulse response $h_0 \in \mathscr{D}'_+(\mathbb{R})$ of the filter (from $\mathscr{D}'_+(\mathbb{R})$ to $\mathscr{D}'_+(\mathbb{R})$) governed by equation (2).

(b) Deduce the impulse response $h \in \mathscr{D}'_+(\mathbb{R})$ of the filter (from $\mathscr{D}'_+(\mathbb{R})$ to $\mathscr{D}'_+(\mathbb{R})$) governed by (1). Compute its step response.

Chapter XI

Sampling and Discrete Filters

Lesson 36

Periodic Distributions

We are going to return to the topics of Lessons 4 and 5 armed with what we now know about tempered distributions and their Fourier transforms. Our objective is to show the connection between Fourier series and the Fourier transform.

36.1 The Fourier series of a locally integrable periodic function

36.1.1 Review of Lesson 4

We saw in Lesson 4 that each periodic function $f \in L^2_{\mathrm{p}}(0,a)$ can represented as a Fourier series

$$f(t) = \sum_{n=-\infty}^{+\infty} c_n e^{2i\pi n \frac{t}{a}} \tag{36.1}$$

and that this series converges to f in the norm of $L^2_{\mathrm{p}}(0,a)$. The Fourier coefficients are given by

$$c_n = \frac{1}{a} \int_0^a f(t) e^{-2i\pi n \frac{t}{a}}\, dt. \tag{36.2}$$

For these coefficients to exist, it is necessary and sufficient that f be integrable on $(0,a)$. Thus if $f \in L^1_{\mathrm{p}}(0,a)$, we can associate with f a trigonometric series that is called its Fourier series:

$$f \mapsto \sum_{n=-\infty}^{+\infty} c_n e^{2i\pi n \frac{t}{a}}.$$

In this general case, we no longer know how to interpret the sum in (36.1). And even if we knew that the series converged in some sense, we would still have to show that its limit is f. We recall a negative result from Section 5.2:

$$f \in L^1_{\mathrm{p}}(0,a) \quad \not\Rightarrow \quad \sum_{n=-N}^{N} c_n e^{2i\pi n \frac{t}{a}} \to f \text{ in } L^1_{\mathrm{p}}(0,a) \text{ as } N \to +\infty.$$

36.1.2 A brief preview

If the Fourier series of f did converge, in a sense to be determined, and if its limit was f, we could (probably) take the Fourier transform of (36.1) and get the formula

$$\widehat{f} = \sum_{n=-\infty}^{+\infty} c_n \delta_{\frac{n}{a}}. \tag{36.3}$$

It is this formula that establishes the connection between the Fourier transform of f and its Fourier series.

Several questions arising from this scenario need to be addressed. Let f be periodic and locally integrable.

Q1: Does the Fourier series of f converge in some sense?

Q2: If yes, does it converge to f?

Q3: Is f tempered?

Q4: Does the Fourier series of f converge to f in $\mathscr{S}\,'$?

We need to answer "yes" to Q4 (thus also to Q1, Q2, Q3) if we are to write (36.3) and expect it to make sense. In fact, this formula requires that f be tempered and that the Fourier transform and the infinite summation can be interchanged.

The answers, all positive, are given in the next result.

36.1.3 Proposition *Let f be a periodic locally integrable function with period $a > 0$. Then we have the following results:*

(i) *f is a tempered distribution.*

(ii) *The equalities*

$$f(t) = \sum_{n=-\infty}^{+\infty} c_n e^{2i\pi n \frac{t}{a}} \tag{36.4}$$

and

$$\widehat{f} \;=\; \sum_{n=-\infty}^{+\infty} c_n \delta_{\frac{n}{a}} \tag{36.5}$$

hold in $\mathscr{S}\,'$ with

$$c_n = \frac{1}{a}\int_0^a f(t) e^{-2i\pi n \frac{t}{a}}\, dt.$$

Proof. Let $f_0 = f \cdot \chi_{[0,a]}$. Note that

$$f = f_0 * \Delta_a. \tag{36.6}$$

The convolution is well-defined, since f_0 is in $\mathscr{E}'$ and Δ_a is in $\mathscr{S}'$ (Proposition 32.3.2). On the other hand,

$$(f_0 * \Delta_a)(t) = \sum_{n=-\infty}^{+\infty} (f_0 * \delta_{na})(t) = \sum_{n=-\infty}^{+\infty} f_0(t - na) = f(t).$$

This proves (i). Next take the Fourier transform of (36.6); applying Proposition 33.2.1 and (31.10), we have

$$\widehat{f} = \widehat{f_0} \cdot \widehat{\Delta_a} = \frac{1}{a} \widehat{f_0} \Delta_{\frac{1}{a}},$$

which by (28.5) is

$$\widehat{f} = \frac{1}{a} \sum_{n=-\infty}^{+\infty} \widehat{f_0}\left(\frac{n}{a}\right) \delta_{\frac{n}{a}}. \tag{36.7}$$

This proves (36.5), since

$$\widehat{f_0}\left(\frac{n}{a}\right) = \int_0^a f(t) e^{-2i\pi n \frac{t}{a}}\, dt = a c_n.$$

We obtained (36.4) by taking the inverse Fourier transform of (36.5). □

36.2 The Fourier series of a periodic distribution

The last proof works equally well if the function f_0 is replaced by a distribution S with compact support.

36.2.1 Proposition *Suppose S is a distribution with compact support and $a > 0$. Then*

$$T = S * \Delta_a$$

is a periodic tempered distribution with period a. It can be decomposed in a Fourier series

$$S * \Delta_a = \frac{1}{a} \sum_{n=-\infty}^{+\infty} \widehat{S}\left(\frac{n}{a}\right) e^{2i\pi n \frac{t}{a}} \tag{36.8}$$

with equality in $\mathscr{S}'$.

Recall that $\widehat{S}$ is a C^∞ function (Theorem 31.5.1). For $S = \delta$ we get the Fourier series representation of Dirac's comb, which was established in Section 29.5.

REMARK: For a periodic function f with period a represented as a Fourier series, we called (Section 7.1) the set of pairs

$$\left(\frac{n}{a}, c_n\right)_{n\in\mathbb{Z}}$$

the spectral lines of f. The representation of these pairs by arrows parallel to the y-axis (Figures 7.1 and 7.2) was just a graphic convenience. Formula (36.5) shows that this representation agrees with that adopted in Section 26.1 for representing Dirac masses (Figure 26.3).

At this point it is natural to ask (and important to answer) the following question: Do all periodic distributions have a Fourier series representation that converges in $\mathscr{S}\,'$? The next theorem provides a sort of converse of Proposition 36.2.1, and also of Theorem 29.4.2.

36.2.2 Theorem *If T is a periodic distribution with period $a > 0$, then the following results hold:*

(i) *T is tempered.*

(ii) *There is a distribution S with compact support such that*

$$T = S * \Delta_a.$$

(iii) *T has a unique Fourier series development that converges to T in $\mathscr{S}\,'$:*

$$T = \sum_{n=-\infty}^{+\infty} \alpha_n e^{2i\pi n\frac{t}{a}}, \tag{36.9}$$

$$\widehat{T} = \sum_{n=-\infty}^{+\infty} \alpha_n \delta_{\frac{n}{a}}. \tag{36.10}$$

(iv) *The sequence of Fourier coefficients (α_n) is slowly increasing, and*

$$\alpha_n = \frac{1}{a}\widehat{S}\Big(\frac{n}{a}\Big).$$

Proof. The idea is the same as that used in Proposition 36.1.3. We would like to define S as the "restriction" of T to $(0, a)$, that is, to write

$$S = \chi_{[0,a]} \cdot T.$$

Unfortunately, as we saw in Section 28.3, such a product is not defined. We get around this difficulty by using a function $\theta \in \mathscr{D}$ that approximates $\chi_{[-1,1]}$ and that also satisfies the relation

$$\sum_{n=-\infty}^{+\infty} \theta(t - na) = 1 \tag{36.11}$$

for all $t \in \mathbb{R}$. Assume for the moment that we have such a function. The distribution $S = \theta T$ will do the job: S has bounded support and

$$S * \Delta_a = \sum_{n=-\infty}^{+\infty} S * \delta_{na} = \sum_{n=-\infty}^{+\infty} \tau_{na} S.$$

For all $\varphi \in \mathscr{D}$,

$$\langle S * \Delta_a, \varphi \rangle = \sum_{n=-\infty}^{+\infty} \langle S, \tau_{-na}\varphi \rangle = \sum_{n=-\infty}^{+\infty} \langle T, \theta\tau_{-na}\varphi \rangle.$$

Since T is periodic,

$$\langle T, \psi \rangle = \langle T, \tau_{na}\psi \rangle$$

for all $\psi \in \mathscr{D}$ and all $n \in \mathbb{Z}$; using (36.11) shows that

$$\langle S * \Delta_a, \varphi \rangle = \sum_{n=-\infty}^{+\infty} \langle T, \tau_{na}\theta\varphi \rangle = \langle T, \varphi \rangle,$$

which proves (ii). Statements (i) and (iii) follow from Proposition 36.2.1. The Fourier coefficients of T are

$$\alpha_n = \frac{1}{a}\widehat{S}\Big(\frac{n}{a}\Big). \tag{36.12}$$

These coefficients are slowly increasing because the function $\widehat{S}$ is slowly increasing (Theorem 31.5.1).

One might think that the coefficients α_n depend on S, that is, on the choice of θ. This is not the case: The Fourier coefficients are unique, as they are for functions in $L^2_{\mathrm{p}}(0, a)$. In view of (36.12) and linearity, it is sufficient to show that

$$S \in \mathscr{E}' \text{ and } S * \Delta_a = 0 \quad \Longrightarrow \quad \widehat{S}\left(\frac{n}{a}\right) = 0$$

for all $n \in \mathbb{Z}$. But $S * \Delta_a = 0$ implies that $\widehat{S} \cdot \widehat{\Delta_a} = 0$, and this in turn implies that

$$\sum_{n=-\infty}^{+\infty} \widehat{S}\Big(\frac{n}{a}\Big)\delta_{\frac{n}{a}} = 0.$$

The only way for this to hold is to have $\widehat{S}\Big(\dfrac{n}{a}\Big) = 0$ for all n.

To finish the proof, we need to show the existence of a function $\theta \in \mathscr{D}$ satisfying (36.11). Let

$$\varphi(t) = \rho\Big(\frac{t}{a}\Big),$$

where ρ is the function in $\mathscr{D}$ defined by (27.3). The sum

$$\widetilde{\varphi}(t) = \sum_{n=-\infty}^{+\infty} \varphi(t - na)$$

exists because for each t there are at most two nonzero terms. For the same reason, $\widetilde{\varphi}$, like φ, is infinitely differentiable. $\widetilde{\varphi}$ is a-periodic and strictly positive; thus a suitable choice for θ is

$$\theta(t) = \frac{\varphi(t)}{\widetilde{\varphi}(t)}.$$

□

REMARK: The Fourier coefficients c_n of a periodic locally integrable function tend to 0 as $|n| \to +\infty$ by the Riemann–Lebesgue theorem. Those of a periodic distribution are only slowly increasing.

36.2.3 Corollary *Let $(y_n)_{n\in\mathbb{Z}}$ be a complex sequence and define the distribution T by*

$$T = \sum_{n=-\infty}^{+\infty} y_n \delta_{na} \tag{36.13}$$

for $a > 0$. T is a tempered distribution if and only if the sequence (y_n) is slowly increasing.

Proof. It was shown in Proposition 31.1.9 that the condition is sufficient. If T is tempered, then the series (36.13) converges in $\mathscr{S}\,'$, and we can take its Fourier transform, which is

$$\widehat{T} = \sum_{n=-\infty}^{+\infty} y_n e^{-2i\pi na\lambda}.$$

$\widehat{T}$ is a periodic tempered distribution with period $\dfrac{1}{a}$. Thus its Fourier coefficients y_{-n} are slowly increasing. □

36.3 The product of a periodic function and a periodic distribution

36.3.1 Theorem *Let f be a periodic C^∞ function with period a and let T be a distribution with the same period. Then the distribution fT is represented by the Fourier series*

$$fT = \sum_{n=-\infty}^{+\infty} \beta_n e^{2i\pi n\frac{t}{a}},$$

and the coefficients are given by

$$\beta_n = \sum_{k=-\infty}^{+\infty} c_k \alpha_{n-k}.$$

The c_k are the Fourier coefficients of f, and the α_n are those of T. The series for β_n is absolutely convergent.

Proof. Write

$$e_n(t) = e^{2i\pi n \frac{t}{a}} \quad \text{and} \quad f_N(t) = \sum_{k=-N}^{N} c_k e_k(t).$$

Then

$$f_N T = \sum_{k=-N}^{N} \sum_{n=-\infty}^{+\infty} c_k \alpha_n e_{n+k} = \sum_{n=-\infty}^{+\infty} \beta_n(N) e_n,$$

where

$$\beta_n(N) = \sum_{k=-N}^{N} c_k \alpha_{n-k}.$$

We will study what happens to $\beta_n(N)$ as $N \to +\infty$.

(1) The series whose general term is $c_k \alpha_{n-k}$ is absolutely convergent for each fixed n. Indeed, the sequence (α_{n-k}) is slowly increasing in k (Theorem 36.2.2), so

$$|\alpha_{n-k}| \leq A|k|^m$$

for sufficiently large $|k|$. By Section 5.3.3(e), we also have

$$|c_k| \leq B|k|^{-m-2}.$$

Thus $|c_k \alpha_{n-k}| \leq C|k|^{-2}$ for sufficiently large $|k|$, and the sequence $\beta_n(N)$ converges as $N \to +\infty$.

(2) We wish to show that $f_N T \to fT$ in $\mathscr{S}'$. If $\varphi \in \mathscr{S}$, then

$$\langle (f - f_N)T, \varphi \rangle = \langle T, (f - f_N)\varphi \rangle,$$

and it is sufficient to show that $\varphi_N = (f - f_N)\varphi$ tends to 0 in $\mathscr{S}$. By Leibniz's formula, this reduces to showing, for arbitrary given integers p, q, and l, that

$$t^p (f - f_N)^{(l)} \varphi^{(q)} \to 0$$

uniformly on $\mathbb{R}$. We know from Theorem 5.3.1 that $(f - f_N)^{(l)}$ tends to 0 uniformly on $\mathbb{R}$. Since $t^p \varphi^{(q)}$ is bounded, the result follows.

(3) Denote the Fourier coefficients of the periodic distribution fT by (β_n). From (1) we know that

$$(f - f_N)T = \sum_{n=-\infty}^{+\infty} \big(\beta_n - \beta_n(N)\big) e_n \to 0$$

in $\mathscr{S}'$. Taking the Fourier transform shows that

$$\sum_{n=-\infty}^{+\infty} (\beta_n - \beta_n(N))\delta_{\frac{n}{a}} \to 0$$

in $\mathscr{S}'$ and hence in $\mathscr{D}'$. This implies that

$$\beta_n(N) \to \beta_n$$

for each $n \in \mathbb{Z}$ as $N \to +\infty$, which means that $\beta_n = \sum_{-\infty}^{+\infty} c_k \alpha_{n-k}$. □

36.4 Exercises

***Exercise 36.1** Let $x = \sum_{n=-\infty}^{+\infty} x_n \delta_{nh}$ be a discrete periodic signal with period $a = Nh$ and grid $h > 0$, and let

$$X_k = \frac{1}{N} \sum_{n=0}^{N-1} x_n \omega_N^{-nk}, \quad k \in \mathbb{Z},$$

be its discrete Fourier transform (see (8.5)).

(a) By writing

$$x = \sum_{j=0}^{N-1} x_j \tau_{jh} \Delta_a$$

(with $\Delta_a = \sum_{n=-\infty}^{\infty} \delta_{na}$), show that

$$\widehat{x}(\lambda) = \frac{1}{a} \sum_{k=0}^{N-1} x_k e^{-2i\pi\lambda kh} \Delta_{\frac{1}{a}}.$$

(b) Deduce from this that

$$\widehat{x} = \frac{1}{h} \sum_{k=-\infty}^{+\infty} X_k \delta_{\frac{k}{a}}.$$

Discuss the relation between the signals x and $\widehat{x}$.

Exercise 36.2 Assume $S \in \mathscr{E}'(\mathbb{R})$. Under what condition does there exist a periodic distribution T with period a such that $T' = S * \Delta_a$?

Exercise 36.3 Let T be a periodic distribution with period a and Fourier coefficients α_n. Show that

$$\langle T, \varphi \rangle = \sum_{n=-\infty}^{+\infty} \alpha_n \widehat{\varphi}\left(-\frac{n}{a}\right)$$

for all $\varphi \in \mathscr{S}(\mathbb{R})$.

Lesson 37

Sampling Signals and Poisson's Formula

We are now going to tackle the problem of sampling analog signals. This operation is a prerequisite of digital signal processing. For example, an analog speech signal must be sampled before it can enter a digital telephone system. A sampler records the level of the signal every a seconds and transforms it into a sequence of impulses (Figure 37.1). An analog-to-digital converter (ADC) codes these impulses as numbers that can be processed digitally.

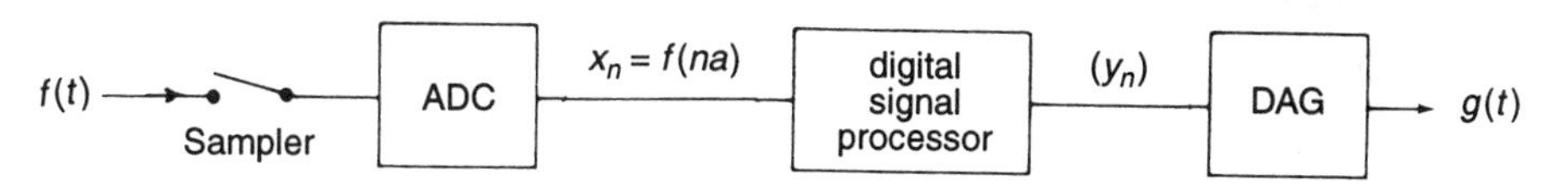

FIGURE 37.1. Processing an analog signal.

Mathematically, sampling has already been defined (Section 29.5.2) as multiplication of the signal by Dirac's comb:

$$f \mapsto af\Delta_a = a \sum_{n=-\infty}^{+\infty} f(na)\delta_{na}.$$

At first glance it would appear that information in the original signal is thereby lost. This is true to some extent, but what we will see in this lesson and the rest of the book is that the sampling rate can be high enough that for all practical purposes the loss of information is not important. On the other hand, digital processing offers many technical and economic advantages. Here is a simplified look at two of them:

(a) During the time between the arrival of two consecutive sample values, a serial processor can be doing calculations based on the value of the last sample to arrive. This enables "real time" processing for applications like automatic control and process control.

(b) Sampling and subsequent digital processing lead to sophisticated ways to compress signals. This means, for example, that speech signals can be compressed, transmitted, and reconstructed digitally without perceptible loss of quality. The economic advantage is that the compressed signal occupies less bandwidth, and so more signals can be transmitted over a given channel.

We are particularly interested in the spectra of signals and thus in spectral analysis. In this context, it is essential to ask what happens to the spectrum of a sampled signal $af\Delta_a$, or, put another way, to ask how the spectra of the sampled and original signals are related. We hasten to note that the spectrum obtained for the sampled signal is not a sampling of the original spectrum.

We will see in the next lesson something that at first seems quite astonishing: For a large (and in practice, important) class of signals, it is possible to sample without losing information. This is the essence of Shannon's theorem, which remains one of the most important theoretical and practical results in the theory of signal processing.

There is, however, an older result that is fundamental for all of the work in this area: It is Poisson's formula, and it establishes the connection between Fourier series and the Fourier transform.

37.1 Poisson's formula in $\mathscr{E}'$

The equation

$$\sum_{n=-\infty}^{+\infty} f(t-na) = \frac{1}{a} \sum_{n=-\infty}^{+\infty} \widehat{f}\left(\frac{n}{a}\right) e^{2i\pi n \frac{t}{a}}, \tag{37.1}$$

where $a > 0$ is arbitrary and fixed, or its dual version

$$\sum_{n=-\infty}^{+\infty} \widehat{g}\left(\lambda - \frac{n}{a}\right) = a \sum_{n=-\infty}^{+\infty} g(na) e^{-2i\pi\lambda na}, \tag{37.2}$$

is called Poisson's summation formula or simply Poisson's formula. We intend to prove this formula for several classes of signals that will be modeled by either functions or distributions.

37.1.1 Preliminary remarks

If formula (37.1) is to make sense, the numbers $\widehat{f}(n/a)$ must be defined; this means that $\widehat{f}$ must be a function that can be evaluated at a given point. This is the case when $\widehat{f}$ is continuous. This is not generally the case if $\widehat{f} \in L^2(\mathbb{R})$—unless, of course, $\widehat{f}$ belongs to an equivalence class that contains a continuous representative.

In spite of appearances, this is not an issue for the left-hand sum; it is to be interpreted as one of the expressions

$$\sum_{n=-\infty}^{+\infty} \tau_{na} f = \sum_{n=-\infty}^{+\infty} f * \delta_{na} = f * \Delta_a, \tag{37.3}$$

where specific values of the function do not appear. Thus f can be any distribution for which the series converges, say, in $\mathscr{D}'$.

Finally, we note once again that the convergence of these series must be interpreted as the (symmetric) limit of $\sum_{n=-N}^{+N}$ as $N \to +\infty$. We know, for example, that the series on the right in (37.1) converges in $\mathscr{D}'$ if $\widehat{f}$ is a slowly increasing function (Theorem 29.4.2). As in the case of the left-hand side, the variable t plays only a symbolic role in the expression on the right. There are, however, cases where the Poisson formula holds for all $t \in \mathbb{R}$. This happens, for example, when $f \in \mathscr{S}$ (see Exercise 37.1).

37.1.2 The case where f is a distribution with compact support

If the distribution f has compact support, $\widehat{f}$ is a slowly increasing C^∞ function (Theorem 31.5.1), and both sides of (37.1) make sense. Furthermore, in view of (37.3), the Poisson formula is just equation (36.8), which was established for periodic distributions. This proves the following result:

The Poisson summation formula (37.1) is true for distributions f with compact support, $f \in \mathscr{E}'$. The dual formula (37.2) is true for all functions g such that $\widehat{g} \in \mathscr{E}'$.

37.2 Poisson's formula in $L^1(\mathbb{R})$

When $f \in L^1(\mathbb{R})$, $\widehat{f}$ is continuous and bounded by Theorem 17.1.3. The right-hand side of (37.1) is then a trigonometric series that converges in $\mathscr{D}'$ to a periodic distribution (Theorem 29.4.2), which, being periodic, is tempered.

We first prove a lemma about the series on the left of (37.1).

37.2.1 Lemma *Assume $f \in L^1(\mathbb{R})$, and for $a > 0$ define*

$$F(t) = \sum_{n=-\infty}^{+\infty} f(t - na). \tag{37.4}$$

Then the following results hold:

(i) *The series (37.4) converges in $L^1(0,a)$, and $F \in L^1_{\mathrm{p}}(0,a)$. The Fourier coefficients of F are*

$$c_k(F) = \frac{1}{a}\widehat{f}\Big(\frac{k}{a}\Big), \quad k \in \mathbb{Z}.$$

(ii) *If in addition, $f' \in L^1(\mathbb{R})$ (where the derivative f' is taken in the sense of distributions), then the series (37.4) converges uniformly on $\mathbb{R}$, and thus F is continuous on $\mathbb{R}$.*

Proof.

(i) We show that the restriction of the series (37.4) to $(0,a)$ converges in the complete space $L^1(0,a)$ by showing that the sequence

$$F_N(t) = \sum_{n=-N}^{N} f(t - na)$$

is a Cauchy sequence. We have

$$\|F_{N+P} - F_N\|_{L^1(0,a)} \leq \sum_{N<|n|\leq N+P} \int_0^a |f(t-na)|\, dt \leq \int_{|x|\geq Na} |f(x)|\, dx.$$

Since f is in $L^1(\mathbb{R})$, the last integral tends to 0 as N and P tend to $+\infty$. Thus the sequence (F_N) converges to F on $(0,a)$, and $F \in L^1_{\mathrm{p}}(0,a)$ by periodicity. The Fourier coefficients of F are the limits of those of F_N:

$$c_k(F) = \lim_{N\to+\infty} c_k(F_N).$$

But

$$c_k(F_N) = \frac{1}{a}\sum_{|n|\leq N} \int_0^a f(t-na) e^{-2i\pi k\frac{t}{a}}\, dt = \frac{1}{a}\int_{-Na}^{(N+1)a} f(x) e^{-2i\pi k\frac{x}{a}}\, dx$$

tends to

$$\frac{1}{a}\int_{-\infty}^{+\infty} f(x) e^{-2i\pi k\frac{x}{a}}\, dx = \frac{1}{a}\widehat{f}\Big(\frac{k}{a}\Big),$$

which proves (i).

(ii) Take $a = 1$ to simplify the notation. We will show that there is a sequence of positive numbers u_n such that

$$|f(t-n)| \leq u_n \tag{37.5}$$

for all $t \in [0,1]$ and all $n \in \mathbb{Z}$, and such that $\sum_{n=-\infty}^{+\infty} u_n < +\infty$.

The assumption in (ii) means that $T'_f = T_{f'}$, and an argument like that given in Section 30.1.2 shows that f is absolutely continuous on all bounded intervals. Thus for $t \in [0, 1]$,

$$f(t-n) - \int_n^{n+1} f(t-x)\,dx = \int_n^{n+1} [f(t-n) - f(t-x)]\,dx,$$

and

$$f(t-n) - f(t-x) = \int_n^x f'(t-y)\,dy.$$

From these relations we see that

$$|f(t-n) - f(t-x)| \le \int_n^{n+1} |f'(t-y)|\,dy$$

and

$$\left| f(t-n) - \int_n^{n+1} f(t-x)\,dx \right| \le \int_n^{n+1} |f'(t-y)|\,dy.$$

The last inequality implies that

$$|f(t-n)| \le \int_n^{n+1} \big[|f(t-x)| + |f'(t-x)|\big]\,dx \le \int_{-(n+1)}^{-(n-1)} \big[|f(y)| + |f'(y)|\big]\,dy.$$

Let $u_n = \int_{-(n+1)}^{-(n-1)} \big[|f(y)| + |f'(y)|\big]\,dy$; then

$$\sum_{n=-\infty}^{+\infty} u_n = 2 \int_{\mathbb{R}} \big[|f(y)| + |f'(y)|\big]\,dy < +\infty. \tag{37.6}$$

This proves that the series (37.4) converges uniformly on $[0, 1]$ (or $[0, a]$ in general). Since F is periodic, the convergence in uniform on $\mathbb{R}$, which proves the result. □

37.2.2 Theorem

(i) *If $f \in L^1(\mathbb{R})$, then Poisson's formula (37.1) expresses the equality in $\mathscr{S}'$ between the function $F \in L^1_{\mathrm{p}}(\mathbb{R})$ and its Fourier series.*

(ii) *If, in addition, $f' \in L^1(\mathbb{R})$ (where the derivative f' is taken in the sense of distributions), equality (37.1) holds for all $t \in \mathbb{R}$. More precisely, the series on the left converges uniformly on $\mathbb{R}$ to a continuous periodic function F, and the Fourier series of F, which is the series on the right, converges uniformly on $\mathbb{R}$ to F.*

Proof. The first result is a restatement of Lemma 37.2.1 and Proposition 36.1.3. For (ii), it is sufficient to show that F is of bounded variation on $[0, a]$ (Theorem 5.2.5). Again, take $a = 1$.

Let $0 = t_0 < t_1 < \cdots < t_p < t_{p+1} = 1$ be any subdivision of $[0, 1]$. Then

$$M_p = \sum_{i=0}^{p} |F(t_{i+1}) - F(t_i)| \leq \sum_{i=0}^{p} \sum_{n=-\infty}^{+\infty} |f(t_{i+1} - n) - f(t_i - n)|$$

$$= \sum_{n=-\infty}^{+\infty} \sum_{i=0}^{p} |f(t_{i+1} - n) - f(t_i - n)| \leq \sum_{n=-\infty}^{+\infty} \sum_{i=0}^{p} \int_{t_i}^{t_{i+1}} |f'(t - n)|\, dt.$$

Finally,

$$M_p \leq \sum_{n=-\infty}^{+\infty} \int_0^1 |f'(t - n)|\, dt = \sum_{n=-\infty}^{+\infty} \int_{-n}^{-(n-1)} |f'(y)|\, dy = \int_{\mathbb{R}} |f'(y)|\, dy.$$

The numbers M_p are bounded independently of the subdivision; thus F is of bounded variation, which proves the theorem. □

37.3 Application to the study of the spectrum of a sampled signal

Let f be a tempered signal whose spectrum contains no frequencies greater than some limiting value λ_c:

$$\operatorname{supp}(\widehat{f}\,) \subset [-\lambda_c, \lambda_c].$$

In this situation, f is said to be *band limited*, which is to say that $\widehat{f} \in \mathscr{E}\,'$. We saw in Section 29.5 that sampling f every a seconds can be expressed by

$$af\Delta_a = a \sum_{n=-\infty}^{+\infty} f(na)\delta_{na}.$$

This makes sense because $\widehat{f} \in \mathscr{E}\,'$ implies $f \in C^{\infty}(\mathbb{R})$.

We wish to determine how sampling modifies the spectrum of f. For this we need to compute the Fourier transform of $af\Delta_a$. By Proposition 33.2.1,

$$\overline{\mathscr{F}}(\widehat{f} * \widehat{\Delta}_a) = f \cdot \Delta_a.$$

Taking the Fourier transform and using (31.10) shows that the spectrum of the sampled function is

$$\widehat{af \cdot \Delta_a} = \widehat{f} * \Delta_{\frac{1}{a}}.$$

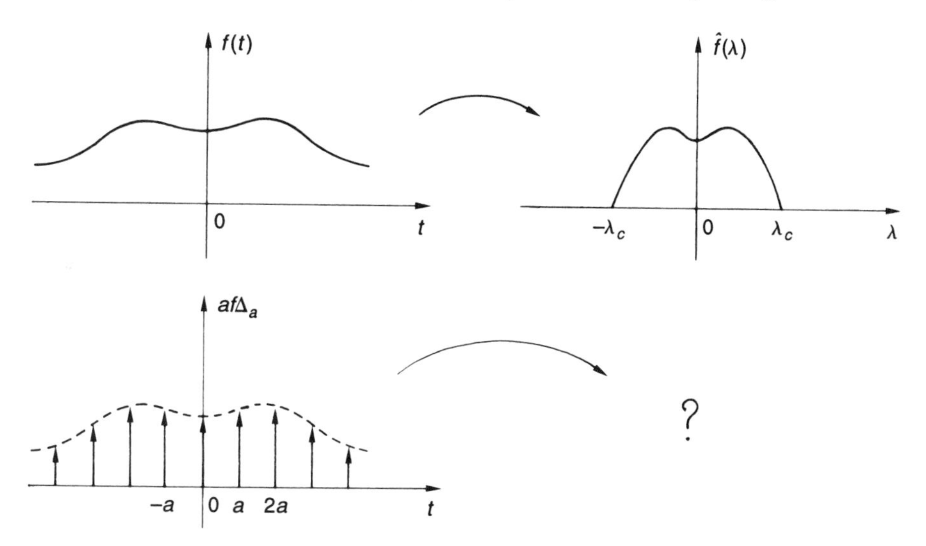

FIGURE 37.2. What is the effect of sampling on the spectrum?

Formula (37.2) with $g = \widehat{f}$ gives the following key result:

$$a\widehat{f\Delta_a} = \sum_{n=-\infty}^{+\infty} \widehat{f}\left(\lambda - \frac{n}{a}\right) = a \sum_{n=-\infty}^{+\infty} f(na)e^{-2i\pi na\lambda}. \tag{37.7}$$

We draw several conclusions:

(a) The spectrum of the sampled signal $a\widehat{f\Delta_a}$ of a band-limited signal f is periodic with period $1/a$.

(b) The spectrum of the sampled signal is obtained by summing all translates of the spectrum of the original signal f, the translations being the integer multiples of $1/a$.

We are now able to answer the question asked in Figure 37.2. There are two cases:

Case 1: $a > 1/(2\lambda_c)$ (Figure 37.3).

The translates of the spectrum of f overlap, and the spectrum of $af\Delta_a$ does not agree with the spectrum of f on the interval $[-\lambda_a, \lambda_a]$.

Case 2: $a \leq 1/(2\lambda_c)$ (Figure 37.4).

The sampling rate is large enough so that the translates of the spectrum of f are separated. The spectrum of the sampled signal is a simple periodic repetition of the initial spectrum $\widehat{f}$, and the two spectra agree on $[-\lambda_a, \lambda_a]$.

The critical sampling rate $a = 1/(2\lambda_c)$ is called the *Nyquist rate.*

If you sample a signal, you periodize its spectrum.

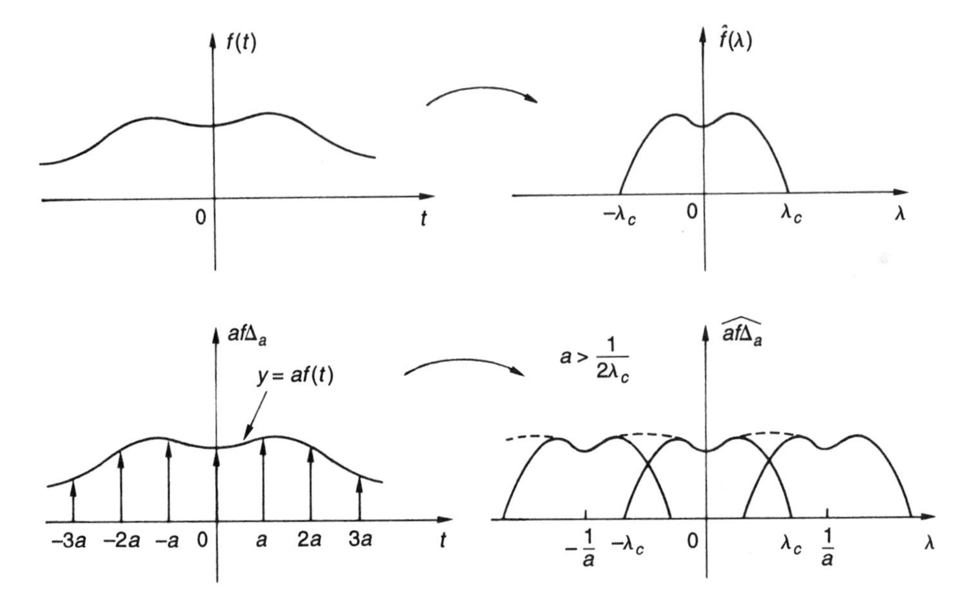

FIGURE 37.3. Sampling a signal periodizes its spectrum.

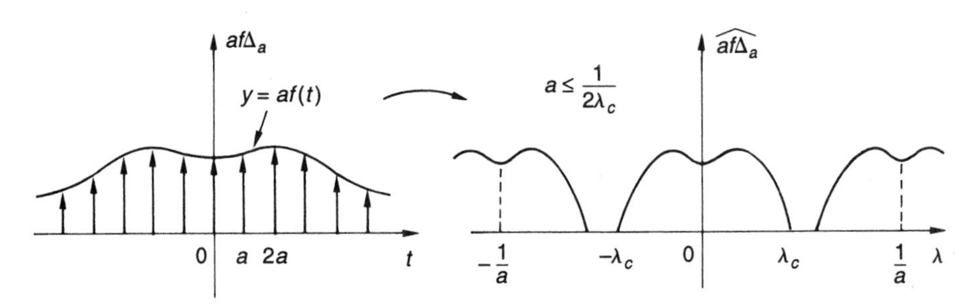

FIGURE 37.4. A high sampling rate separates the components of the spectrum.

37.4 Application to accelerating the convergence of a Fourier series

We illustrate the idea with an example. Consider the function F defined by its Fourier series:

$$F(t) = \sum_{n=-\infty}^{+\infty} \frac{1}{n^2 + b^2} e^{2i\pi nt}. \tag{37.8}$$

This series converges uniformly on $\mathbb{R}$, and F is continuous with period 1. If f is the function defined by

$$\widehat{f}(\lambda) = \frac{1}{\lambda^2 + b^2}, \quad b > 0,$$

then we see that the right-hand side of (37.8) is exactly the right-hand side of (37.1) with $a = 1$. We will compute F using f. From Section 18.2.2 we know that

$$f(t) = \frac{\pi}{b} e^{-2\pi b|t|},$$

and clearly f and f' are integrable. Thus

$$F(t) = \frac{\pi}{b} \sum_{n=-\infty}^{+\infty} e^{-2\pi b|t-n|} \tag{37.9}$$

for all $t \in \mathbb{R}$. This series converges much faster than (37.8), and in this case we can compute the sum explicitly. For $t \in [0, 1]$,

$$\frac{b}{\pi} F(t) = \sum_{n=-\infty}^{0} e^{-2\pi b(t-n)} + \sum_{n=1}^{+\infty} e^{-2\pi b(n-t)} = \frac{e^{-2\pi bt} + e^{-2\pi b(1-t)}}{1 - e^{-2\pi b}},$$

and hence, for $t \in [0, 1]$,

$$\sum_{n=-\infty}^{+\infty} \frac{1}{n^2 + b^2} e^{2i\pi nt} = \frac{\pi}{b \sinh \pi b} \cosh \left[2\pi b \left(t - \frac{1}{2} \right) \right].$$

Taking $t = 0$, we see that

$$\sum_{n=-\infty}^{+\infty} \frac{1}{n^2 + b^2} = \frac{\pi}{b} \coth(\pi b).$$

37.5 Exercises

Exercise 37.1 (Poisson's formula in $\mathscr{S}(\mathbb{R})$)

(a) Use equation (31.10) to show that

$$\sum_{n=-\infty}^{\infty} \varphi\left(\frac{n}{a}\right) = a \sum_{n=-\infty}^{\infty} \widehat{\varphi}(na), \quad a \neq 0,$$

for all $\varphi \in \mathscr{S}(\mathbb{R})$

(b) Deduce from this that

$$\sum_{n=-\infty}^{\infty} e^{-\pi n^2 t} = \frac{1}{\sqrt{t}} \sum_{n=-\infty}^{\infty} e^{-\pi \frac{n^2}{t}}, \quad t > 0.$$

Exercise 37.2 (sampling a sinusoidal signal)

Consider the signal $g(t) = \cos(2\pi\lambda t + \varphi)$ that is sampled with a frequency τ. Let g_k denote the values of g at the times $t_k = k/\tau$, $k \in \mathbb{N}$.

(a) Show that there exists a frequency $f \in [-\tau/2, \tau/2)$ such that the signal

$$h(t) = \cos(2\pi f t + \varphi)$$

gives the same samples as g at the times t_k.

(b) Deduce from this result a necessary condition (involving λ and τ) for g to be completely determined by its samples.

Exercise 37.2 Use the method illustrated in Section 37.4 to transform the Fourier series

$$F(t) = \sum_{n=-\infty}^{+\infty} e^{-bn^2} e^{2i\pi nt}, \quad b > 0.$$

Does the transformed series converge faster than the original series?

Lesson 38

The Sampling Theorem and Shannon's Formula

Shannon's formula is an interpolation formula that expresses the value $f(t)$ of a signal at any time t in terms of its values $f(na)$ at the discrete points na, $n \in \mathbb{Z}$. The signal f is thus completely determined by the sampled signal $af\Delta_a$. This is what we had in mind when we mentioned in the last lesson that sampling does not destroy information. Since this property is patently false for arbitrary signals, some restrictive assumptions must be made about f. Our point of departure will be Poisson's formula (37.7); thus the first assumption is that f is band limited.

Let f be a band-limited signal:

$$\operatorname{supp}(\widehat{f}) \subset [-\lambda_c, \lambda_c]. \tag{38.1}$$

Then f is infinitely differentiable and slowly increasing (Theorem 31.5.1), and we have Poisson's formula (34.4)

$$\sum_{n=-\infty}^{+\infty} \widehat{f}\left(\lambda - \frac{n}{a}\right) = a \sum_{n=-\infty}^{+\infty} f(na)e^{-2i\pi\lambda na} \tag{38.2}$$

with equality in $\mathscr{S}\,'$. When the sampling rate is high enough, which is when

$$a \leq \frac{1}{2\lambda_c},$$

the translates of the spectrum $\widehat{f}$ in the left-hand side of (38.2) do not overlap; they are separated by $a^{-1} - 2\lambda_c \geq 0$.

The idea behind Shannon's formula is to isolate the central copy of $\widehat{f}$ and use it to reconstruct f, which will then be expressed in terms of its values $f(na)$ (see Figure 38.1). The next assumption is that $\widehat{f}$, and hence f, is square integrable:

$$\widehat{f} \in L^2(\mathbb{R}). \tag{38.3}$$

The left-hand side of (38.2) is then a periodic function $F(\lambda)$ with period

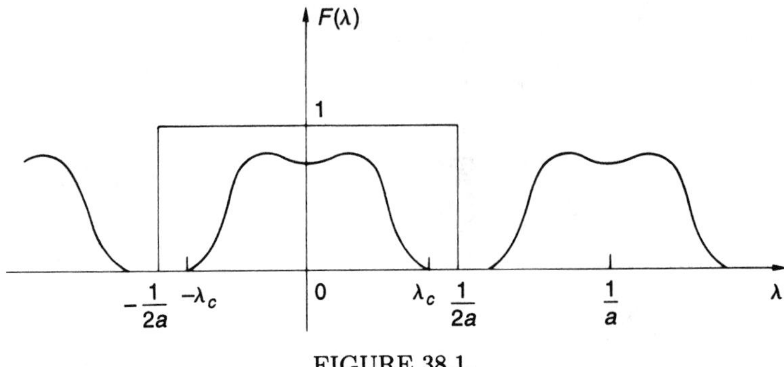

FIGURE 38.1.

$1/a$ that is square integrable over one period. Thus F can be expanded in a Fourier series

$$F(\lambda) = \sum_{n=-\infty}^{+\infty} c_n e^{2i\pi\lambda na}. \tag{38.4}$$

The sequence (c_{-n}) is square integrable, and the equality (38.4) holds in $L^2(0, 1/a)$. Since Fourier expansions of periodic tempered distributions are unique (Theorem 36.2.2),

$$c_{-n} = af(na).$$

First conclusion: Under the assumptions (38.1) and (38.3) the equality (38.2) holds in $L^2_{\mathrm{p}}(0, 1/a)$ and

$$\sum_{n=-\infty}^{+\infty} |f(na)|^2 < +\infty.$$

If we multiply $F(\lambda)$ by the characteristic function

$$r(\lambda) = \chi_{[-\frac{1}{2a}, \frac{1}{2a}]}(\lambda),$$

we see that

$$\widehat{f}(\lambda) = a \sum_{n=-\infty}^{+\infty} f(na) r(\lambda) e^{-2i\pi\lambda na}, \tag{38.5}$$

which holds in $L^2(\mathbb{R})$. From the continuity of $\overline{\mathscr{F}}$ on $L^2(\mathbb{R})$ and from the relation

$$\overline{\mathscr{F}}\left[r(\lambda)e^{-2i\pi\lambda na}\right] = (\overline{\mathscr{F}} r)(t - na) = \frac{\sin \frac{\pi}{a}(t - na)}{\pi(t - na)},$$

we finally obtain the interpolation formula

$$f(t) = \sum_{n=-\infty}^{+\infty} f(na) \frac{\sin \frac{\pi}{a}(t - na)}{\frac{\pi}{a}(t - na)}. \tag{38.6}$$

The series on the right converges to f in $L^2(\mathbb{R})$. If in addition,

$$\sum_{n=-\infty}^{+\infty} |f(na)| < +\infty,$$

then the series in (38.6) converges uniformly on $\mathbb{R}$ to a continuous function g. But this implies that the series converges to g in $L^2(J)$, where J is any bounded interval. The conclusion is that $f = g$ a.e. on $\mathbb{R}$; hence $f(t) = g(t)$ for all $t \in \mathbb{R}$ because f and g are continuous.

38.1 Shannon's theorem

Shannon's theorem *Let f be a signal that contains no frequencies greater than some value λ_c and assume that f has finite energy:*

$$\operatorname{supp}(\widehat{f}\,) \subset [-\lambda_c, \lambda_c] \quad \text{and} \quad \widehat{f} \in L^2(\mathbb{R}).$$

Then for all $a > 0$,

$$\sum_{n=-\infty}^{+\infty} |f(na)|^2 < +\infty, \tag{38.7}$$

and for all $a \leq \dfrac{1}{2\lambda_c}$,

$$f(t) = \sum_{n=-\infty}^{+\infty} f(na) \frac{\sin \frac{\pi}{a}(t - na)}{\frac{\pi}{a}(t - na)}. \tag{38.8}$$

This equality holds in $L^2(\mathbb{R})$. If in addition,

$$\sum_{n=-\infty}^{+\infty} |f(na)| < +\infty, \tag{38.9}$$

then the series converges uniformly on $\mathbb{R}$ and equality holds for all $t \in \mathbb{R}$.

REMARK: Shannon's formula can also be written as

$$f(t) = \frac{a}{\pi} \sin \frac{\pi}{a} t \sum_{n=-\infty}^{+\infty} f(na) \frac{(-1)^n}{t - na}. \tag{38.10}$$

This causes poles, which do not belong to f, to appear in the series at the points $t_n = na$.

38.2 The case of a function $f(t) = \sum_{n=-N}^{N} c_n e^{2i\pi\lambda_n t}$

The spectrum of this function is $\sum_{n=-N}^{N} c_n \delta_{\lambda_n}$, so it has bounded support. Although this function is not square integrable, it is easy to see using the theory of Fourier series that Shannon's formula is true for trigonometric series. By linearity it is sufficient to prove (38.8) for the function

$$f(t) = e^{2i\pi\lambda t}, \quad \lambda \in \mathbb{R}.$$

Let g be the $1/a$-periodic function that agrees with f on $(-1/(2a), 1/(2a))$. For a real and fixed, the Fourier coefficients of g are

$$c_n = \frac{a \sin \frac{\pi}{a}(\lambda - na)}{\pi(\lambda - na)}, \quad n \in \mathbb{Z},$$

and

$$g(t) = \sum_{n=-\infty}^{+\infty} \frac{\sin \frac{\pi}{a}(\lambda - na)}{\frac{\pi}{a}(\lambda - na)} e^{2i\pi nat}. \tag{38.11}$$

This equality holds in $L^2(0, 1/a)$, but in view of Theorem 5.2.4, it is also true for all t in the open interval $(-1/(2a), 1/(2a))$. Hence, for all $\lambda \in \mathbb{R}$,

$$e^{2i\pi\lambda t} = \sum_{n=-\infty}^{+\infty} e^{2i\pi nat} \frac{\sin \frac{\pi}{a}(\lambda - na)}{\frac{\pi}{a}(\lambda - na)}, \quad |t| < \frac{1}{2a}. \tag{38.12}$$

Shannon's formula (38.8) is obtained by interchanging t and λ.

REMARK: We note that here it is not possible to take a to be one of the extreme values $\pm 1/(2\lambda)$. The relation would clearly be false in this case.

38.2.1 Theorem *For a trigonometric signal*

$$f(t) = \sum_{n=-N}^{N} c_n e^{2i\pi\lambda_n t}, \quad \lambda_n \in \mathbb{R},$$

Shannon's formula

$$f(t) = \sum_{n=-\infty}^{+\infty} f(na) \frac{\sin \frac{\pi}{a}(t - na)}{\frac{\pi}{a}(t - na)} \tag{38.13}$$

holds for all $a \in (0, 1/(2\lambda_c))$ and all $t \in \mathbb{R}$ with $\lambda_c = \max_{|n| \le N} |\lambda_n|$, and the convergence is pointwise on $\mathbb{R}$.

38.3 Shannon's formula fails in $\mathscr{S}'$

It is natural to ask whether Shannon's formula (38.8) is true in $\mathscr{S}'$ for all signals f such that $\widehat{f} \in \mathscr{E}'$ (f is band limited). If this equality were true in $\mathscr{S}'$ for all such signals, then we would have

$$\sum_{n=-N}^{N} f(na) \int_{\mathbb{R}} \frac{\sin \frac{\pi}{a}(t-na)}{\frac{\pi}{a}(t-na)} \varphi(t)\, dt \to \int_{\mathbb{R}} f(t)\varphi(t)\, dt$$

for all $\varphi \in \mathscr{S}$. This, however, is not the case: One can find such signals f and functions φ for which the sum does not converge to the integral on the right (see Exercise 38.5).

38.4 The cardinal sine functions

The function $s_a(t) = \sin \frac{\pi}{a}t \Big/ \left(\frac{\pi}{a}t\right)$, whose translates $s_{an}(t) = s_a(t-na)$ appear in Shannon's formula

$$f = \sum_{n=-\infty}^{+\infty} f(na) s_{an}, \tag{38.14}$$

is called the cardinal sine. We have $s_n(na) = 1$, and $s_n(ma) = 0$ for $m \in \mathbb{Z}$ and $m \neq n$. The cardinal sine and its translates are in $L^2(\mathbb{R})$, and (38.14) suggests that the s_n might form a basis for the Hilbert space $L^2(\mathbb{R})$. One must remember, however, that (38.14) was developed only for functions in $L^2(\mathbb{R})$ that have bounded spectra contained in $[-1/(2a), 1/(2a)]$. This leads us to introduce the following definition:

$$V_a = \left\{ v \in L^2(\mathbb{R}) \,\middle|\, \operatorname{supp}(\widehat{v}) \subset \left[-\frac{1}{2a}, \frac{1}{2a}\right] \right\}.$$

It is easy to show that V_a is a closed subspace of $L^2(\mathbb{R})$.

38.4.1 Proposition

(i) *The family of functions $(s_{an})_{n\in\mathbb{Z}}$ is an orthogonal basis for the Hilbert space V_a.*

(ii) *If $(a_j)_{j\in\mathbb{N}}$ is any sequence such that $\lim_{j\to+\infty} a_j = 0$, then $\bigcup_{j\in\mathbb{N}} V_{a_j}$ is dense in $L^2(\mathbb{R})$.*

Proof. We first prove orthogonality. By Proposition 22.1.2,

$$\int_{\mathbb{R}} s_{an}\overline{s_{ap}} = \int_{\mathbb{R}} \widehat{s_{an}}\, \overline{\widehat{s_{ap}}}.$$

We know how to compute the Fourier transform of s_{an}:

$$\widehat{s_{an}}(\lambda) = \widehat{\tau_{na} s_a}(\lambda) = \widehat{s_a}(\lambda) e^{-2i\pi\lambda na} = ar(\lambda) e^{-2i\pi\lambda na},$$

where r is the characteristic function of $[-1/(2a), 1/(2a)]$. Consequently,

$$\int_{\mathbb{R}} s_{an} \overline{s_{ap}} = a^2 \int_{-\frac{1}{2a}}^{\frac{1}{2a}} e^{-2i\pi\lambda(n-p)a}\, d\lambda = \begin{cases} a & \text{if} \quad n = p, \\ 0 & \text{if} \quad n \neq p, \end{cases} \tag{38.15}$$

which proves orthogonality. Next we show that linear combinations of the s_{an} are dense in V_a.

Take $g \in V_a$ and $\varepsilon > 0$. By (38.14) and (38.15),

$$\|g - \sum_{n=-N}^{N} g(na) s_{an}\|_2^2 = \| \sum_{|n|>N} g(na) s_{an}\|_2^2 = a \sum_{|n|>N} |g(na)|^2,$$

and hence by (38.7) there is an $N_0 \in \mathbb{N}$ such that

$$\|g - \sum_{n=-N_0}^{N_0} g(na) s_{an}\|_2 < \varepsilon.$$

This proves density and completes the proof of (i).

To prove (ii), take $f \in L^2(\mathbb{R})$, $\varepsilon > 0$, and define g_n by

$$\widehat{g_n}(\lambda) = \begin{cases} \widehat{f}(\lambda) & \text{if} \quad |\lambda| \leq n, \\ 0 & \text{otherwise.} \end{cases}$$

There exists an $n_0 \in \mathbb{N}$ such that for all $n \geq n_0$,

$$\|f - g_n\|_2^2 = \int_{|\lambda| \geq n} |\widehat{f}(\lambda)|^2\, d\lambda < \varepsilon,$$

and $g_{n_0} \in V_{a_j}$ for sufficiently large j. □

38.4.2 Remark It happens that the decomposition (38.8) is not particularly useful in practice for numerical computation. The cardinal sine tends to zero too slowly. Figure 38.2 illustrates a representation using (38.8).

It is nevertheless true that the function

$$f_N(t) = \sum_{n=-N}^{N} f(na) s_{an}(t),$$

which interpolates f at the points $t_n = na$, $-N \leq n \leq N$, and is zero at the other subdivision points, is the best approximation of f in the subspace of $L^2(\mathbb{R})$ spanned by $\{s_{-aN}, \dots, s_0, \dots, s_{aN}\}$.

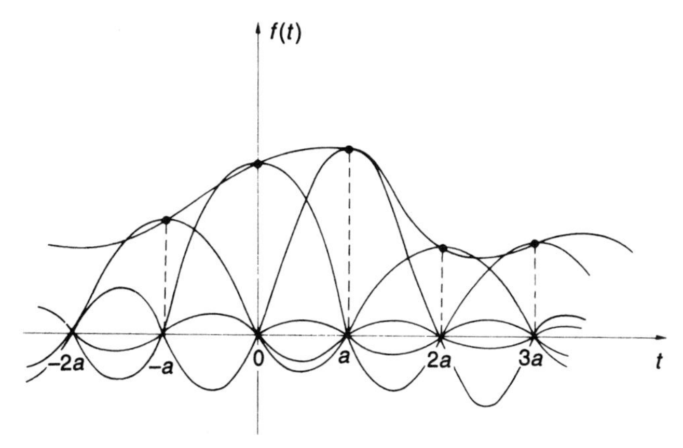

FIGURE 38.2. The cardinal sine basis.

We have just encountered the problem of looking for a "good" orthogonal basis for representing a signal $f \in L^2(\mathbb{R})$, where "good" is related to the kind of signal processing we have in mind. We will see in Lesson 42 how this question is being dealt with today in view of results on wavelets that began to appear in the 1980s.

38.5 Sampling and the numerical evaluation of a spectrum

38.5.1 The sampling problem

Suppose we wish to compute the spectrum $\widehat{f}(\lambda)$ of a signal f that is presented to us in some form—for example, as an analog recording—where we have access to the function values at "all times" t. If we have no explicit formula for the function or other information, the best we can do is sample f and try to compute its spectrum from the sampled function. But what sampling rate should be used? Without more information, there is no answer. Experts in signal processing can tell us in concrete cases what frequencies are essential for carrying information; this means that in a given, well known situation, an expert can specify a limit λ_c above which the higher frequencies are considered to be noise. The simplest example is perhaps the case of sound in the human audio range. We know, in general, that humans do not hear frequencies beyond about 20,000 Hz. Thus frequencies higher than this can be suppressed in transmission and reproduction systems without perceptible loss of quality. The limit λ_c =20,000 Hz corresponds to a basic sampling rate of 40,000 times per second, which is approximately what is used in digital recordings. In fact, four times this rate, or 160,000 Hz, is used for the production of compact discs. In cases where the signal varies slowly, it is possible to sample at a lower rate. It is the specific situation with its specific definition of "quality" that determines the sampling rate.

While it is up to the expert to define what is considered to be the bound λ_c of the spectrum, one must always keep in mind that this assumption of a limited spectrum implies that the signal itself is an analytic function. In particular, we cannot assume without contradiction that both the signal and its spectrum have bounded support, since an analytic function that vanishes on an interval must vanish identically (Theorem 31.5.2). It is important to keep these facts in mind.

To assume that the signal f is band limited implies that f is an analytic function and that $\mathrm{supp}(f) = \mathbb{R}$. *(In particular, f cannot be causal.) Conversely, to assume that f has bounded support implies that its spectrum cannot have bounded support.*

38.5.2 The phenomenon of aliasing

If one is not careful, computing the spectrum from samples taken directly from a recorded signal can lead to unpleasant surprises. Any recorded physical signal is going to be contaminated by noise. In addition to the "real" signal f, the recorded signal will typically look like $g = f + r$, where r has relatively small amplitude but contains relatively high frequencies, and the spectrum of g will be broader than the spectrum of f. This means that even though one has a priori an idea about the band width of f, a sampling rate based on this knowledge will have a good chance of being too low and will lead to the situation illustrated in Figure 37.3. This is phenomenon is called *aliasing*. When this happens, the computed spectrum will not be the one that is sought. To avoid this problem, it is necessary to filter the signal before it is sampled. By passing the signal through a well-designed low-pass filter, one gains two advantages: High-frequency noise is eliminated, and one has a better idea about the appropriate sampling rate.

To compute the spectrum of a physical signal numerically, it is necessary to filter the signal before it is sampled. This is to avoid the problem of aliasing.

Aliasing appears when processing a sampled signal in formula (8.7),

$$c_n^N = \cdots c_{n-2N} + c_{n-N} + c_n + c_{n+N} + c_{n+2N} + \cdots,$$

which we saw in connection with the discrete Fourier transform. Here the approximate spectrum c_n^N is "contaminated" with extra copies of the real spectrum c_n that appear as the terms c_{n+pN}, $p \neq 0$. Prefiltering eliminates these terms, which can be too large for practical computations, even though they eventually tend to zero.

38.5.3 Computation using the FFT

Assume that the signal f has been filtered and that

$$\operatorname{supp}(\widehat{f}\,) \subset [-\lambda_c, \lambda_c].$$

Then by (38.5), for $a < \dfrac{1}{2\lambda_c}$,

$$\widehat{f}(\lambda) = a \sum_{n=-\infty}^{+\infty} f(na)e^{-2i\pi na\lambda}$$

for all $\lambda \in [-\lambda_c, \lambda_c]$.

Suppose the signal is observed during the time $t \in [-Na, (N-1)a]$. The approximation of the spectrum that is based the samples $x_n = f(na)$, $n = -N, \dots, N-1$, will be

$$S_N(\lambda) = a \sum_{n=-N}^{N-1} x_n e^{-2i\pi\lambda na},$$

and its values at the points $\lambda_k = \dfrac{k}{2Na} = \dfrac{k}{T}$,

$$S_N\left(\frac{k}{T}\right) = a \sum_{n=-N}^{N-1} x_n \omega_{2N}^{-nk},$$

are easily computed using the FFT as described in Lesson 9.

We see that the mesh of the grid on which the spectrum is computed is $1/T$, where T is the length of observation.

In practice, one avoids cutting the signal abruptly at the two extremes, since this operation, which amounts to multiplying the function by some characteristic function $\chi_{[a,b]}$, introduces perturbations on the spectrum. Replacing $\chi_{[a,b]}$ by a smooth window lessens these effects. We will return to this question in Lesson 41.

38.6 Exercises

Exercise 38.1 Let f be an element of $\mathscr{S}\,'(\mathbb{R})$ such that

$$\widehat{f}(\lambda) = \frac{\pi}{2} \cos \frac{\pi}{2}\lambda \cdot \chi_{[-1,1]}(\lambda).$$

(a) Show that

$$f(t) = \frac{-2\cos 2\pi t}{16t^2 - 1}$$

and verify that f is infinitely differentiable.

(b) Use Shannon's formula with $a = 1/2$ to show that

$$\pi \cot 2\pi t = (16t^2 - 1) \sum_{n=-\infty}^{+\infty} \frac{1}{(4n^2 - 1)(2t - n)}$$

when $2t$ is not an integer.

(c) Write the general term of the last series in partial fractions (in the variable n) and show that

$$\cot x = \frac{1}{x} + \sum_{n=1}^{+\infty} \left(\frac{1}{x - n\pi} + \frac{1}{x + n\pi} \right)$$

when x is not a multiple of π.

Exercise 38.2

Apply Shannon's formula to the function $f(t) = \cos 2\pi t$ with $a = 1/2$ and verify that one obtains directly the expression for $\cot x$ found in the last exercise.

Hint: Use the proof of Theorem 38.2.1 with

$$f(t) = \cos 2\pi t = \frac{e^{2i\pi t} + e^{-2i\pi t}}{2}$$

and notice that in this case one can apply Theorem 5.2.4 for all $t \in \mathbb{R}$.

Exercise 38.3

Suppose $f \in L^2(\mathbb{R})$ and $\operatorname{supp}(\widehat{f}) \subset [\alpha - \lambda_0, \alpha + \lambda_0]$ with $\alpha \in \mathbb{R}$ and $\lambda_0 > 0$. Show that f is determined by a sampling $(f(na))$ with $0 < a \leq 1/2\lambda_0$.

Exercise 38.4

Write equality (38.12) at the points $t = \pm 1/(2a)$ using Theorem 5.2.4.

Exercise 38.5 (Shannon's formula fails in $\mathscr{S}'$)

We use the notation of Section 38.3 with $a = 1$. For $\varphi \in \mathscr{S}$, define

$$S_N(\varphi) = \sum_{n=-N}^{N} f(n) I_n(\varphi)$$

with

$$I_n(\varphi) = \int_{\mathbb{R}} \widehat{r}(t)\varphi(t + n)\, dt$$

and $r = \chi_{[-\frac{1}{2}, \frac{1}{2}]}$.

(a) Show that

$$I_n(\varphi) = \int_{-\frac{1}{2}}^{\frac{1}{2}} (\overline{\mathscr{F}}\varphi)(x) e^{-2i\pi n x}\, dx.$$

(b) Take $f(t) = t$. Show that

$$S_N(\varphi) = \sum_{n=-N}^{N} nc_n(\psi),$$

where ψ is the function with period 1 that agrees with $\overline{\mathscr{F}}\varphi$ on the interval $(-1/2, 1/2)$.

(c) Take $\varphi = \widehat{g}$, where g is an element of $\mathscr{D}(\mathbb{R})$ such that

$$g(x) = x \quad \text{if} \quad |x| < \frac{1}{2}$$

(Exercise 27.4). Compute $S_N(\varphi)$ and conclude that Shannon's formula is not generally true in $\mathscr{S}'$.

Lesson 39

Discrete Filters and Convolution

We are going to study several specific questions about discrete signals and filters in this and the following lesson. The current lesson concentrates on the convolution of discrete signals and its application to discrete filters.

39.1 Discrete signals and filters

39.1.1 Discrete signals

Let a be a positive real number. Any distribution of the form

$$x = \sum_{n=-\infty}^{+\infty} x_n \delta_{na}, \quad x_n \in \mathbb{C},$$

will be called a *discrete signal*; we denote the set of discrete signals by X_a:

$$X_a = \left\{ x \in \mathscr{D}' \;\middle|\; x = \sum_{n=-\infty}^{+\infty} x_n \delta_{na} \right\}.$$

This is a vector space that is usually endowed with the topology induced by that of $\mathscr{D}'$, which is the topology of pointwise convergence:

$$\lim_{N\to+\infty} x_N = x \text{ in } \mathscr{D}' \quad \Longleftrightarrow \quad \lim_{N\to+\infty} x_{Nn} = x_n \text{ for all } n \in \mathbb{Z}.$$

39.1.2 Definition (discrete filter) Any mapping $D : X \mapsto X_a$ that is linear, continuous, and commutes with the translations τ_{ka}, $k \in \mathbb{Z}$, (see Section 2.1.3) will be called a discrete filter whenever the space X satisfies the following conditions:

(i) X is a subspace of X_a that contains δ and that is invariant under the translations τ_{ka}, $k \in \mathbb{Z}$.

(ii) X is endowed with a topology that is at least as fine as the topology induced by $\mathscr{D}'$.

Unless otherwise indicated, the topology of X will be that induced by $\mathscr{D}'$. This definition is modeled on the one given for analog filters (Definition 34.1.1). Here, however, the translations must be limited to integer multiples of the step a, which is fixed once and for all. The spaces most frequently encountered in practice are the following:

$X = X_a$	all of the discrete signals.
$X = X_a \cap \mathscr{D}'_+$	the discrete causal signals.
$X = X_a \cap \mathscr{E}' = \mathscr{L}_a$	the discrete signals with finite support.
$X = X_a \cap \mathscr{S}'$	the discrete tempered signals.

39.1.3 Examples of discrete filters

(a) A delay (or shift): $X = X_a$, $Dx = y$ with $y_n = x_{n-k}$.

(b) An average: $X = X_a$, $Dx = y$ with, for example,

$$y_n = \frac{1}{2}(x_n + x_{n-1}) \quad \text{or} \quad y_n = \frac{1}{4}x_n + \frac{1}{2}x_{n-1} + \frac{1}{4}x_{n-2}.$$

(c) The recursive system defined by (1.1): $y_k = x_k + \alpha y_{k-1}$.

(d) Any convolution system: $Dx = h * x$.

Naturally, one assumes that the convolution is defined on the space of input signals X. It is clear that D is linear and invariant; it will generally be continuous, but this depends on the space X (see Lesson 32). As in the case of analog filters, most discrete filters are convolution systems.

39.1.4 Proposition *Let $D : X \to X_a$ be a discrete filter and let $h = D\delta$. Then D is a convolution system*

$$Dx = h * x, \quad x \in X,$$

in the following two cases:

(i) *$X = X_a$ and h is finite.*

(ii) *$X = X_a \cap \mathscr{D}'_+$ and h is a causal signal.*

Proof. The result follows immediately in case (i), and it is obtained by the density of $\mathscr{L}_a$ in X for case (ii). □

The next step is to examine several frequently encountered cases where the convolution of two discrete signals is defined and to determine how to compute the convolution in these cases.

39.2 The convolution of two discrete signals

We first look at two simple cases:

(a) $h \in \mathscr{L}_a$; that is, h is finite.

(b) The supports of h and x are limited on the left.

We have

$$h = \sum_{m=-\infty}^{+\infty} h_m \delta_{ma}, \quad x = \sum_{k=-\infty}^{+\infty} x_k \delta_{ka}, \quad y = h * x = \sum_{n=-\infty}^{+\infty} y_n \delta_{na}.$$

Operating formally, we compute y_n in terms of (h_m) and (x_k):

$$h * x = \Big(\sum_{m=-\infty}^{+\infty} h_m \delta_{ma}\Big) * \Big(\sum_{k=-\infty}^{+\infty} x_k \delta_{ka}\Big) = \sum_{m,k} h_m x_k \delta_{(m+k)a}.$$

By regrouping the terms $m + k = n$, we see that

$$y_n = \sum_{k=-\infty}^{+\infty} h_k x_{n-k}. \tag{39.1}$$

We are going to investigate the validity of this computation.

39.2.1 Proposition *If h is finite, or if the supports of h and x are limited on the left, then $y = h * x$ is given by equation (39.1), which is a finite sum.*

Proof. The proof in the finite case follows directly from the formal computation using the distributivity of the convolution. For the other case, we have

$$\langle h, \varphi \rangle = \sum_{k=k_0}^{+\infty} h_k \varphi(ka), \quad \langle x, \varphi \rangle = \sum_{m=m_0}^{+\infty} x_m \varphi(ma)$$

for all $\varphi \in \mathscr{D}$, and these sums have only a finite number of terms. Hence,

$$\langle h * x, \varphi \rangle = \big\langle h_k, \langle x_m, \varphi\big((m+k)a\big)\rangle\big\rangle = \sum_{k=k_0}^{+\infty} h_k \psi_k,$$

where

$$\psi_k = \sum_{m=m_0}^{+\infty} x_m \varphi\big((m+k)a\big).$$

The result follows by interchanging the (finite) sums and by the change of variable $n = m + k$:

$$\langle h * x, \varphi \rangle = \sum_{m=m_0}^{+\infty} \sum_{k=k_0}^{+\infty} h_k x_m \varphi\big((m+k)a\big) = \sum_{n=m_0+k_0}^{+\infty} \Big(\sum_{k=k_0}^{+\infty} h_k x_{n-k}\Big) \varphi(na).$$

This proves equation (39.1). In addition, $y_n = 0$ if $n < m_0 + k_0$, and the series for y_n is indeed a finite sum. □

39.3 Cases where the two supports are not bounded

Here we will see the first of several cases where $h * x$ exists when the supports of h and x extend from $-\infty$ to $+\infty$. This case is the discrete version of the continuous convolution $\mathscr{S} * \mathscr{S}'$ in the same way that the last two cases were the discrete versions of the convolutions $\mathscr{E}' * \mathscr{D}'$ and $\mathscr{D}'_+ * \mathscr{D}'_+$. The condition that h have finite support is replaced by the condition that the sequence (h_n) be rapidly decreasing; x must be tempered, which means that (x_n) is slowly increasing (Corollary 36.2.3).

39.3.1 Proposition *Suppose that the two discrete signals*

$$h = \sum_{k=-\infty}^{+\infty} h_k \delta_{ka} \quad \text{and} \quad x = \sum_{n=-\infty}^{+\infty} x_n \delta_{na}$$

are such that (h_k) is rapidly decreasing and (x_n) is slowly increasing. Then the convolution $h * x$ is well defined. Furthermore,indexdiscrete signals!convolution of

(i) *$h * x$ is a tempered distribution.*

(ii) *$h * x = \sum_{n=-\infty}^{+\infty} y_n \delta_{na}$ with $y_n = \sum_{k=-\infty}^{+\infty} h_k x_{n-k}$, and the series for y_n converges absolutely.*

Proof. The proof is based on Fubini's theorem for the discrete measure space $(\mathbb{Z}\times\mathbb{Z}, \mathscr{T}, \mu)$ where $\mathscr{T}$ is the σ-algebra generated by the finite subsets of $\mathbb{Z}\times\mathbb{Z}$ and μ is the measure defined by $\mu(S) =$ the number of point in S. A function $u : \mathbb{Z} \times \mathbb{Z} \to \mathbb{C}$ is integrable (or summable) if and only if $|u|$ is integrable. One part of Fubini's theorem states that the condition

$$\sum_{n=-\infty}^{+\infty} \left(\sum_{k=-\infty}^{+\infty} |u_{n,k}| \right) < +\infty$$

implies that u is integrable (summable) and that

$$\int_{\mathbb{Z}^2} u = \sum_{(n,k)\in\mathbb{Z}^2} u_{n,k} = \sum_{n=-\infty}^{+\infty} \left(\sum_{k=-\infty}^{+\infty} u_{n,k} \right) = \sum_{k=-\infty}^{+\infty} \left(\sum_{n=-\infty}^{+\infty} u_{n,k} \right).$$

This is the discrete version of Theorem 14.3.1.

Now let φ be an element of $\mathscr{S}$. The same formal computation that was done in Section 39.2 shows that if $h*x$ is to exist as a tempered distribution, we must have

$$\langle h * x, \varphi \rangle = \sum_{k=-\infty}^{+\infty} h_k \psi_k$$

with

$$\psi_k = \sum_{m=-\infty}^{+\infty} x_m \varphi\big((m+k)a\big).$$

Thus, for the convolution to exist, it is sufficient that the function

$$h_k x_m \varphi\big((m+k)a\big) \tag{39.2}$$

be summable on $\mathbb{Z} \times \mathbb{Z}$, or, after renaming the indices, that the function

$$h_k x_{n-k} \varphi(na)$$

be summable. This leads us to examine the double sum

$$\sum_{n=-\infty}^{+\infty} \Big(\sum_{k=-\infty}^{+\infty} |h_k||x_{n-k}| \Big) |\varphi(na)|. \tag{39.3}$$

If (39.3) is finite, Fubini's theorem tells us that all of the series involved in the formal computations are absolutely convergent and summation in any order gives the same answer.

The sequence $|h_k|$ is rapidly decreasing and $|x_n|$ is slowly increasing. If we define

$$f(t) = \sum_{k=-\infty}^{+\infty} |h_k| e^{2i\pi k \frac{t}{a}} \quad \text{and} \quad T = \sum_{n=-\infty}^{+\infty} |x_n| e^{2i\pi n \frac{t}{a}},$$

then f is an infinitely differentiable function (Proposition 5.3.4) and T is a tempered distribution. By Theorem 36.3.1, the Fourier coefficients β_n of fT are

$$\beta_n = \sum_{k=-\infty}^{+\infty} |h_k||x_{n-k}|,$$

and they are slowly increasing (Theorem 36.2.2(iv)). Thus there exist $A > 0$ and $\alpha > 0$ such that

$$\beta_n \leq A(1 + |n|^\alpha)$$

for all $n \in \mathbb{Z}$. On the other hand, since $\varphi \in \mathscr{S}$, there is a $B > 0$ such that

$$|\varphi(na)| \leq \frac{B}{1 + |n|^{\alpha+2}}$$

for all n. These two inequalities imply that

$$\sum_{n=-\infty}^{+\infty} \beta_n |\varphi(na)| \leq AB \sum_{n=-\infty}^{+\infty} \frac{1+|n|^\alpha}{1+|n|^{\alpha+2}} < +\infty.$$

This shows that the sum (39.3) is finite and hence that (39.2) is summable. We thus can sum (39.2) in any order, and in particular,

$$\langle h * x, \varphi \rangle = \sum_{n=-\infty}^{+\infty} \left(\sum_{k=-\infty}^{+\infty} h_k x_{n-k} \right) \varphi(na),$$

which proves that $h * x$ makes sense and is given by (ii). The estimate

$$|y_n| \leq \beta_n$$

shows that (y_n) is slowly increasing, so (ii) follows from Corollary 36.2.3.in-dexdiscrete signals!convolution of □

39.3.2 Corollary (periodic convolution and $\mathscr{F}$)

(i) *If f is a periodic C^∞ function with period $a > 0$ and if T is a periodic distribution with the same period, then*

$$\widehat{fT} = \widehat{f} * \widehat{T}.$$

(ii) *Let (h_n) be a rapidly decreasing complex sequence, let (x_n) be a slowly increasing sequence, and let*

$$h = \sum_{n=-\infty}^{+\infty} h_n \delta_{na} \quad \text{and} \quad x = \sum_{n=-\infty}^{+\infty} x_n \delta_{na}$$

be the associated distributions. Then

$$\widehat{h * x} = \widehat{h} \cdot \widehat{x}.$$

Proof. T is tempered, $\widehat{T} = \sum_{n=-\infty}^{\infty} \alpha_n \delta_{\frac{n}{a}}$, and the sequence (α_n) is slowly increasing (Theorem 36.2.2). f is tempered, $\widehat{f} = \sum_{n=-\infty}^{\infty} c_n \delta_{\frac{n}{a}}$, and the coefficients c_n are rapidly decreasing (Proposition 36.1.3 and Section 5.3.3). From Proposition 39.3.1 we know that $\widehat{f} * \widehat{T}$ is tempered and that $\widehat{f} * \widehat{T} = \sum_{n=-\infty}^{\infty} y_n \delta_{\frac{n}{a}}$, where the y_n are equal to $\sum_{k=-\infty}^{\infty} c_k \alpha_{n-k}$. This and Theorem 36.3.1 imply that $\widehat{f} * \widehat{T} = \widehat{fT}$, which proves (i). To prove (ii), first observe that (i) is true if we replace $\mathscr{F}$ by $\overline{\mathscr{F}}$. The result follows by applying (i) to $f = \widehat{h}$ and $T = \widehat{x}$ with $\mathscr{F}$ replaced by $\overline{\mathscr{F}}$. □

39.3.3 The convolution $l_a^1 * l_a^\infty$

If we define

$$l_a^p = \left\{ x = \sum_{n=-\infty}^{+\infty} x_n \delta_{na} \;\middle|\; \sum_{n=-\infty}^{+\infty} |x_n|^p < +\infty \right\}$$

and

$$l_a^\infty = \left\{ x = \sum_{n=-\infty}^{+\infty} x_n \delta_{na} \;\middle|\; \sup_n |x_n| < +\infty \right\},$$

then the convolution $l_a^1 * l_a^\infty$ is well-defined in the same way the convolution $L^1 * L^\infty$ is well-defined in the continuous case. In fact, going back to the proof of Proposition 39.3.1, for $h \in l_a^1$ and $x \in l_a^\infty$ we have

$$\sum_{n=-\infty}^{+\infty} \left(\sum_{k=-\infty}^{+\infty} |h_k||x_{n-k}| \right) |\varphi(na)| \leq \|\varphi\|_\infty \cdot \sum_{k=-\infty}^{+\infty} |h_k| \cdot \sup_n |x_n|,$$

and thus we have formula (39.1) with $h * x \in l_a^\infty$.

39.3.4 The convolution $l_a^2 * l_a^2$

From Schwarz's inequality

$$\sum_{k=-\infty}^{+\infty} |h_k||x_{n-k}| \leq \left(\sum_{k=-\infty}^{+\infty} |h_k|^2 \right)^{1/2} \left(\sum_{n=-\infty}^{+\infty} |x_n|^2 \right)^{1/2},$$

we see (by a computation similar to the one above) that $h * x$ exists for all $h, x \in l_a^2$ and that $h * x \in l_a^2$. (The convolution $l_a^1 * l_a^2$ does not need to be studied as a special case since $l_a^2 \subset l_a^\infty$.)

39.4 Summary

The convolution $h * x$ is defined for the distributions

$$h = \sum_{n=-\infty}^{+\infty} h_n \delta_{na} \quad \text{and} \quad x = \sum_{n=-\infty}^{+\infty} x_n \delta_{na}$$

in the following cases:

(a) h (or x) is finite.

(b) h and x have their supports bounded on the left (or on the right).

(c) (h_n) is rapidly decreasing and (x_n) is slowly increasing.

(d) $h \in l_a^1$ and $x \in l_a^\infty$ $(h * x \in l_a^\infty)$.

(e) $h \in l_a^2$ and $x \in l_a^2$ $(h * x \in l_a^\infty)$.

In all of these cases,

$$h * x = \sum_{n=-\infty}^{+\infty} y_n \delta_{na} \quad \text{with} \quad y_n = \sum_{k=-\infty}^{+\infty} h_k \, x_{n-k}.$$

In cases (a) and (b), the series for y_n is a finite sum; in the other cases, the series is absolutely convergent. In case (c),

$$\widehat{h * x} = \widehat{h} \cdot \widehat{x}.$$

These results show that the mapping

$$\begin{aligned} D : X &\to X_a, \\ x &\mapsto D(x) = h * x \end{aligned}$$

is a discrete filter in the following cases:
Case 1: h is finite, and $X = X_a$.
Case 2: h is causal, and $X = X_a \cap \mathscr{D}'_+$.
Case 3: h is rapidly decreasing, and $X = X_a \cap \mathscr{S}'$ (slowly increasing).
Case 4: $h \in l^1_a$, and $X = l^\infty_a$.
Case 5: $h \in l^2_a$, and $X = l^2_a$.
Case 6: $h \in l^\infty_a$, and $X = l^1_a$.
Case 7: $h \in X_a$, and $X = X_a \cap \mathscr{E}' = \mathscr{L}_a$ (finite inputs).

In Cases 1, 2, and 3, the topology on X is that induced by $\mathscr{D}'$. In Cases 4, 5, and 6, one can take the topologies of the l^p_a spaces. In Case 7, one has many choices.

39.5 Causality and stability of a discrete filter

The general definition of causality of a system was given in Section 2.1.2. As in the analog case, linearity and invariance reduce the definition to the following:

$$\begin{bmatrix} \text{The filter} \\ D : X \to X_a \\ \text{is realizable} \\ \text{(or causal).} \end{bmatrix} \iff \left[x_n = 0 \text{ for all } n < 0 \Rightarrow y_n = 0 \text{ for all } n < 0. \right]$$

We define stability as follows:

$$\begin{bmatrix} \text{The filter} \\ D : X \to X_a \\ \text{is stable.} \end{bmatrix} \iff \begin{bmatrix} \text{There is an } A > 0 \text{ such that} \\ \|Dx\|_\infty \le A \|x\|_\infty \\ \text{for all } x \in X \cap l^\infty_a. \end{bmatrix}$$

In particular, a bounded input produces a bounded output.

The next result characterizes these two properties in terms of the impulse response.

39.5.1 Theorem *Let $D : X \to X_a$ belong to one of the 7 cases listed above and let h be its impulse response. Then the following hold:*

(i) *D is stable if and only if $\sum_{n=-\infty}^{+\infty} |h_n| < +\infty$.*

(ii) *D is realizable if and only if $h_n = 0$ for all $n < 0$.*

Proof. If $h \in l_a^1$, then from (39.1),

$$|y_n| \leq \sum_{k=-\infty}^{+\infty} |h_k||x_{n-k}| \leq \sup_n |x_n| \sum_{k=-\infty}^{+\infty} |h_k|$$

and

$$\|Dx\|_\infty \leq \left(\sum_{k=-\infty}^{+\infty} |h_k| \right) \|x\|_\infty.$$

Hence D is stable. To prove the converse, assume that D is stable. In Cases 1, 3, and 4, there is nothing to prove, since $h \in l_a^1$. For the other cases, consider the sequence of signals x^p, $p \in \mathbb{N}$, defined by

$$x_n^p = \begin{cases} \text{sign}(h_{p-n}) & \text{if } 0 \leq n \leq 2p \text{ and } h_{p-n} \neq 0, \\ 0 & \text{otherwise.} \end{cases}$$

(For $c = |c|e^{i\theta}$, $\text{sign}(c) = e^{-i\theta}$.) The signals x^p are finite, so they are in $X \cap l_a^\infty$ for Cases 2, 5, 6, and 7, and $\|x^p\|_\infty \leq 1$. Then

$$y_n^p = \sum_{k=-\infty}^{\infty} h_k x_{n-k}^p = \sum_{k=-2p+n}^{n} h_k \text{sign}(h_{p-n+k}),$$

and

$$y_p^p = \sum_{k=-p}^{p} |h_k|$$

for all $p \geq 0$. From the definition of stability we conclude that

$$|y_p^p| = \sum_{k=-p}^{p} |h_k| \leq A$$

for all $p \geq 0$; hence

$$\sum_{k=-\infty}^{+\infty} |h_k| < +\infty,$$

which proves (i).

If D is realizable, then $h = D\delta$, and the definition shows that $h_n = 0$ for all $n < 0$. Conversely, if this property holds, then formula (39.1) shows that

$$x_n = 0 \text{ for all } n < 0 \implies y_n = 0 \text{ for all } n < 0,$$

and this proves (ii). □

39.6 Exercises

Exercise 39.1 Let $x = \sum_{n=-\infty}^{\infty} x_n \delta_{na}$, $a > 0$, be a discrete signal. Compute the impulse responses of the following filters $y = Dx$.

(a) $y_n = x_{n-1}$.

(b) $y_n = \frac{1}{2}(x_n + x_{n-1})$.

(c) $y_n = \frac{1}{3}(x_{n+1} + x_n + x_{n-1})$.

Which of these filters are realizable?

Exercise 39.2 Show that $\widehat{h * x} = \widehat{h} \cdot \widehat{x}$ when h and x are in l_a^2 (use the result in Section 39.3.4).

Exercise 39.3 Consider the discrete filter whose impulse response $h = (h_n)$ is given by

$$h_n = \begin{cases} 0 & \text{if } n \leq 0, \\ \dfrac{1}{n} & \text{if } n > 0, \end{cases}$$

and that belongs to Case 7 in Section 39.4. Show that the response of every finite signal (which is necessarily bounded) is bounded but that the filter is not stable.

Exercise 39.4 Show that the sequence

$$y_n = \sum_{k=1}^{n-1} \frac{1}{k(n-k)}, \quad n \geq 2,$$

is bounded.

Exercise 39.5 Can the proof of Proposition 39.1.4 be adapted to Cases 4, 5, 6, and 7 of Section 39.4?

Hint: Yes for 5, 6, and 7; no for 4, since $\mathscr{L}_a$ is not dense in l_a^∞ in the topology induced by the sup norm $\|\cdot\|_\infty$.

Lesson 40

The z-Transform and Discrete Filters

40.1 The z-transform of a discrete signal

The spectrum of a discrete tempered signal $x = \sum_{n=-\infty}^{+\infty} x_n \delta_{na}$ is the periodic distribution

$$\widehat{x}(\lambda) = \sum_{n=-\infty}^{+\infty} x_n e^{-2i\pi\lambda na}. \tag{40.1}$$

The change of variable $z = e^{2i\pi\lambda a}$ transforms $\widehat{x}$ into the function

$$X(z) = \sum_{n=-\infty}^{+\infty} x_n z^{-n}, \tag{40.2}$$

which is represented as a Laurent series in the complex variable z. By freeing this variable from the constraint $|x| = 1$, we obtain what is called the *z-transform* of the discrete signal x. We know from elementary results on power series that this Laurent expansion defines a function X that is holomorphic in an annulus (which is possibly empty)

$$r < |z| < R$$

with $0 \leq r \leq R \leq +\infty$ (Figure 40.1). The series diverges outside this annulus, and the behavior on $|z| = r$ or $|z| = R$ is uncertain. It is clear, however, that $R = +\infty$ for causal signals.

For discrete signals, it is customary to study the complex function $X(z)$ rather than the Fourier transform $\widehat{x}(\lambda)$. These two functions are related through the equation

$$\widehat{x}(\lambda) = X(e^{2i\pi\lambda a}). \tag{40.3}$$

The z-transform of a discrete signal does not always exist. For example, there is no z-transform for Dirac's comb.

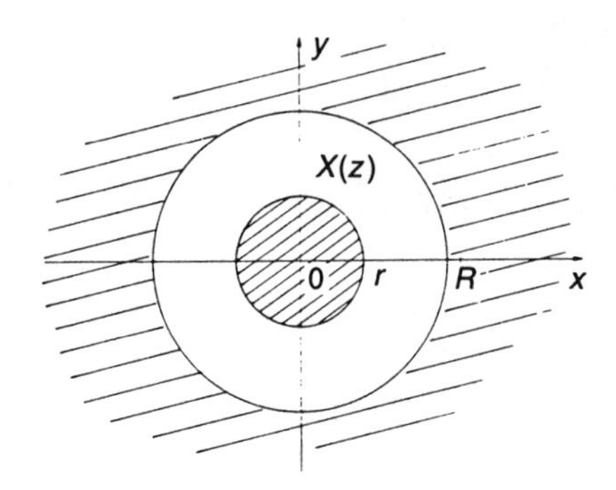

FIGURE 40.1. Annulus of convergence for $X(z)$.

EXAMPLES:

(a) For $\alpha > 0$ and $\beta > 0$, define

$$x_n = \begin{cases} \beta^n & \text{if } n < 0, \\ \alpha^n & \text{if } n \geq 0. \end{cases}$$

Then

$$X(z) = \sum_{n=-\infty}^{-1} \beta^n z^{-n} + \sum_{n=0}^{+\infty} \alpha^n z^{-n} = \frac{z}{\beta - z} + \frac{z}{z - \alpha}$$

for values of z satisfying $|z/\beta| < 1$ and $|\alpha/z| < 1$. Thus the z-transform exists if $\alpha < \beta$. It is defined and holomorphic in the annulus $\alpha < |z| < \beta$.

(b) The discrete version of the unit step function (Heaviside function) is defined by

$$x_n = u_n = \begin{cases} 0 & \text{if } n < 0, \\ 1 & \text{if } n \geq 0, \end{cases}$$

and

$$U(z) = \sum_{n=0}^{+\infty} z^{-n} = \frac{1}{1 - z^{-1}}$$

if $|z| > 1$. Here the annulus of convergence is the exterior of the unit disk: $r = 1$, $R = +\infty$.

40.1.1 Elementary properties of the z-transform

(a) *Linearity*

The transform $x \mapsto X$ is clearly linear.

(b) *Effect of a delay*

If the z-transform of $x = \sum_{n=-\infty}^{\infty} x_n \delta_{na}$ is $X(z)$, then $z^{-1}X(z)$ is the transform of $\tau_a x$ and $z^{-k}X(z)$ is the transform of $\tau_{ka} x$.

(c) *Transform of a convolution*

Assume that $h = \sum_{n=-\infty}^{+\infty} h_n\delta_{na}$ and $x = \sum_{n=-\infty}^{+\infty} x_n\delta_{na}$ are discrete signals that belong to one of the cases in Section 39.4 where the convolution $h * x$ exists. Their respective z-transforms H and X exist in the annuli A_1 and A_2. In the annulus $A = A_1 \cap A_2$, assumed to be nonempty, we have

$$\sum_{n,k\in\mathbb{Z}} |h_k||x_{n-k}||z|^{-n} \leq \left(\sum_{k=-\infty}^{+\infty} |h_k||z|^{-k}\right)\left(\sum_{n=-\infty}^{+\infty} |x_n||z|^{-n}\right) < +\infty.$$

The function $(n,k) \mapsto h_k x_{n-k} z^{-k}$ is summable (integrable) on $\mathbb{Z}^2$, and by Fubini's theorem, it can be summed in any order. Thus for $y = h * x$,

$$Y(z) = \sum_{n=-\infty}^{+\infty}\left(\sum_{k=-\infty}^{+\infty} h_k x_{n-k}\right) z^{-n} = \sum_{k=-\infty}^{+\infty} h_k z^{-k} \sum_{n=-\infty}^{+\infty} x_{n-k} z^{-(n-k)}.$$

It follows that

$$Y(z) = H(z) \cdot X(z)$$

for all $z \in A$. One should not be surprised that the z-transform of a convolution of two signals is the product of their z-transforms!

40.1.2 Inverting the z-transform

Given the z-transform of a signal x, one can recover x by either of two methods: (a) by expanding $X(z)$ in a Laurent series, or (b) by using the residue theorem to compute

$$x_n = \frac{1}{2i\pi}\int_\Gamma X(z) \cdot z^{n-1}\, dz, \tag{40.4}$$

where Γ is a contour around the origin situated in the annulus of convergence and taken in the positive direction (Figure 40.2).

EXAMPLE: Let $X(z) = z(z-r)^{-1}$, $r > 0$, and take the annulus of convergence to be the exterior of the disk $|z| \leq r$.

The first method gives

$$X(z) = \frac{1}{1 - \dfrac{r}{z}} = \sum_{n=0}^{+\infty} r^n z^{-n},$$

so

$$x = \sum_{n=0}^{+\infty} r^n \delta_{na}.$$

By the second method,

$$x_n = \frac{1}{2i\pi}\int_\Gamma \frac{z^n}{z-r}\, dz.$$

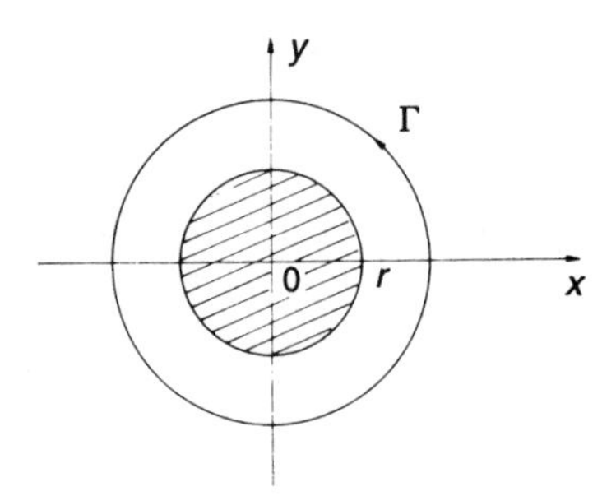

FIGURE 40.2.

If $n \geq 0$, the residue of $f(z) = z^n(z-r)^{-1}$ at $z = r$ is r^n. If $n < 0$, another pole appears at $z = 0$. The residue at $z = 0$ is obtained by expanding $f(z)$ around $z = 0$:

$$f(z) = -\frac{z^n}{r}\frac{1}{1-\dfrac{z}{r}} = -\sum_{p=0}^{+\infty} \frac{z^{n+p}}{r^{p+1}}.$$

The residue at $z = 0$ is the coefficient of z^{-1}, which is equal to $-r^n$. The two residues cancel each other, and we have $x_n = 0$ for $n < 0$. Thus

$$x = \sum_{n=0}^{+\infty} r^n \delta_{na}.$$

40.2 Applications to discrete filters

In most applications, a discrete filter $D : x \mapsto y$ will be a convolution system; thus $Dx = h * x$ for some $h \in X_a$. This is established either by applying one of the results from Lesson 39 or by direct verification.

When this is the case, the z-transform $H(z)$ of the impulse response h is called the *transfer function* of the discrete filter D. The next result relates the stability and realizability of D to properties of H.

40.2.1 Theorem *Assume that the filter D is a convolution system with transfer function $H(z)$ that converges in a nonempty annulus A.*

(i) *D is stable if and only if the unit circle $|z| = 1$ is in A.*

(ii) *If D is realizable, then it is stable if and only if the poles of $H(z)$ are in the interior of the unit disk.*

Proof. The filter D is stable if and only if (Theorem 39.5.1)

$$\sum_{n=-\infty}^{+\infty} |h_n| < +\infty.$$

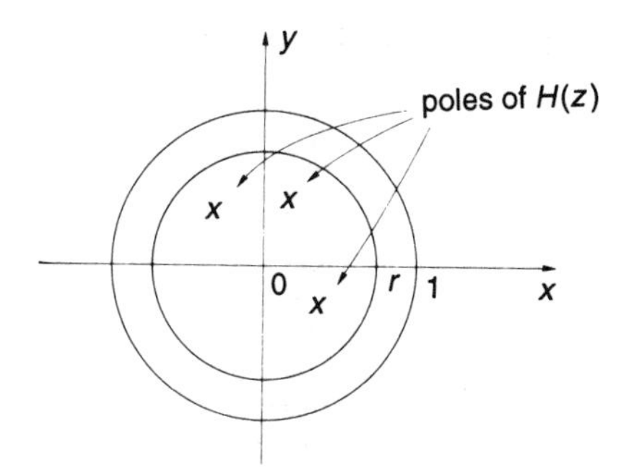

FIGURE 40.3. A realizable and stable filter.

This is equivalent to saying that the series

$$H(z) = \sum_{n=-\infty}^{+\infty} h_n z^{-n}$$

is absolutely convergent for $|z| = 1$. This proves (i).

If D is realizable, the annulus of convergence is the exterior ($|z| > r$) of some disk $|z| \leq r$ (Figure 40.3). If the poles p_k of $H(z)$ are in the interior of the unit disk, that is, if $|p_k| < 1$, then $r < 1$. Conversely, if D is realizable and stable, then $H(z)$ converges absolutely on $|z| = 1$, so the poles must be in the interior of the disk $|z| \leq 1$. □

EXAMPLE: Let $H(z) = z(z-r)^{-1}$ with the annulus of convergence $|z| > r$. This corresponds to a realizable filter. The pole is r. Thus the filter is stable if $r < 1$.

40.2.2 Filters governed by linear difference equations with constant coefficients

In the same way that analog filters are often governed by differential equations, discrete filters can be governed by linear difference equations with constant coefficients:

$$\sum_{k=0}^{q} b_k y_{n-k} = \sum_{j=0}^{p} a_j x_{n-j}, \quad b_0 = 1. \tag{40.5}$$

The output y is completely determined by some additional condition, for example, that the filter is realizable.

COMPUTING THE TRANSFER FUNCTION: By taking the z-transform of both sides of (40.5) and using Section 40.1.1(b), we have

$$\left(\sum_{k=0}^{q} b_k z^{-k}\right) Y(z) = \left(\sum_{j=0}^{p} a_j z^{-j}\right) X(z).$$

The transfer function is the rational function

$$H(z) = \frac{\sum_{j=0}^{p} a_j z^{-j}}{\sum_{k=0}^{q} b_k z^{-k}}. \tag{40.6}$$

COMPUTING THE IMPULSE RESPONSE: This is the inversion problem for the z-transform that we examined in Section 40.1.2. We can obtain the Laurent expansion of $H(z)$ from the relation

$$\left(\sum_{n=-\infty}^{+\infty} h_n z^{-n}\right)\left(\sum_{k=0}^{q} b_k z^{-k}\right) = \sum_{j=0}^{p} a_j z^{-j}.$$

When the filter is realizable, $h_n = 0$ for all $n < 0$. In this case, the h_n are obtained from the recurrence

$$\begin{aligned} h_0 &= a_0, \\ h_n &= a_n - \sum_{k=1}^{n} b_k h_{n-k}, \quad n = 1, 2, \dots, \end{aligned}$$

where we define $a_n = 0$ if $n > p$ and $b_k = 0$ if $k > q$, and we have

$$y_n = \sum_{k=0}^{+\infty} h_k x_{n-k}$$

for all $n \in \mathbb{Z}$.

40.2.3 Example

The discrete form of the realizable RC filter, $RCv' + v = f$, is

$$RC\,\frac{y_n - y_{n-1}}{a} + y_n = x_n, \quad n \in \mathbb{Z}.$$

In this case, the annulus of convergence is the exterior of the unit disk: $r = 1$, $R = +\infty$. The discrete filter has the form

$$y_n - by_{n-1} = cx_n$$

with

$$b = \frac{RC}{RC + a} \quad \text{and} \quad c = \frac{a}{RC + a}.$$

The transfer function is

$$H(z) = \frac{c}{1 - bz^{-1}} = \frac{cz}{z - b}, \quad |z| > b.$$

The series expansion is

$$H(z) = c \sum_{n=0}^{+\infty} b^n z^{-n},$$

and the impulse response is

$$h = c \sum_{n=0}^{+\infty} b^n \delta_{na}.$$

The filter is stable, since $|b| < 1$.

40.3 Exercises

Exercise 40.1 Invert the z-transform defined by $X(z) = z(z-r)^{-1}$ in the annulus $0 < |z| < r$ and compare the result with the example in §40.1.2.

Exercise 40.2 Invert the z-transform defined by

$$H(z) = \frac{z^2+1}{z^2-1},$$

knowing that the associated filter is realizable.

Exercise 40.3 Let $X(z)$ be the z-transform of the signal

$$x = (x_n) = \sum_{n=-\infty}^{\infty} x_n \delta_{na}.$$

(a) Compute the z-transforms of the signals

$$(x_{-n}), \quad (\alpha^n x_n)\ \alpha \neq 0, \quad (\bar{x}_n), \quad (nx_n).$$

(b) Use (a) to compute the z-transforms of the signals

$$x = (nu_n) \quad \text{and} \quad y = (n2^n u_n),$$

where $u_n = u(n)$.

Exercise 40.4 Give an example of a noncausal signal for which $R = +\infty$.

Chapter XII

Current Trends: Time–Frequency Analysis

Lesson 41

The Windowed Fourier Transform

41.1 Limitations of standard Fourier analysis

Current research is to a large extent motivated by industrial applications of mathematical analysis and signal processing. Seismic exploration, the analysis and synthesis of sound, medical imaging, and the digital telephone are a few of the applications that come to mind. In all cases, one wishes to extract from the signal the pertinent information as discrete numerical values. This set of digital information must be rich enough to characterize the signal, but it should be no larger than necessary for the task at hand. If, for example, it is a question of speech and the digital telephone, one wants enough numerical information at the receiver to reconstruct a recognizable voice, but economy dictates the need to minimize the amount of information that must be transmitted.

Fourier analysis is the oldest of the various techniques avaliable for signal analysis and synthesis. Since the invention of the fast Fourier transform (FFT), it has become an efficient tool, particularly for analyzing sufficiently smooth periodic signals (Lesson 9). In these cases, the Fourier coefficients c_n decrease rapidly as $|n| \to +\infty$, and relatively few numerical coefficients are needed to reconstruct the signal for most practical purposes. Unfortunately, as soon as the signal becomes irregular, like, for example, a transient, the number of coefficients necessary to reconstruct the signal (and hence the amount of data that must either be stored or transmitted) becomes large and often economically impractical.

Before the advent of the FFT, Fourier analysis was mainly a theoretical tool—indeed, one of the most important and pervasive. This quickly changed with the arrival of the FFT and efficient digital computing, and these twin techniques have had widespread applications in the last third of the twentieth century. Nevertheless, even with the FFT and modern computing, Fourier analysis does not provide a satisfactory analysis for all kinds of signals. Although the Fourier transform $\widehat{f}$ contains all of the information about f, much of this information in "hidden." For example, none of the temporal aspects of f are revealed by $\widehat{f}$. If f is a finite signal, the spectrum does not indicate the beginning and the end of the signal, and if there is a singularity, the time of occurrence is hidden throughout $\widehat{f}$.

Faced with these kinds of issues, one would like to have an analytic tool that provides information both in time and in frequency. The model that is often cited is musical notation: the horizontal position of a note (its "start time," its duration, and its frequency are all represented.

There is another problem that has surely not escaped the reader's notice: To compute the spectrum $\widehat{f}(\lambda)$ it is necessary to know $f(t)$ for all real values of t. This is impossible in the case of analysis in "real time" where the signal must be processed as it arrives. One cannot know the spectrum, even approximately, of a signal when one knows nothing of its future; the interesting information may be yet to arrive. We should not despair, however; the previous eleven chapters retain their value today both theoretically and numerically in spite of the cited problems. These technical constraints simply motivate us to refine existing tools and to develop new ones.

41.2 Opening windows

One of the first ideas was to truncate the signal and to analyze only what happens on a finite interval $[-A, A]$. One is forced to do this when making numerical computations. Mathematically, this amounts to multiplying the signal $f(t)$ by a characteristic function $\chi_{[-A,A]} = r_A$ (or a translate) and taking the Fourier transform of the product. The result is

$$\widehat{g}(\lambda) = \widehat{r_A \cdot f}(\lambda) = \left(\frac{\sin 2\pi A\lambda}{\pi\lambda}\right) * \widehat{f}(\lambda) = (s_A * \widehat{f})(\lambda).$$

Thus truncating the signal results in convolving its spectrum with the cardinal sine (Figure 41.1).

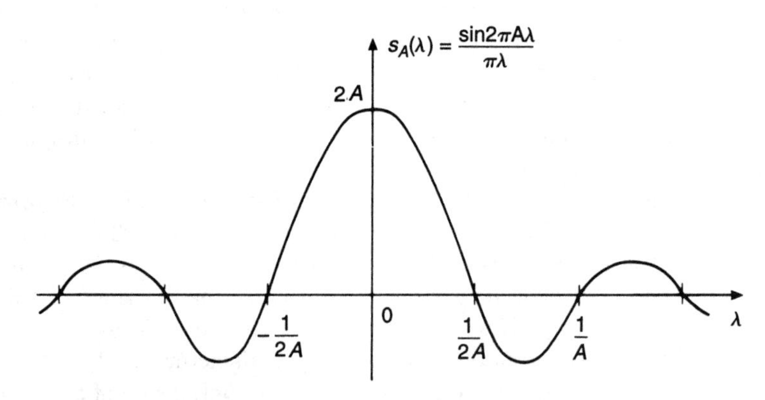

FIGURE 41.1. The cardinal sine.

The approximation of $\widehat{f}$ by $\widehat{g}$ becomes better as A increases, that is, as s_A better approximates the Dirac impulse. Unfortunately, the computations for this process quickly become very voluminous. The cardinal sine decays slowly and has important lobes near the origin. To avoid these problems, one replaces $\chi_{[-A,A]}$ with a more regular function. These functions are all called windows, and they are concentrated around the origin.

EXAMPLES:

(a) *Triangular window* (Figure 41.2)

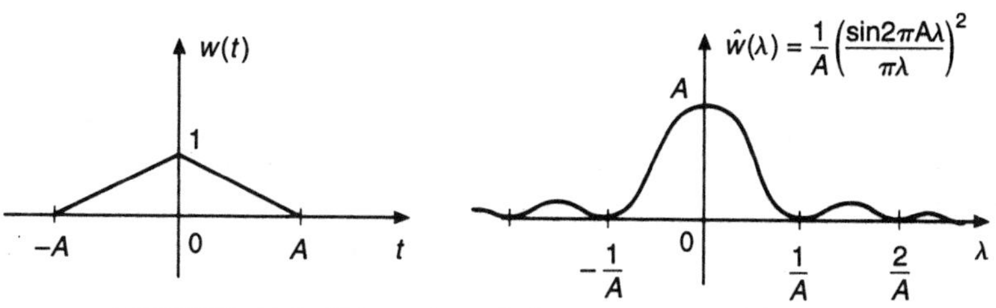

FIGURE 41.2. Triangular window in time and frequency.

(b) *Hamming and Hanning windows* (Figure 41.3)

These are of the form $\omega(t) = [\alpha + (1-\alpha)\cos(2\pi t/A)]r(t)$. For $\alpha = 0.54$ we have Hamming's window and for $\alpha = 0.50$ Hanning's window. These coefficients have been computed to minimize certain criteria (see [Kun84]).

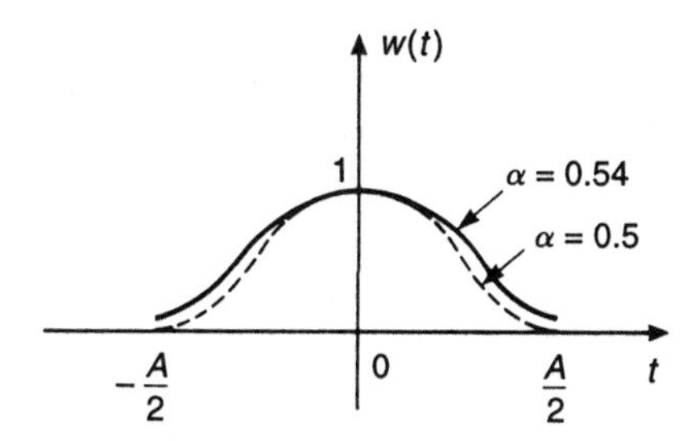

FIGURE 41.3. Hamming and Hanning windows.

(c) *Gaussian window* $\omega(t) = Ae^{-\alpha t^2}$ $(\alpha, A > 0)$ (Figure 41.4)

These windows are used in practice, and they significantly improve the computation of the spectrum.

One is led naturally to slide this window along the graph of the function and thereby analyze the whole function. One then obtains a family of coefficients depending on two real variables λ and b given by

$$W_f(\lambda, b) = \int_{-\infty}^{+\infty} f(t)\overline{w}(t-b)e^{-2i\pi\lambda t}\,dt. \tag{41.1}$$

$W_f(\lambda, b)$ replaces $\widehat{f}(\lambda)$. The mapping $f \mapsto W_f$ is called the *sliding window Fourier transform* or simply the *windowed Fourier transform*.

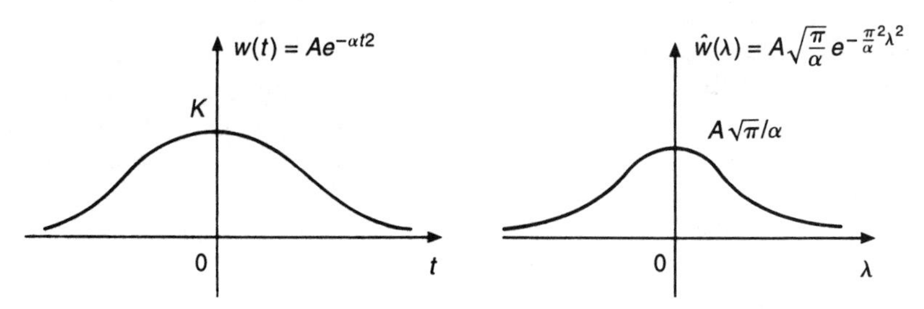

FIGURE 41.4. Gaussian window in time and frequency.

The parameter λ plays the role of a frequency, localized around the abscissa b of the temporal signal. $W_f(\lambda, b)$ thus provides an indication of how the signal behaves at time $t = b$ for the frequency λ. We use the function $\overline{\omega}$ rather than ω in (41.1) for reasons of convenience and because we wish to allow complex-valued windows. Thus, W_f becomes a scalar product in L^2:

$$\begin{aligned} W_f(\lambda, b) &= (f, w_{\lambda b}), \\ w_{\lambda b}(t) &= w(t-b)e^{2i\pi\lambda t}. \end{aligned} \tag{41.2}$$

41.3 Dennis Gabor's formulas

Intuitively, one might expect that knowing $W_f(\lambda, b)$ for all values of λ and b completely determines the signal f. One could even conjecture that the information contained in $W_f(\lambda, b)$ is redundant, since we have replaced a one-parameter family $\widehat{f}$ with a two-parameter family. We will see below that these speculations are well founded.

In his 1946 paper [Gab46], Dennis Gabor used a window that was essentially the Gaussian $w(t) = \pi^{-1/4}e^{-t^2/2}$. Such a function has the advantage of approximating a square window while avoiding the disadvantage of introducing abrupt discontinuities. One of Gabor's important contributions was to show that $W_f(\lambda, b)$ can be inverted to recover f.

41.3.1 Theorem *Suppose that $w \in L^1 \cap L^2$ is a window such that $|\widehat{w}|$ is even and $\|w\|_2 = 1$. Write*

$$w_{\lambda b}(t) = w(t-b)e^{2i\pi\lambda t}, \quad \lambda, b \in \mathbb{R}.$$

For all signals $f \in L^2$ we define the coefficients

$$W_f(\lambda, b) = \int_{-\infty}^{+\infty} f(t)\overline{w}_{\lambda b}(t)\, dt.$$

Under these conditions, we have the following two results:

(a) *Conservation of energy:*

$$\iint_{\mathbb{R}^2} |W_f(\lambda, b)|^2 \, d\lambda \, db = \int_{-\infty}^{+\infty} |f(t)|^2 \, dt. \tag{41.3}$$

(b) *Reconstruction formula:*

$$f(x) = \iint_{\mathbb{R}^2} W_f(\lambda, b) w_{\lambda b}(x) \, d\lambda \, db \tag{41.4}$$

in the sense that if

$$g_A(x) = \iint_{\substack{|\lambda| \leq A \\ b \in \mathbb{R}}} W_f(\lambda, b) w_{\lambda b}(x) \, d\lambda \, db,$$

then $g_A \to f$ *in* L^2 as $A \to +\infty$.

Proof. We first give another expression for $W_f(\lambda, b)$:

$$W_f(\lambda, b) = \int_{-\infty}^{+\infty} f(t) \overline{w}_{\lambda b}(t) \, dt = \int_{-\infty}^{+\infty} \widehat{f}(\xi) \overline{\widehat{w}}_{\lambda b}(\xi) \, d\xi.$$

Since

$$\widehat{w}_{\lambda b}(\xi) = e^{-2i\pi(\xi - \lambda)b} \widehat{w}(\xi - \lambda), \tag{41.5}$$

this becomes

$$W_f(\lambda, b) = e^{-2i\pi\lambda b} \int_{-\infty}^{+\infty} \widehat{f}(\xi) \overline{\widehat{w}}(\xi - \lambda) e^{2i\pi\xi b} \, d\xi,$$

so

$$W_f(\lambda, b) = e^{-2i\pi\lambda b} \overline{\mathscr{F}}_\xi [\widehat{f}(\xi) \overline{\widehat{w}}(\xi - \lambda)](b). \tag{41.6}$$

The function of ξ in brackets is in L^1, since it is the product of two functions in L^2. It is also in L^2 because, w being in L^1, $\widehat{w}$ is bounded. Thus we have

$$\begin{aligned}
\iint_{\mathbb{R}^2} |W_f(\lambda, b)|^2 \, d\lambda \, db &= \int_{-\infty}^{+\infty} \left(\int_{-\infty}^{+\infty} |\overline{\mathscr{F}}_\xi [\widehat{f}(\xi) \overline{\widehat{w}}(\xi - \lambda)](b)|^2 \, db \right) d\lambda \\
&= \int_{-\infty}^{+\infty} \left(\int_{-\infty}^{+\infty} |\widehat{f}(\xi) \overline{\widehat{w}}(\xi - \lambda)|^2 \, d\xi \right) d\lambda \quad \text{(Parseval)} \\
&= \int_{-\infty}^{+\infty} \left(|\widehat{f}(\xi)|^2 \int_{-\infty}^{+\infty} |\widehat{w}(\xi - \lambda)|^2 \, d, \lambda \right) d\xi \\
&= \|f\|_2^2 \|\widehat{w}\|_2^2 = \|f\|_2^2.
\end{aligned}$$

This establishes (a).

To prove (b), we first show that g_A is well-defined for all $A > 0$ by showing that $(\lambda, b) \mapsto W_f(\lambda, b) w_{\lambda b}(x)$ is integrable on the strip $[-A, A] \times \mathbb{R}$. Let

$$J_A(x) = \int_{-A}^{A} \left(\int_{-\infty}^{+\infty} |\overline{\mathscr{F}}_\xi[\widehat{f}(\xi)\overline{\widehat{w}}(\xi - \lambda)](b)|\, |w(x-b)|\, db \right) d\lambda.$$

By Schwarz's inequality and Parseval's relation, we have (Theorem 22.1.4)

$$\begin{aligned} J_A(x) &\leq \int_{-A}^{A} \|\overline{\mathscr{F}}_\xi[\widehat{f}(\xi)\overline{\widehat{w}}(\xi - \lambda)](b)\|_2 \|w\|_2 \, d\lambda \\ &= \int_{-A}^{A} \|\widehat{f}(\xi)\overline{\widehat{w}}(\xi - \lambda)\|_2 \, d\lambda. \end{aligned}$$

The function $h(\lambda)$ under the last integral sign satisfies

$$h^2(\lambda) = \int_{-\infty}^{+\infty} |\widehat{f}(\xi)|^2 |\widehat{w}(\xi - \lambda)|^2 \, d\xi = (|\widehat{f}|^2 * |\widehat{w}|^2)(\lambda).$$

Since $L^1 * L^1 \subset L^1$, it follows that $|h|^2 \in L^1$ and hence that $h \in L^2$. Finally,

$$J_A(x) \leq \int_{-A}^{A} h(\lambda)\, d\lambda \leq \sqrt{2A}\|h\|_2 < +\infty$$

for all $x \in \mathbb{R}$ and $A > 0$. Integrability allows us to choose the order of integration in the definition of g_A, so in view of (41.6), we have

$$g_A(x) = \int_{-A}^{A} g(\lambda)\, d\lambda$$

with

$$g(\lambda) = \int_{-\infty}^{+\infty} \overline{\mathscr{F}}_\xi[\widehat{f}(\xi)\overline{\widehat{w}}(\xi - \lambda)](b) w(x-b) e^{2i\pi\lambda(x-b)}\, db,$$

which by Proposition 22.1.5 is

$$g(\lambda) = \int_{-\infty}^{+\infty} \widehat{f}(\xi)\overline{\widehat{w}}(\xi - \lambda) \overline{\mathscr{F}}_b[w(x-b)e^{2i\pi\lambda(x-b)}](\xi)\, d\xi.$$

After computing the Fourier transform $\overline{\mathscr{F}}_b[w(x-b)e^{2i\pi\lambda(x-b)}]$, we see that

$$g(\lambda) = \int_{-\infty}^{+\infty} \widehat{f}(\xi)\overline{\widehat{w}}(\xi - \lambda)\widehat{w}(\xi - \lambda) e^{2i\pi\xi x}\, d\xi,$$

so

$$g_A(x) = \int_{-A}^{A} \left(\int_{-\infty}^{+\infty} \widehat{f}(\xi)|\widehat{w}(\xi - \lambda)|^2 e^{2i\pi\xi x}\, d\xi \right) d\lambda. \tag{41.7}$$

The next step is to verify that the function of (λ, ξ) under the double integral (41.7) is integrable on $[-A, A] \times \mathbb{R}$. Since $|\widehat{w}|$ is even,

$$\int_{-A}^{A} \left(\int_{-\infty}^{+\infty} |\widehat{f}(\xi)| |\widehat{w}(\xi - \lambda)|^2 \, d\xi \right) d\lambda = \int_{-A}^{A} (|\widehat{f}| * |\widehat{w}|^2)(\lambda) \, d\lambda.$$

Since $|\widehat{f}| \in L^2$ and $|\widehat{w}|^2 \in L^1$, it follows (Proposition 20.3.2) that $h = |\widehat{f}| * |\widehat{w}|^2 \in L^2(\mathbb{R})$ and hence that $h \in L^1[-A, A]$. Thus the integral is well-defined, and we can interchange the order of integration in (41.7):

$$g_A(x) = \int_{-\infty}^{+\infty} \widehat{f}(\xi) e^{2i\pi\xi x} \left(\int_{-A}^{A} |\widehat{w}(\xi - \lambda)|^2 \, d\lambda \right) d\xi.$$

Denote the second integral by $\varphi_A(\xi)$. Then $0 \le \varphi_A(\xi) \le 1$, since $\|\widehat{w}\|_2 = 1$. Since φ_A is bounded, $\widehat{f}\varphi_A$ is in L^2 and $g_A = \overline{\mathscr{F}}(\widehat{f} \cdot \varphi_A)$. The last step is to show that g_A tends to f in L^2 as $A \to +\infty$. For this we evaluate the norm of the difference:

$$\|f - \overline{\mathscr{F}}(\widehat{f} \cdot \varphi_A)\|_2^2 = \|\overline{\mathscr{F}}[(1 - \varphi_A)\widehat{f}\,]\|_2^2 = \|(1 - \varphi_A)\widehat{f}\,\|_2^2 = \varepsilon(A).$$

We estimate the integral

$$\varepsilon(A) = \int_{-\infty}^{+\infty} [1 - \varphi_A(\xi)]^2 |\widehat{f}(\xi)|^2 \, d\xi \tag{41.8}$$

in two parts. If $|\xi| \le A/2$, then

$$1 - \varphi_A(\xi) = \int_{|\lambda| \ge A} |\widehat{w}(\xi - \lambda)|^2 \, d\lambda = \int_{-\infty}^{\xi - A} |\widehat{w}(y)|^2 \, dy + \int_{\xi + A}^{+\infty} |\widehat{w}(y)|^2 \, dy,$$

so

$$0 \le 1 - \varphi_A(\xi) \le \int_{-\infty}^{-\frac{A}{2}} |\widehat{w}(y)|^2 \, dy + \int_{\frac{A}{2}}^{+\infty} |\widehat{w}(y)|^2 \, dy = \varepsilon_1(A),$$

which tends to 0 as $A \to +\infty$. As a consequence,

$$\int_{-\frac{A}{2}}^{\frac{A}{2}} [1 - \varphi_A(\xi)]^2 |\widehat{f}(\xi)|^2 \, d\xi \le \varepsilon_1^2(A) \|f\|_2^2.$$

If $|\xi| \ge A/2$, then

$$\int_{|\xi| \ge \frac{A}{2}} [1 - \varphi_A(\xi)]^2 |\widehat{f}(\xi)|^2 \, d\xi \le \int_{|\xi| \ge \frac{A}{2}} |\widehat{f}(\xi)|^2 \, d\xi,$$

which also tends to 0 as $A \to +\infty$. These two estimates show that $\varepsilon(A)$ (41.8) tends to 0 as A tends to infinity, and this proves (b). $\square$

This result shows that for the windowed Fourier transform in L^2 we have formulas analogous to those for the ordinary Fourier transform in L^2: conservation of energy (Parseval's formula) and an inversion formula. There is a nice harmony in these formulas; this will also appear in the theory of wavelets.

In practice, one generally uses a function w that is well localized around the origin $t = 0$, for example, a Gaussian. The function $w_{\lambda b}$ is then localized around the point $t = b$, while $\widehat{w}_{\lambda b}$, given by (41.5), is localized around the point $\xi = \lambda$. This means that

$$W_f(\lambda, b) = (f, w_{\lambda b}) = (\widehat{f}, \widehat{w}_{\lambda b})$$

contains information in both time and frequency around the point (b, λ).

For numerical computations, the coefficients $W_f(\lambda, b)$ are evaluated on a grid $(m\lambda_0, nb_0)$ with $m, n \in \mathbb{Z}$ and $\lambda_0, b_0 > 0$. One thus obtains a double sequence $W_{m,n}(f) = W_f(m\lambda_0, nb_0)$, which is a discretized version of the function of the two real variables λ and b.

41.4 Comparing the methods of Fourier and Gabor

The transforms of Fourier and Gabor, which we can write formally as

$$f(x) = \int_{-\infty}^{+\infty} \widehat{f}(\xi) e^{2i\pi x\xi}\, d\xi,$$

$$f(x) = \int_{\mathbb{R}^2} W_f(\lambda, b) w_{\lambda b}(x)\, d\lambda\, db,$$

can be interpreted as decomposing the signal f in terms of functions that play the role of basis functions, except that sums are replaced by integrals.

In the Fourier transform, these functions are sinusoids; in the Gabor transform, they are strongly attenuated sinusoids, or looked at the other way, modulated Gaussians (Figure 41.5). In the frequency space, we have the representations illustrated in Figure 41.6.

With Fourier's method, the "basis functions" are completely concentrated in frequency (Dirac impulses) and totally distributed in time (unattenuated sinusoids extending from $-\infty$ to $+\infty$). This is another way to explain that taking the Fourier transform gives the maximum amount of information about the distribution of the frequencies but completely loses information relative to time.

With Gabor's method, the figures show that time–frequency information remains coupled, although there is always a compromise: The uncertainty principle limits the simultaneous localization in time and frequency. In spite

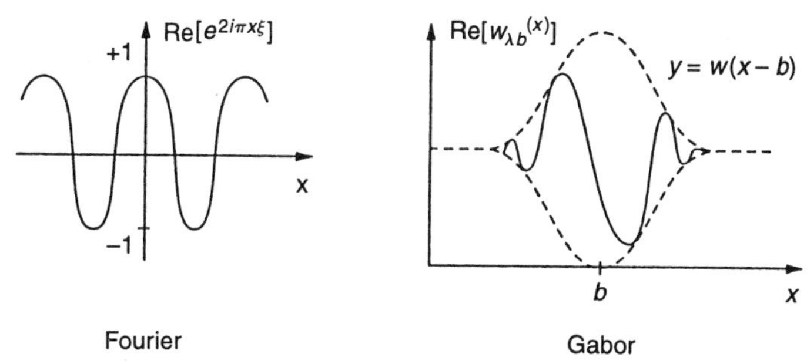

FIGURE 41.5. Basis functions for Fourier and Gabor decompositions.

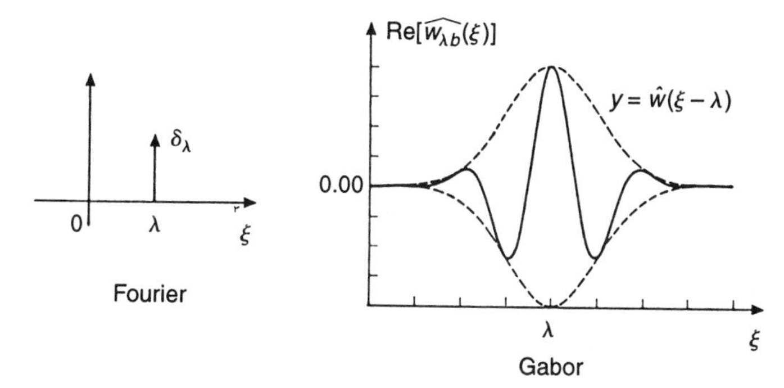

FIGURE 41.6. Basis functions for Fourier and Gabor in frequency space.

of this—which is a fact of life for any time–frequency analysis—Gabor's method has advantages over Fourier analysis for certain applications.

A signal f of finite duration provides one of the best illustrations of the difference between the two methods. The reconstruction of f using the inverse Fourier formula necessitates knowing the values of $\widehat{f}(\xi)$ with considerable precision over a very large range of values, for although $\widehat{f}(\xi)$ tends to zero, it can do so frustratingly slowly (consider the transform of $\chi_{[a,b]}$). The effects of all the sinusoids must come together to give zero outside the support of f.

The situation is quite different for Gabor analysis. It f vanishes on a long enough interval $(b_0 - \alpha, b_0 + \alpha)$ and if $w(t)$ is small for $|t| \geq 1$, then the coefficients $W_f(\lambda, b)$ will be negligible for b in a neighborhood of b_0, since

$$W_f(\lambda, b) \approx \int_{b-1}^{b+1} f(t)\overline{w}_{\lambda b}(t)\, dt = 0.$$

On the other hand, if f oscillates strongly at $t = b_0$, the value of $W_f(\lambda, b)$ will be large for b near b_0 when the values of λ "match" the frequency of f near b_0. This gives an idea about the "local frequency" of f.

In spite of its advantages for certain applications, the Gabor method has the major disadvantage that the size of the window is fixed. In terms of the uncertainty principle, this means that Δt is fixed (Section 22.3), and this limits the ability to localize events in time. Problems arise when one wishes to analyze signals that contain features on scales that range over several orders of magnitude. This is the case, for example, with speech. Consider the word "school." It begins with a short high-frequency attack followed by a longer relatively lower-frequency component. Fluid mechanics provides another important example. In fully developed turbulence, one observes events on scales that range from the macroscopic to the microscopic.

The geophysicist Jean Morlet encountered these kinds of problems in connection with seismic exploration for oil. Here it is necessary to analyze signals that result from a pulse being reflected (and delayed and compressed) from various layers in the earth. This led Morlet to introduce a new method where the window is not only translated but is also dilated and contracted. This was the beginning of the use of wavelets for numerical signal processing.

41.5 Exercises

Exercise 41.1 With the notation and hypotheses of Theorem 41.3.1, show that for f and $g \in L^2(\mathbb{R})$,

$$\iint_{\mathbb{R}^2} W_f(\lambda, b)\overline{W}_g(\lambda, b)\, d\lambda\, db = \int_{\mathbb{R}} f(t)\overline{g}(t)\, dt.$$

Exercise 41.2 Consider the signal $f(t) = e^{2i\pi\alpha t}$, $\alpha \in \mathbb{R}$, and the Gaussian window $w(t) = e^{-\pi t^2}$.

(a) Verify that

$$W_f(\lambda, b) = \int_{\mathbb{R}} f(t)\overline{w}(t - b)e^{-2i\pi\lambda t}\, dt$$

is well-defined (even though $f \notin L^2(\mathbb{R})$).

(b) Compute $W_f(\lambda, b)$ using the following result:

$$\text{For } a > 0 \text{ and } x \in \mathbb{R}, \quad \int_{\mathbb{R}} e^{-\pi a(t+ix)^2}\, dt = a^{-\frac{1}{2}}.$$

(c) Show that $|W_f(\lambda, b)|^2$ attains its maximum when $\lambda = \alpha$.

Exercise 41.3 Consider the Gaussian window $w(t) = Ae^{-\alpha t^2}$ with $A, \alpha > 0$ and the signal $f(t) = Be^{-\beta t^2}$ with $B, \beta > 0$. Use the result in Exercise 41.2(b) to compute

$$W_f(\lambda, b) = \int_{\mathbb{R}} f(t)\overline{w}(t - b)e^{-2i\pi\lambda t}\, dt.$$

Lesson 42

Wavelet Analysis

Gabor's method dates from the 1940s. With wavelets we enter a dynamic contemporary research environment; what is now known as the modern theory of wavelets emerged in the 1980s, notably with the article [GM84] by Alex Grossmann and Jean Morlet. We say "modern" wavelet theory because looking back over the mathematical landscape from a late twentieth century perspective we can identify many earlier ideas and techniques that are now logically included in this theory. Work by Haar in 1909; work in the late 1920s by Strömberg; results from the 1930s by Littlewood and Paley, Lusin, and Franklin; and later work in the 1960s, particularly the result of Calderón on operators with singular kernels—all these efforts and others are now interpreted in the language of wavelets.

What happened in the 1980s was qualitatively different; there occurred a conjunction of requirement and solution. Jean Morlet, a geophysicist, wished to analyze a particular class of signals associated with seismic exploration, and he had an idea about how this should be done. He sought the collaboration of Alex Grossmann, who, being a theoretical physicist, had command of certain mathematical tools, particularly those associated with coherent states and group representations from quantum theory. The immediate result was their celebrated 1984 paper; it was also the beginning of a productive collaboration between mathematics and other sectors of science and technology. We will say more about contemporary research at the end of the lesson, once some basic results have been established.

42.1 The basic idea: the accordion

Starting with a function ψ, called the analyzing wavelet or "mother" wavelet, we construct the family of functions

$$\psi_{ab}(t) = \frac{1}{\sqrt{a}}\psi\left(\frac{t-b}{a}\right), \quad b \in \mathbb{R},\ a > 0.$$

The wavelet coefficients of a signal f are the numbers

$$C_f(a,b) = (f, \psi_{a,b}) = \int_{-\infty}^{+\infty} f(t)\overline{\psi}_{a,b}(t)\,dt.$$

The properties of ψ are quite different from those of a window, which has more or less the aspect of a characteristic function, while ψ, on the other hand, oscillates and its integral is zero. We also want ψ and $\widehat{\psi}$ to be well localized, which means that they both converge to zero at infinity fairly rapidly. In this way one obtains a function that looks like a wave: It oscillates and quickly decays. This is the source of its name. Morlet used the function

$$\psi(t) = e^{-\frac{t^2}{2}} \cos 5t,$$

which is now known as Morlet's wavelet; derivatives of the Gaussian are widely used in practice. Figures 42.1–42.4 illustrate differences in the behavior of the Gabor functions $w_{\lambda b}(t)$, which have a ridged envelope, and wavelets, which are dilated and contracted. With wavelets one sees the action of an accordion. (The factor $a^{-1/2}$ has not been used in the figures.) Unlike Gabor functions, wavelets do not have a rigid envelope.

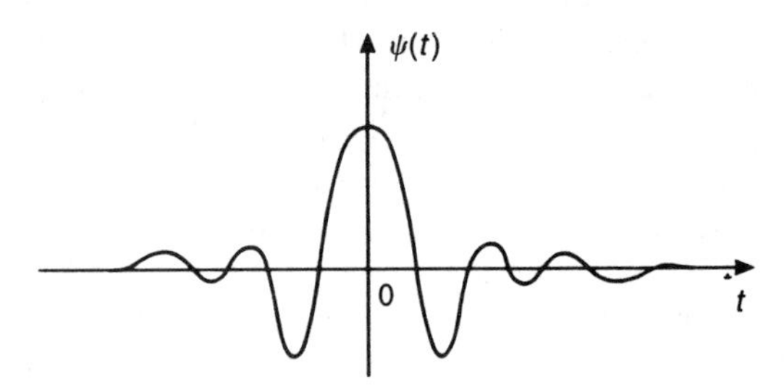

FIGURE 42.1. A wavelet oscillates and decays.

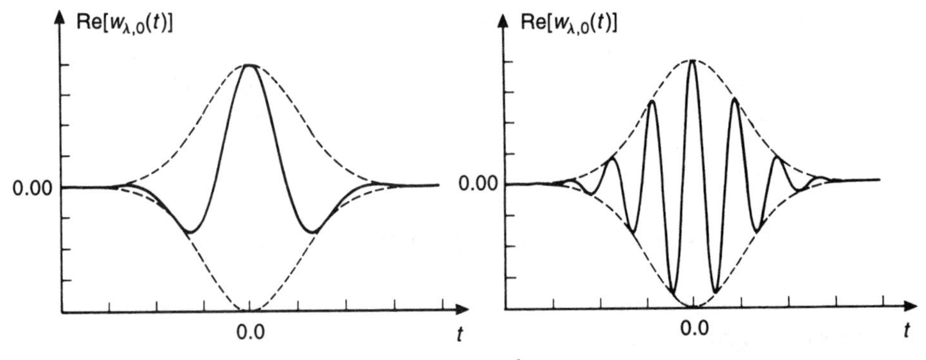

FIGURE 42.2. Gabor functions $w_{\lambda b}(t) = e^{-\frac{1}{2}(t-b)^2} e^{2i\pi\lambda t}$: The envelope is rigid, and the number of oscillations varies with frequency.

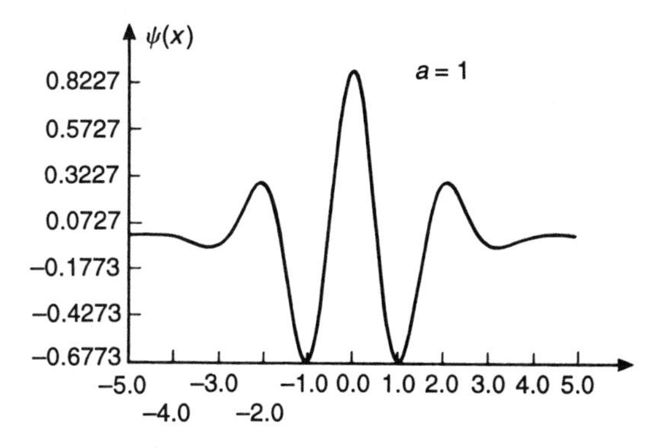

FIGURE 42.3. A mother wavelet (8th derivative of a Gaussian).

42.2 The wavelet transform

42.2.1 Theorem *Suppose that the function $\psi \in L^1(\mathbb{R}) \cap L^2(\mathbb{R})$ satisfies the following conditions:*

(i) $\displaystyle\int_{-\infty}^{+\infty} \frac{|\widehat{\psi}(\lambda)|^2}{|\lambda|}\, d\lambda = K < +\infty.$

(ii) $\|\psi\|_2 = 1.$

Construct the family of wavelets

$$\psi_{ab}(t) = \frac{1}{\sqrt{|a|}}\psi\left(\frac{t-b}{a}\right), \quad a, b \in \mathbb{R},\ a \neq 0,$$

and for any signal $f \in L^2(\mathbb{R})$ consider the wavelet coefficients

$$C_f(a,b) = \int_{-\infty}^{+\infty} f(t)\overline{\psi}_{ab}(t)\, dt.$$

Under these conditions we have the following results:

(a) *Conservation of energy:*

$$\frac{1}{K}\iint_{\mathbb{R}^2} |C_f(a,b)|^2 \frac{da\, db}{a^2} = \int_{-\infty}^{+\infty} |f(t)|^2\, dt.$$

(b) *Reconstruction formula:*

$$f(x) = \frac{1}{K}\iint_{\mathbb{R}^2} C_f(a,b)\psi_{ab}(x)\frac{da\, db}{a^2}$$

in the sense that if

$$f_\varepsilon(x) = \frac{1}{K}\iint_{\substack{|a|\geq\varepsilon \\ b\in\mathbb{R}}} C_f(a,b)\psi_{ab}(x)\frac{da\, db}{a^2},$$

then $f_\varepsilon \to f$ in $L^2(\mathbb{R})$ as $\varepsilon \to 0^+$.

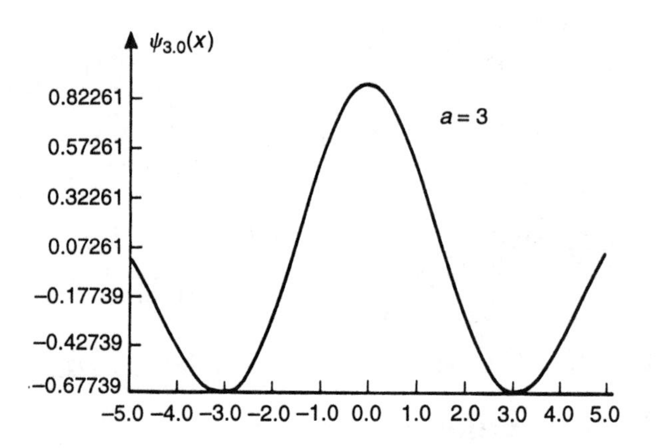

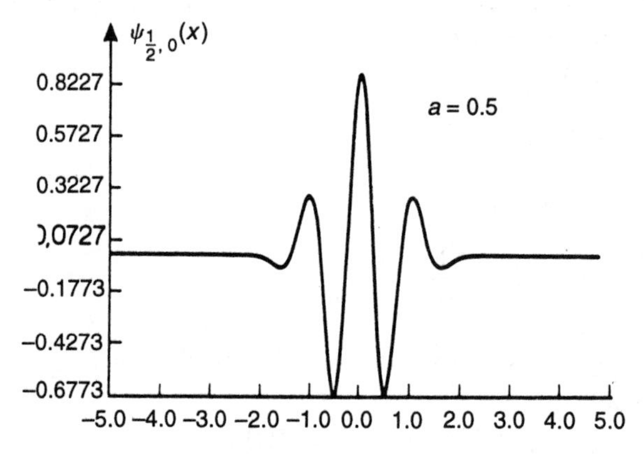

FIGURE 42.4. Wavelets at low and high frequency: They have the same form and the same number of oscillations; they are dilated for large a and contracted for small a.

Proof. First two observations: The ψ_{ab} are normalized so that $\|\psi_{ab}\|_2 = 1$, and the proof is similar to that of Theorem 41.3.1. Thus, as before, we find another expression for $C_f(a,b)$:

$$C_f(a,b) = \int_{-\infty}^{+\infty} f(t)\overline{\psi}_{ab}(t)\,dt = \int_{-\infty}^{+\infty} \widehat{f}(\lambda)\overline{\widehat{\psi}}_{ab}(\lambda)\,d\lambda,$$

and since

$$\widehat{\psi}_{ab}(\lambda) = \sqrt{|a|}e^{-2i\pi\lambda b}\widehat{\psi}(a\lambda), \tag{42.1}$$

we have

$$C_f(a,b) = \sqrt{|a|}\,\overline{\mathscr{F}}_{\lambda}\big[\widehat{f}(\lambda)\overline{\widehat{\psi}}(a\lambda)\big](b). \tag{42.2}$$

The function of λ in brackets is in $L^1(\mathbb{R})$ because it is a product of two functions in $L^2(\mathbb{R})$; it is also in $L^2(\mathbb{R})$, since $\psi \in L^1(\mathbb{R})$ implies $\widehat{\psi}$ is bounded.

To prove (a), we compute the double integral using (42.2). Hence,

$$\begin{aligned} I &= \int_{-\infty}^{+\infty} \left(\int_{-\infty}^{+\infty} |C_f(a,b)|^2 \, db \right) \frac{da}{a^2} \\ &= \int_{-\infty}^{+\infty} \left(\int_{-\infty}^{+\infty} |\overline{\mathscr{F}}_\lambda[\widehat{f}(\lambda)\overline{\widehat{\psi}}(a\lambda)](b)|^2 \, db \right) \frac{da}{|a|}. \end{aligned}$$

Using this and Parseval's relation, we obtain

$$\begin{aligned} I &= \int_{-\infty}^{+\infty} \left(\int_{-\infty}^{+\infty} |\widehat{f}(\lambda)|^2 |\widehat{\psi}(a\lambda)|^2 \, d\lambda \right) \frac{da}{|a|} \\ &= \int_{-\infty}^{+\infty} |\widehat{f}(\lambda)|^2 \left(\int_{-\infty}^{+\infty} \frac{|\widehat{\psi}(a\lambda)|^2}{|a|} \, da \right) d\lambda. \end{aligned}$$

By the change of variable $\xi = a\lambda$, we see that the last integral is constant and equal to K, which proves the result.

To prove (b), we first compute

$$J(a) = \int_{-\infty}^{+\infty} C_f(a,b)\psi_{ab}(x) \, db = \sqrt{|a|} \int_{-\infty}^{+\infty} \overline{\mathscr{F}}_\lambda[\widehat{f}(\lambda)\overline{\widehat{\psi}}(a\lambda)](b)\psi_{ab}(x) \, db.$$

Using Parseval's relation again, we have

$$J(a) = \sqrt{|a|} \int_{-\infty}^{+\infty} \widehat{f}(\lambda)\overline{\widehat{\psi}}(a\lambda)\overline{\mathscr{F}}_b[\psi_{ab}(x)](\lambda) \, d\lambda,$$

and since

$$\overline{\mathscr{F}}_b[\psi_{ab}(x)](\lambda) = \sqrt{|a|}\,\widehat{\psi}\,(a\lambda)e^{2i\pi\lambda x},$$

it follows that

$$J(a) = |a| \int_{-\infty}^{+\infty} \widehat{f}(\lambda)|\widehat{\psi}(a\lambda)|^2 e^{2i\pi\lambda x} \, d\lambda. \tag{42.3}$$

Define

$$g_\varepsilon(x) = \int_{|a|\geq\varepsilon} J(a)\frac{da}{a^2} = \int_{|a|\geq\varepsilon} \left(\int_{-\infty}^{+\infty} \widehat{f}(\lambda)|\widehat{\psi}(a\lambda)|^2 e^{2i\pi\lambda x} \, d\lambda \right) \frac{da}{|a|}. \tag{42.4}$$

The next step is to show that the function of (a,λ) under the integral signs is integrable on $(|a| \geq \varepsilon) \times \mathbb{R}$.

By the change of variable $\xi = a\lambda$, we see that

$$\begin{aligned} A &= \int_{-\infty}^{+\infty} |\widehat{f}(\lambda)| \left(\int_{|a|\geq\varepsilon} \frac{|\widehat{\psi}(a\lambda)|^2}{|a|} \, da \right) d\lambda \\ &= \int_{-\infty}^{+\infty} |\widehat{f}(\lambda)| \left(\int_{|\xi|\geq\varepsilon|\lambda|} \frac{|\widehat{\psi}(\xi)|^2}{|\xi|} \, d\xi \right) d\lambda. \end{aligned}$$

A is estimated in two parts. For $|\lambda| \le 1$,

$$A_1 = \int_{-1}^{1} |\widehat{f}(\lambda)| \left(\int_{|\xi| \ge \varepsilon|\lambda|} \frac{|\widehat{\psi}(\xi)|^2}{|\xi|} \, d\xi \right) d\lambda$$

$$\le K \int_{-1}^{1} |\widehat{f}(\lambda)| \, d\lambda \le K\sqrt{2} \, \|f\|_2.$$

For $|\lambda| \ge 1$,

$$A_2 = \int_{|\lambda| \ge 1} |\widehat{f}(\lambda)| \left(\int_{|\xi| \ge \varepsilon|\lambda|} \frac{|\widehat{\psi}(\xi)|^2}{|\xi|} \, d\xi \right) d\lambda$$

$$\le \int_{|\lambda| \ge 1} |\widehat{f}(\lambda)| \left(\frac{1}{\varepsilon|\lambda|} \int_{|\xi| \ge \varepsilon|\lambda|} |\widehat{\psi}(\xi)|^2 \, d\xi \right) d\lambda,$$

so

$$A_2 \le \frac{1}{\varepsilon} \|\psi\|_2^2 \int_{|\lambda| \ge 1} \frac{|\widehat{f}(\lambda)|}{|\lambda|} \, d\lambda \le \frac{1}{\varepsilon} \|\psi\|_2^2 \, \|f\|_2 \left(\int_{|\lambda| \ge 1} \frac{d\lambda}{\lambda^2} \right)^{\frac{1}{2}} < +\infty.$$

This means that we can interchange the order of integration in (42.4); thus

$$g_\varepsilon(x) = \int_{-\infty}^{+\infty} \widehat{f}(\lambda) e^{2i\pi\lambda x} \left(\int_{|a| \ge \varepsilon} \frac{|\widehat{\psi}(a\lambda)|^2}{|a|} \, da \right) d\lambda = \overline{\mathscr{F}} [\widehat{f} \cdot \theta_\varepsilon](x)$$

with

$$\theta_\varepsilon(\lambda) = \int_{|a| \ge \varepsilon} \frac{|\widehat{\psi}(a\lambda)|^2}{|a|} \, da.$$

To show that $g_\varepsilon \to Kf$ in L^2, we evaluate the norm of the difference:

$$\|Kf - g_\varepsilon\|_2^2 = \|\overline{\mathscr{F}} (K\widehat{f} - \widehat{f} \cdot \theta_\varepsilon)\|_2^2 = \|\widehat{f}(K - \theta_\varepsilon)\|_2^2,$$

or

$$\|Kf - g_\varepsilon\|_2^2 = \int_{-\infty}^{+\infty} [K - \theta_\varepsilon(\lambda)]^2 \, |\widehat{f}(\lambda)|^2 \, d\lambda. \tag{42.5}$$

Again we examine two cases depending on the relation of λ to $\varepsilon^{-1/2}$. If $|\lambda| \le \varepsilon^{-1/2}$, then

$$\theta_\varepsilon(\lambda) = \int_{|\xi| \ge \varepsilon|\lambda|} \frac{|\widehat{\psi}(\xi)|^2}{|\xi|} \, d\xi \ge \int_{|\xi| \ge \sqrt{\varepsilon}} \frac{|\widehat{\psi}(\xi)|^2}{|\xi|} \, d\xi = K(\varepsilon).$$

Thus $0 \le K - \theta_\varepsilon(\lambda) \le K - K(\varepsilon)$, and $K(\varepsilon) \to K$ as $\varepsilon \to 0^+$ by (i). If $|\lambda| \ge \varepsilon^{-1/2}$, it is sufficient to note that $0 \le \theta_\varepsilon(\lambda) \le K$. Then from (42.5) we have

$$\|Kf - g_\varepsilon\|_2^2 \le [K - K(\varepsilon)]^2 \|f\|_2^2 + K^2 \int_{|\lambda| \ge \varepsilon^{-\frac{1}{2}}} |\widehat{f}(\lambda)|^2 \, d\lambda,$$

and these two terms tend to zero as $\varepsilon \to 0$. □

42.2.2 Remarks

(a) Hypothesis (i) implies that $\widehat{\psi}(0) = \int_{\mathbb{R}} \psi(t)\, dt = 0$, since $\widehat{\psi}$ is continuous. In all practical cases this condition is also sufficient. For example, if ψ and $x\psi$ are integrable, then $\widehat{\psi} \in C^1(\mathbb{R})$ and $\xi \to |\xi|^{-1}|\widehat{\psi}(\xi)|^2$ is continuous at $\xi = 0$. There is no problem with the integral at infinity, since $\psi \in L^2(\mathbb{R})$.

(b) For signals f belonging to $L^1(\mathbb{R}) \cap L^2(\mathbb{R})$ such that $\widehat{f}$ is also in $L^1(\mathbb{R})$, the proof of the theorem is simplified because all of the integrals exist in the usual sense when $\varepsilon = 0$. From (42.3) we deduce that

$$\begin{aligned}
\int_{-\infty}^{+\infty} J(a) \frac{da}{a^2} &= \iint_{\mathbb{R}^2} \widehat{f}(\lambda) |\widehat{\psi}(a\lambda)|^2 e^{2i\pi\lambda x}\, d\lambda\, \frac{da}{|a|} \\
&= \int_{-\infty}^{+\infty} \widehat{f}(\lambda) e^{2i\pi\lambda x} \left(\int_{-\infty}^{+\infty} \frac{|\widehat{\psi}(a\lambda)|^2}{|a|}\, da \right) d\lambda = K f(x)
\end{aligned}$$

by the Fourier inversion formula (Theorem 18.1.1). The reconstruction formula then holds for almost all $x \in \mathbb{R}$, or for all x if f is the continuous representative of its class.

42.2.3 Examples

(a) The wavelet first used by Morlet (Figures 42.5 and 42.6),

$$\psi(t) = e^{-\frac{t^2}{2}} \cos 5t, \tag{42.6}$$

is not normalized, but this is not a problem. On the other hand, the hypothesis (i) is not satisfied, since

$$\widehat{\psi}(0) = \sqrt{2\pi}\, e^{-\frac{25}{2}} > 0.$$

Thus $K = +\infty$! However, the value of $\widehat{\psi}(0)$ is on the order of 10^{-5}. For numerical computations this is essentially zero, and in practice things work well. Nevertheless, the theorem does not apply to Morlet's wavelet.

(b) The simplest example of a wavelet is the piecewise constant function ψ defined by

$$\psi(x) = \begin{cases} 1 & \text{if } \ 0 < x < \frac{1}{2}, \\ -1 & \text{if } \ \frac{1}{2} < x < 1, \\ 0 & \text{elsewhere.} \end{cases}$$

This is the Haar wavelet (Figure 42.7), and

$$\widehat{\psi}(\xi) = i e^{-i\pi\xi} \frac{1 - \cos \pi\xi}{\pi\xi}.$$

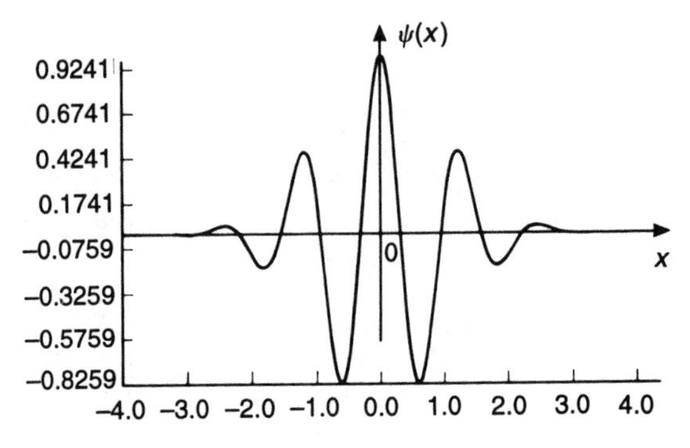

FIGURE 42.5. The Morlet wavelet.

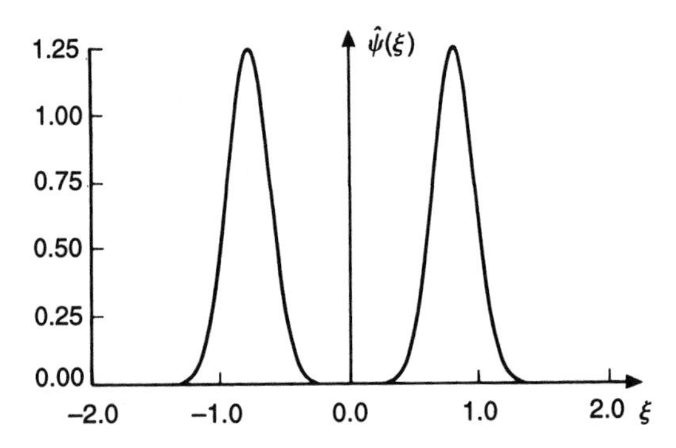

FIGURE 42.6. Spectrum of Morlet's wavelet.

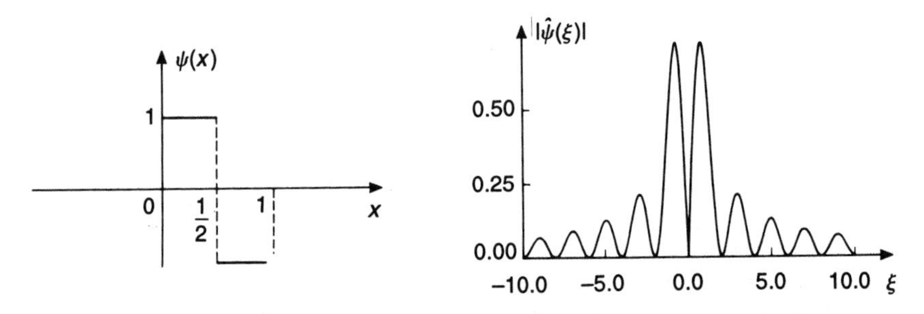

FIGURE 42.7. The Haar wavelet and the modulus of its spectrum.

The convergence of $\widehat{\psi}(\lambda)$ to 0 at infinity is very slow due to the irregularity of ψ, and this is a considerable problem for applications.

(c) Almost any function ψ that oscillates and has a zero integral and is such that both ψ and $\widehat{\psi}$ are well localized can be used as a mother wavelet. Examples include the derivatives of the Gaussian. The second derivative

$$\psi(x) = \frac{2}{\sqrt{3}\pi^{\frac{1}{4}}}(1 - x^2)e^{-\frac{1}{2}x^2} \tag{42.7}$$

is called the Mexican hat (Figure 42.8). Its spectrum is

$$\widehat{\psi}(\xi) = K\xi^2 e^{-2\pi^2\xi^2}.$$

Both ψ and $\widehat{\psi}$ belong to $\mathscr{S}$ and are well localized. Figure 42.9 illustrates the 8th derivative of the Gaussian and its spectrum.

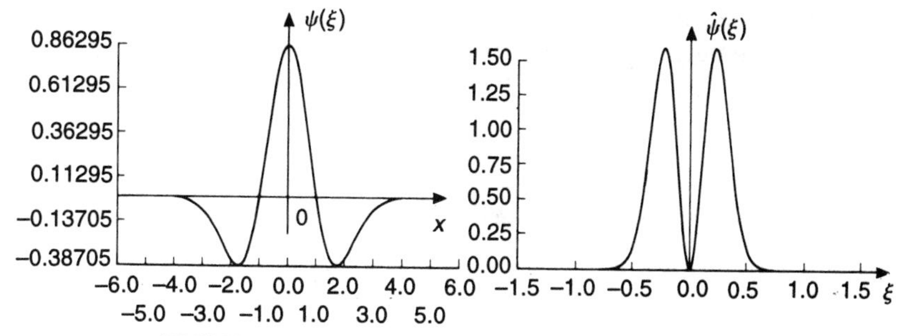

FIGURE 42.8. The Mexican hat and its spectrum.

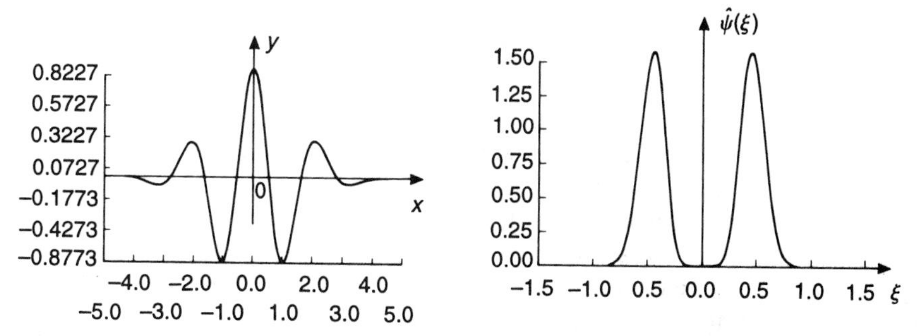

FIGURE 42.9. The 8th derivative of the Gaussian and its spectrum.

42.2.4 The wavelet transform as an analytic tool

Like the Fourier transform, the wavelet transform is both a theoretical and practical tool, but unlike the Fourier transform, there is the opportunity to choose different analyzing wavelets depending on the job at hand. There are, however, common properties shared by large classes of analyzing wavelets.

The wavelet coefficients of the signal $f = e^{2i\pi\lambda_0 t}$ are

$$C_f(a,b) = \sqrt{|a|}\,\overline{\widehat{\psi}}(a\lambda_0)e^{2i\pi\lambda_0 b}.$$

In this case, $|C_f(a,b)|$ depends only on a, and when $\widehat{\psi}$ is real, the argument of $C_f(a,b)$ is proportional to b modulo 2π. Here we see that the wavelet coefficients tell us something about the behavior of the function f, and this provides a simple illustration of how the wavelet transform can be used as an analytic tool. It will have not gone unnoticed that $f = e^{2i\pi\lambda_0 t}$ is not in $L^2(\mathbb{R})$. In general, the extension of Theorem 42.2.1 to functions not in $L^2(\mathbb{R})$, or to distributions, is a difficult problem, but this does not mean that the wavelet transform has no applications outside the context of $L^2(\mathbb{R})$. On the contrary, the wavelet transform has proved to be a powerful tool for investigating the local behavior of functions that are, for example, assumed to be only in $L^\infty(\mathbb{R})$. Here are a few other important properties.

Consider the function $f(t) = 1$. Its wavelet coefficients are all zero,

$$f(t) = 1 \quad \Longrightarrow \quad C_f(a,b) = 0,$$

which means that the wavelet transform "ignores" constants. For $f(t) = t$,

$$C_f(a,b) = a|a|^{1/2} \int_{-\infty}^{+\infty} x\overline{\psi}(x)\,dx$$

if $x\psi$ is integrable, which is most often the case. We then have

$$f(t) = t \quad \Longrightarrow \quad C_f(a,b) = -\frac{a|a|^{1/2}}{2i\pi}\,\widehat{\overline{\psi}}'(0).$$

If ψ is a derivative of order $m \geq 2$ of a Gaussian, then the coefficients of f are again all zero. The general result is this: The wavelet transform $C_f(a,b)$ will vanish for all polynomials f of degree $\leq p$ when $m \geq p+1$. One implication is that by "ignoring" the "smooth" part of a function, the wavelet transform "sees" only the "rough" part. In particular, by choosing an analyzing wavelet with sufficiently many vanishing moments, the wavelet transform will ignore the polynomial "trends" in a signal. Finally, we note that the wavelet transform is invariant with respect to translations of the signal,

$$g(t) = f(t - t_0) \quad \Longrightarrow \quad C_g(a,b) = C_f(a, b - t_0),$$

and that for $k \neq 0$,

$$g(t) = f(kt) \quad \Longrightarrow \quad C_g(a,b) = \frac{1}{|k|} C_f(ka, kb).$$

This last property plays an important role when the wavelet transform is used to analyze the singularities of a function.

42.2.5 Numerical computation

Throughout the text we have computed explicitly the Fourier transform of functions and distributions. In contrast, the wavelet transform is hardly ever computed explicitly, even for simple functions f and ψ. Naturally, for numerical computations it is necessary to restrict the parameters a and b to a discrete (indeed finite) subset of $\mathbb{R}$. For example, with

$$a_m = 2^{-m} \quad \text{and} \quad b_n = n2^{-m}, \quad m, n \in \mathbb{Z},$$

we have

$$\psi_{a_m,b_n}(x) = 2^{m/2}\psi(2^m x - n);$$

or more generally, with $\alpha > 1$ and $\beta > 0$,

$$\psi_{a_m,b_n}(x) = \alpha^{m/2}\psi(\alpha^m x - n\beta).$$

The information in this set becomes more redundant—and the computations become more voluminous—the closer α is to 1 and β is to 0. The choice $\alpha = 2$ corresponds to different octaves in music.

Ingrid Daubechies studied under what conditions the mapping

$$C : f \mapsto C_f = (C_f(a_m, b_n))_{(m,n)\in\mathbb{Z}^2}$$

from $L^2(\mathbb{R})$ into $l^2(\mathbb{Z}^2)$ is 1-to-1, which means that the wavelet coefficients characterize the signal. She also asked when the inverse of this mapping is continuous on its domain, which is important for numerical stability. For any reasonable analyzing wavelet ψ (good decay in both time and frequency and $\int \psi(x)\,dx = 0$), these requirements are equivalent to the existence of two positive constants A and B such that

$$A\|f\|_2^2 \le \sum_{m,n\in\mathbb{Z}} |(\psi_{a_m,b_n}, f)|^2 \le B\|f\|_2^2.$$

Daubechies showed that in this case the reconstruction formula can be written

$$f = \frac{2}{A+B} \sum_{m,n\in\mathbb{Z}} C_f(a_m, b_n)\psi_{a_m,b_n} + Rf.$$

If the remainder, or error term, Rf is small enough, it can be neglected. If not, then Daubechies has a reconstruction algorithm that converges exponentially (see [Dau92] for a complete discussion).

42.3 Orthogonal wavelets

Since the information contained in the coefficients $C_f(a, b)$ (which are determined in terms of the "basis" functions ψ_{ab}) is redundant, a natural challenge for the early researchers was to find a family of orthogonal wavelets $\{\psi_{jk}\}$, $j, k \in \mathbb{Z}$, on which every signal $f \in L^2(\mathbb{R})$ could be decomposed in a double series

$$f(x) = \sum_{j,k\in\mathbb{Z}} (f, \psi_{jk})\psi_{jk}(x)$$

with

$$\psi_{jk}(x) = 2^{j/2}\psi(2^j x - k).$$

One would thus have an orthogonal basis, in the usual sense, for the Hilbert space $L^2(\mathbb{R})$, where the coefficients $c_{jk} = C_f(j,k)$ are independent of one another. What is special about this basis is that the functions are all determined from one wavelet ψ by dilations and translations.

42.3.1 The Haar system

Such a family of functions ψ_{jk} has been known since the beginning of the century. It is the *Haar system* introduced by Alfred Haar in 1910 [Haa10]. The mother wavelet for this system is defined in Section 42.2.3(b) and shown in Figure 42.7; a graph of ψ_{jk} is shown in Figure 42.10.

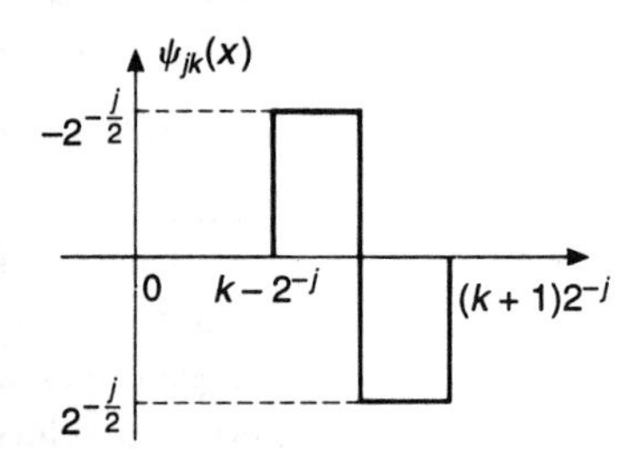

FIGURE 42.10. The orthogonal Haar system.

The Haar system is an orthonormal basis for $L^2(\mathbb{R})$. We have seen, however, that the absence of regularity of ψ causes $\widehat{\psi}$ to be poorly localized. More to the point, we will see below that the Haar coefficients c_{jk} converge to 0 very slowly as $j \to +\infty$ even for C^∞ signals.

42.3.2 The problem of moments for wavelets

Assume that $f \in C^\infty \cap L^2(\mathbb{R})$ and to simplify the notation, write $n = 2^j$. We wish to study the rate at which

$$u_n = \int_{-\infty}^{+\infty} f(x)\overline{\psi}(nx)\,dx = \frac{1}{\sqrt{n}} c_{j,0}$$

converges to 0 as $n \to +\infty$. Taylor's formula with integral remainder applied to f at $x = 0$ shows that at order q,

$$u_n = \sum_{l=0}^{q} f^{(l)}(0) \int_{-\infty}^{+\infty} \frac{x^l}{l!}\overline{\psi}(nx)\,dx + \int_{-\infty}^{+\infty} R(x)\overline{\psi}(nx)\,dx$$

with

$$R(x) = \int_0^x \frac{(x-t)^q}{q!} f^{(q+1)}(t)\,dt.$$

Denoting the moments of $\overline{\psi}$ by M_l and the remainder by r_n, we have

$$M_l = \int_{-\infty}^{+\infty} x^l\, \overline{\psi}(x)\, dx, \quad l \in \mathbb{N},$$

$$u_n = \sum_{l=0}^{q} \frac{f^{(l)}(0) M_l}{l!\, n^{l+1}} + r_n.$$

An easy computation shows that $|r_n| \leq Cn^{-(q+2)}$ for some constant C. Thus

$$u_n = \frac{f'(0)}{n^2}\frac{M_1}{1!} + \frac{f''(0)}{n^3}\frac{M_2}{2!} + \cdots + \frac{f^{(q)}(0)}{n^{q+1}}\frac{M_q}{q!} + O\Big(\frac{1}{n^{q+2}}\Big),$$

and we see that the rate of convergence u_n to 0 is controlled by the first nonzero moment of $\overline{\psi}$. For the Haar wavelet, $M_1 \neq 0$, and this is the source of the numerical problems related to the lack of concentration of the coefficients. These considerations lead to the definition of a wavelet with a certain amount of regularity and localization [Mey90].

42.3.3 Definition Suppose $r \in \mathbb{N}$. A wavelet of order r is any function $\psi : \mathbb{R} \to \mathbb{C}$ such that ψ and its derivatives up to order r belong to $L^\infty(\mathbb{R})$ and that satisfies the following two conditions:

(a) ψ and its derivatives up to order r decrease rapidly. (42.8)

(b) $\displaystyle\int_{-\infty}^{+\infty} x^q \psi(x)\, dx = 0 \quad \text{for} \quad 0 \leq q \leq r.$ (42.9)

42.3.4 Definition We say that the family $\{\psi_{jk}\}_{j,k\in\mathbb{Z}}$ is an orthonormal wavelet basis for $L^2(\mathbb{R})$ if the ψ_{jk} are of the form

$$\psi_{jk}(x) = 2^{j/2}\psi(2^j x - k) \tag{42.10}$$

for $\psi \in L^2(\mathbb{R})$ with $\|\psi\|_2 = 1$, and ψ is a wavelet of some order $r \geq 0$.

Later in the lesson we will see how to construct orthonormal bases of the form (42.10) where all we know about ψ is that it is in $L^2(\mathbb{R})$. This is not particularly interesting for applications, as we have seen in the case of the Haar wavelet (which is of order 0). For efficient numerical computations it is necessary to use higher-order wavelets, which means that the wavelet and its Fourier transform have reasonably good localization and regularity.

42.3.5 Yves Meyer's C^∞ wavelet

We know from Proposition 17.2.1 that saying that the first few moments of ψ vanish is the same as saying that

$$\widehat{\psi}^{(q)}(0) = 0, \quad q = 0, 1, 2, \ldots .$$

For a wavelet of order r, condition (b) of Definition 42.3.3 on the moments is equivalent to $\xi = 0$ being a zero of order $r+1$ of $\widehat{\psi}$. Thus if one wishes to find a wavelet of infinite order, one could start with a function $\widehat{\psi}$ that vanishes in a neighborhood of the origin. But this is easier said than done.

In 1985 Yves Meyer was able to produce such a wavelet, which is in $\mathscr{S}$, by first constructing $\widehat{\psi}$ belonging to $\mathscr{D}$. This construction is rather subtle; the details can be fond in [Dau92, p. 116]. What are we to think when Professor Meyer confesses modestly to have made this discovery "by accident"? In fact, Meyer was quite dissatisfied that the construction of ψ did not fit into a general framework; this was created a little later with the advent of multiresolution analysis. Here are the steps in the construction of this C^∞ wavelet.

One starts with a real, even function $\omega \in \mathscr{D}$ having the shape shown in Figure 42.11.

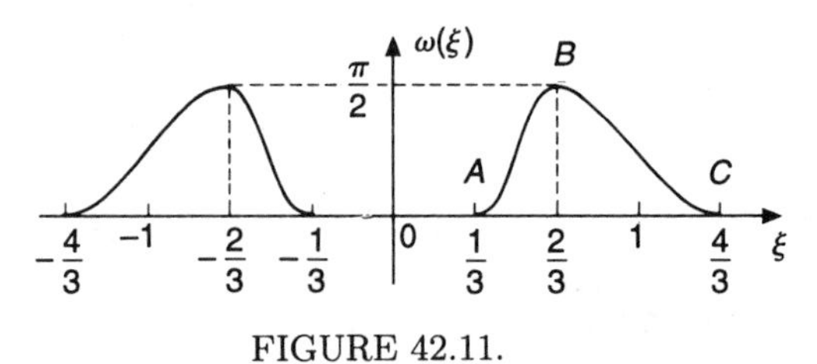

FIGURE 42.11.

The curve AB is required to have a certain symmetry:

$$\omega(1-\xi) = \frac{\pi}{2} - \omega(\xi), \quad \xi \in \left[\frac{1}{3}, \frac{2}{3}\right].$$

The curve BC is required to have the same form as AB, but reversed and stretched:

$$\omega(2\xi) = \frac{\pi}{2} - \omega(2(1-\xi)), \quad \xi \in \left[\frac{1}{3}, \frac{2}{3}\right].$$

Then $\widehat{\psi}$ is defined by

$$\widehat{\psi}(\xi) = e^{-i\pi\xi} \sin[\omega(\xi)], \quad \xi \in \mathbb{R}. \tag{42.11}$$

It is easy to see that $\widehat{\psi}$ is in $\mathscr{D}$. From this it follows that $\psi \in \mathscr{S}$ and is given by

$$\psi(t) = 2\int_0^{+\infty} \sin[\omega(\xi)] \cos\left[2\pi\left(t - \frac{1}{2}\right)\xi\right] d\xi. \tag{42.12}$$

Observe thatψ is real and its graph is symmetric with respect to $t = 1/2$.

It turns out that there is not much leeway in the choice of ω, and the wavelets constructed all have about the same appearance as the one shown in Figure 42.12. We note that although this function decreases rapidly,

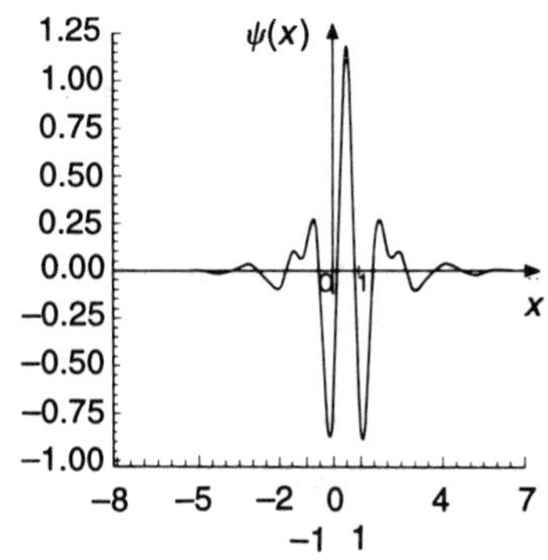

FIGURE 42.12. Meyer's C^∞ wavelet.

it has a rather large "numerical" support. Meyer proved (and this is not simple [LM86]) that the ψ_{jk} form an orthonormal basis for $L^2(\mathbb{R})$. Thus for all $f \in L^2(\mathbb{R})$,

$$f = \sum_{j=-\infty}^{+\infty} \sum_{k=-\infty}^{+\infty} (f, \psi_{jk})\psi_{jk}, \tag{42.13}$$

and

$$\sum_{j=-\infty}^{+\infty} \sum_{k=-\infty}^{+\infty} |(f, \psi_{jk})|^2 = \int_{-\infty}^{+\infty} |f(t)|^2 \, dt.$$

This wavelet basis has been tested numerically by Stéphane Jaffard by approximating the curve AB with a polynomial. Equation (42.13) is a series expansion similar to a Fourier series, except that here the series is double and f is not required to be periodic.

With an orthonormal wavelet basis, we obtain a decomposition of a signal f in "voices" f_j:

$$f = \sum_{j=-\infty}^{+\infty} f_j \quad \text{with} \quad f_j = \sum_{k=-\infty}^{+\infty} (f, \psi_{jk})\psi_{jk}.$$

Here we have a chorus with infinitely many voices. The approximation

$$F_n = \sum_{j=-\infty}^{n-1} \sum_{k=-\infty}^{+\infty} (f, \psi_{jk})\psi_{jk},$$

which is a projection of f on a certain subspace V_n, tends to f as $n \to +\infty$. The voice f_j represents exactly the detail that must be added to F_j to obtain the finer approximation F_{j+1}. These ideas led to the notion of a multiresolution analysis of the space $L^2(\mathbb{R})$. This concept was introduced by Stéphane Mallat and Yves Meyer in 1987 [Mal89].

42.4 Multiresolution analysis of $L^2(\mathbb{R})$

42.4.1 An introductory example

We begin with a uniform subdivision of the real line defined, for simplicity, by $t_k = k$ for all $k \in \mathbb{Z}$. (In approximation theory, the points t_k are known as knots.) An approximation F_0 of a signal $f \in L^2(\mathbb{R})$ can be defined in terms of an orthogonal projection on a subspace V_0 of approximations that are defined with respect to the given subdivision. For example, let V_0 be the subspace of $L^2(\mathbb{R})$ consisting of the continuous functions in $L^2(\mathbb{R})$ whose restrictions to the intervals $[k, k+1]$ are polynomials of degree ≤ 1 (see Figure 42.13).

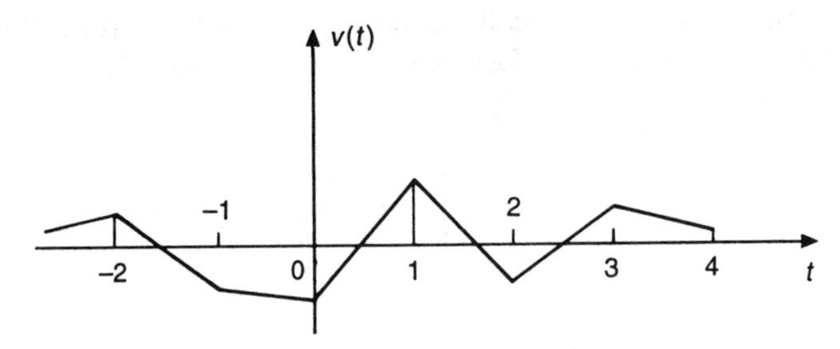

FIGURE 42.13. A cardinal spline of degree 1.

Any such function, square integrable or not, is called a cardinal spline of degree 1. It is not difficult to show that V_0 is isomorphic to the Hilbert space $l^2(\mathbb{Z})$. Since $l^2(\mathbb{Z})$ is complete, V_0 is a closed subspace of $L^2(\mathbb{R})$ on which the orthogonal projection F_0 of f is well-defined. Having defined such a function, we can improve the approximation of f by projecting f onto a larger subspace V_1 that contains V_0. Then V_1 is defined the same way we defined V_0, except that this time we refine the subdivision by adding all of the mid-points of the original intervals $[k, k+1]$. Then V_1 is easily characterized in terms of V_0, namely,

$$v(t) \in V_0 \iff v(2t) \in V_1.$$

In the same way we define the spaces $V_2, V_3, \ldots$ by taking finer subdivisions, and the spaces $V_{-1}, V_{-2}, \ldots$ by taking coarser subdivisions. In the latter construction, the knots for V_{-1} are the points $2k$, $k \in \mathbb{Z}$; those for V_{-2} are 2^2k, and so on. In this way we obtain a sequence of closed, nested subspaces of $L^2(\mathbb{R})$

$$\cdots \subset V_{-2} \subset V_{-1} \subset V_0 \subset V_1 \subset V_2 \subset \cdots$$

such that for all $j \in \mathbb{Z}$,

$$v(t) \in V_0 \iff v(2^j t) \in V_j.$$

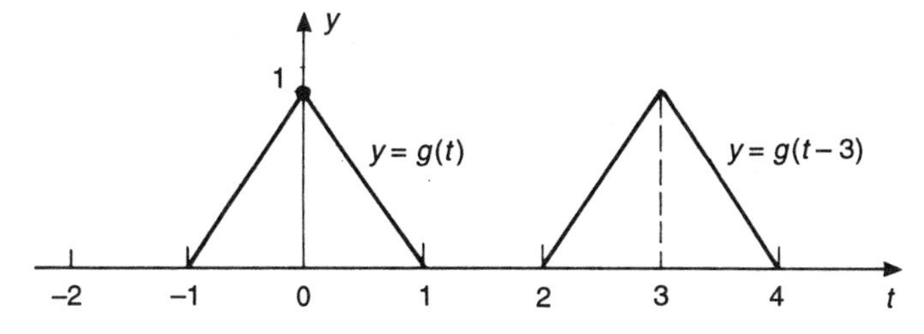

FIGURE 42.14. The hat-function basis.

V_0 is invariant under integer translations of the variable, and one can show that the translates $\tau_k g$ of the hat function g (Figure 42.14) form a basis for the Hilbert space V_0. Thus every $v \in V_0$ can be expressed as

$$v = \sum_{k=-\infty}^{+\infty} v(k)\tau_k g.$$

This example serves as a model for the definition of a multiresolution analysis of $L^2(\mathbb{R})$.

42.4.2 Definition A multiresolution analysis of $L^2(\mathbb{R})$ is a increasing sequence $\{V_j\}_{j\in\mathbb{Z}}$ of closed subspaces of $L^2(\mathbb{R})$ that have the following properties:

(i) $v(t) \in V_j \Leftrightarrow v(2t) \in V_{j+1}$ for all $j \in \mathbb{Z}$.

(ii) V_0 is invariant under integer translations of the variable: $v \in V_0$ implies that $\tau_k v \in V_0$ for all $k \in \mathbb{Z}$.

(iii) $\bigcup_{j\in\mathbb{Z}} V_j$ is dense in $L^2(\mathbb{R})$ and $\bigcap_{j\in\mathbb{Z}} V_j = \{0\}$.

(iv) There is a function g in V_0 such that the family $\{\tau_k g\}_{k\in\mathbb{Z}}$ is an unconditional basis for V_0.

The subspace V_j can be interpreted, as in the example, as the space of all possible approximations at the scale 2^{-j}. Property (iii) means that the sequence of orthogonal projections F_j of f tends f in $L^2(\mathbb{R})$ as $j \to +\infty$ and that $F_j \to 0$ as $j \to -\infty$.

The example of the spline functions of degree 1 clearly satisfies properties (i) and (ii). If v is in V_j for all j, then v would have to be a linear function, and being in L^2, this means that it must vanish identically. For the density, it is sufficient to show that the union of the V_j is dense in $\mathscr{D}$, since $\mathscr{D}$ itself is dense in $L^2(\mathbb{R})$. Suppose f is in $\mathscr{D}$, and consider the function $v_j \in V_j$ that agrees with f at the points $k2^{-j}$, $k \in \mathbb{Z}$. We know that v_j converges to f uniformly on $\mathbb{R}$ as $j \to +\infty$. Since the supports of f and the v_j are all contained in some bounded interval, the v_j also converge to f in $L^2(\mathbb{R})$.

To understand point (iv) it is necessary to define an unconditional basis (or Riesz basis) for a Hilbert space, since the definition of a topological basis given in Lesson 16 was only for the case where the basis elements were orthogonal. We do not assume that the vectors $\tau_k g$ in Definition 42.4.2 are orthogonal, and in important cases they are not.

42.4.3 Definition A sequence of elements $\{e_k\}_{k\in\mathbb{Z}}$ in a Hilbert space H is called an unconditional basis for H if the following conditions are satisfied:

(i) For each $f \in H$ there exists a unique complex sequence $(c_k)_{k\in\mathbb{Z}}$ in $l^2(\mathbb{Z})$ such that

$$\|f - \sum_{k=-N}^{N} c_k e_k\| \to 0 \quad \text{as} \quad N \to +\infty. \tag{42.14}$$

(ii) There are two positive constants A and B such that

$$A\|f\|^2 \leq \sum_{k=-\infty}^{+\infty} |c_k|^2 \leq B\|f\|^2, \tag{42.15}$$

which means that $f \mapsto \left(\sum_{k\in\mathbb{Z}} |c_k|^2\right)^{1/2}$ defines a norm on H that is equivalent to the original norm on H.

Having an unconditional basis for H is equivalent to having an isomorphism T between the two Hilbert spaces $l^2(\mathbb{Z})$ and H. If $A = B = 1$, we have the definition of a Hilbert basis. It is left as an exercise to show that the hat functions $\tau_k g$ in Section 42.4.1 form an unconditional basis for V_0.

42.4.4 Cardinal spline functions

The example in Section 42.4.1 is easily generalized by simultaneously increasing the degree of the polynomials and the global regularity of the approximations. Thus V_0 can be expanded to the subspace of continuously differentiable functions in $L^2(\mathbb{R})$ whose restrictions to the intervals $[k, k+1]$ are polynomials of degree less than or equal to 2. Note that V_0 is not trivial, since it is easy to exhibit nonzero functions of this sort. These are the cardinal splines of degree 2.

If r denotes the function that equals 1 on $[1/2, 1/2]$ and zero elsewhere, then the function g in Figure 42.16 is equal to $r * r$. Similarly, one can show that $g = r * r * r$ is in V_0 and that the sequence of translates of g forms an unconditional basis for the new space V_0.

On can continue this process and consider the spaces V_0 of cardinal splines of degrees $3, 4, \ldots, n, \ldots$ that are in $C^{n-1} \cap L^2$, and in this way create a family of multiresolution analyses of $L^2(\mathbb{R})$.

42.4.5 A different multiresolution analysis

Here is an example of a multiresolution analysis that is not based on spline functions. Let

$$V_0 = \{v \in L^2(\mathbb{R}) \mid \operatorname{supp}(\widehat{v}) \subset [-1, 1]\}.$$

V_0 is closed in $L^2(\mathbb{R})$ and invariant under translation. We have

$$V_j = \{v \in L^2(\mathbb{R}) \mid \operatorname{supp}(\widehat{v}) \subset [-2^j, 2^j]\},$$

and the spaces V_j are closed and nested. The density of $\bigcup V_j$ in $L^2(\mathbb{R})$ was proved in Section 38.4, and it is clear that the intersection reduces to $\{0\}$.

For the function $g \in V_0$ we can take the cardinal sine

$$g(t) = \frac{\sin 2\pi t}{\pi t},$$

and we have seen with Shannon's formula that the translates $\tau_k g$ are a basis for V_0: For all $v \in V_0$,

$$v(t) = \sum_{k=-\infty}^{+\infty} v(k) g(t-k).$$

It happens in this case that the $g(t-k)$ are orthonormal. On the other hand, g converges slowly at infinity and is not integrable. In the sense of Definitions 42.3.3 and 42.3.4, the $g(t-k)$ do not form a wavelet basis.

42.5 Multiresolution analysis and wavelet bases

We are going to see how it is possible, given a multiresolution analysis of $L^2(\mathbb{R})$, to construct an orthonormal basis for $L^2(\mathbb{R})$ of the form

$$\psi_{j,k}(t) = 2^{j/2}\psi(2^j t - k), \quad j, k \in \mathbb{Z}.$$

To have a wavelet basis in the sense of Definition 42.3.4, it is then sufficient to verify that the wavelet ψ satisfies Definition 42.3.3 for some r.

Finding such a function ψ is not an easy problem, as we have seen in the case of Meyer's C^∞ wavelet. We will first look for orthonormal bases of the form $\{\varphi_j(t-k)\}_{k\in\mathbb{Z}}$ for the subspaces V_j. This will not solve the problem directly, but it is an important initial step. Based on the relations

$$v(t) \in V_0 \iff v(2^j t) \in V_j,$$

it is sufficient to find an orthonormal basis of the form $\{\varphi(t-k)\}_{k\in\mathbb{Z}}$ for V_0. The corresponding bases for the V_j will be

$$\varphi_j(t-k) = 2^{j/2}\varphi(2^j t - k), \quad j, k \in \mathbb{Z}.$$

We are assuming that we have a multiresolution analysis, so by definition we have an unconditional basis for V_0, namely, the functions $g_k(t) = g(t-k)$. One way to proceed is to transform $\{g_k\}$ into an orthonormal basis using the Gram–Schmidt method. Meyer used instead a method, attributed to Henri Poincaré, that does not depend on the way the g_k are indexed and that transforms a basis of the form $g_k(x) = g(x-k)$ into one of the same form [Mey90].

Let $\{g_k\}_{k\in\mathbb{Z}}$ be an unconditional basis for a Hilbert space H. Define the linear operator $T : H \to H$ by

$$Tx = \sum_{k=-\infty}^{+\infty} (x, g_k) g_k.$$

This operator is continuous and self-adjoint, and there is an $a > 0$ such that for all $x \in H$,

$$(Tx, x) \geq a\|x\|.$$

Thus T has an inverse that is also positive and self-adjoint. By a result of Schur, it is possible to define a positive self-adjoint operator U such that $U^2 = T^{-1}$. It is then easy to verify that the vectors $\{Ug_k\}_{k\in\mathbb{Z}}$ form the desired orthogonal basis.

Although this is an abstract result, it can be used as a guide to attack the problem directly to produce an orthonormal basis expressed in terms of the g_k. This is the approach we now take.

42.5.1 Proposition *The set of functions $\{\tau_k\varphi\}_{k\in\mathbb{Z}}$ is an orthonormal family in V_0 if and only if*

$$\sum_{k=-\infty}^{+\infty} |\widehat{\varphi}(\lambda + k)|^2 = 1 \tag{42.16}$$

for almost every $\lambda \in \mathbb{R}$.

Proof. By definition, the functions $\tau_k\varphi$ are orthonormal if and only if

$$\int_{-\infty}^{+\infty} \varphi(t-p)\overline{\varphi}(t-q)\, dt = \begin{cases} 0 & \text{if } p \neq q, \\ 1 & \text{if } p = q \end{cases}$$

for all $p, q \in \mathbb{Z}$. This is equivalent to

$$\int_{-\infty}^{+\infty} \widehat{\varphi}(\lambda)\overline{\widehat{\varphi}}(\lambda) e^{-2i\pi(p-q)\lambda}\, d\lambda = \begin{cases} 0 & \text{if } p \neq q, \\ 1 & \text{if } p = q, \end{cases}$$

which, by letting $n = p - q$, is equivalent to having

$$\int_{-\infty}^{+\infty} |\widehat{\varphi}(\lambda)|^2 e^{-2i\pi n\lambda}\, d\lambda = \begin{cases} 0 & \text{if } n \neq 0, \\ 1 & \text{if } n = 0. \end{cases} \tag{42.17}$$

Finally, we show that (42.17) is equivalent to (42.16). Since $|\widehat{\varphi}(\lambda)|^2 \in L^1$, we know from Theorem 37.2.2 that $\sum_{k=-\infty}^{+\infty} |\widehat{\varphi}(\lambda+k)|^2 \in L^1_{\mathrm{p}}(0,1)$ and that

$$\sum_{k=-\infty}^{+\infty} |\widehat{\varphi}(\lambda+k)|^2 = \sum_{k=-\infty}^{+\infty} \mathscr{F}\ |\widehat{\varphi}|^2(n)e^{2i\pi n\lambda},$$

in the sense of $\mathscr{S}\,'$. If (42.16) holds (in the sense of L^1), then the Fourier coefficients of $|\widehat{\varphi}|^2$ satisfy condition (42.17). On the other hand, if we have (42.17), then

$$\sum_{k=-\infty}^{+\infty} |\widehat{\varphi}(\lambda+k)|^2 = 1$$

in the sense of $\mathscr{S}\,'$. But by Exercise 21.6, this implies that the relation holds for almost every $\lambda \in \mathbb{R}$. □

42.5.2 Proposition *If the family $\{\tau_k\varphi\}_{k\in\mathbb{Z}}$ is an orthonormal basis for V_0, then there exists a function M in $L^2_{\mathrm{p}}(0,1)$ such that*

$$\widehat{\varphi}(\lambda) = M(\lambda)\widehat{g}(\lambda)$$

and for a.e. $\lambda \in \mathbb{R}$,

$$|M(\lambda)| = \left(\sum_{k=-\infty}^{+\infty} |\widehat{g}(\lambda+k)|^2\right)^{-\frac{1}{2}}. \tag{42.18}$$

Proof. Since $\varphi \in V_0$, there exists a sequence (m_k) in $l^2(\mathbb{Z})$ such that

$$\varphi(t) = \sum_{k=-\infty}^{+\infty} m_k g(t-k)$$

in $L^2(\mathbb{R})$. By taking the Fourier transform of both sides, we see that

$$\widehat{\varphi}(\lambda) = \sum_{k=-\infty}^{+\infty} m_k e^{-2i\pi k\lambda}\widehat{g}(\lambda) = M(\lambda)\widehat{g}(\lambda),$$

where

$$M(\lambda) = \sum_{k=-\infty}^{+\infty} m_k e^{-2i\pi k\lambda}.$$

Clearly, $M(\lambda)$ is in $L^2_{\mathrm{p}}(0,1)$. Using Proposition 42.5.1 we have

$$\sum_{k=-\infty}^{+\infty} |\widehat{\varphi}(\lambda+k)|^2 = |M(\lambda)|^2 \sum_{k=-\infty}^{+\infty} |\widehat{g}(\lambda+k)|^2 = 1,$$

which proves the result. □

One can show that the Poincaré process leads essentially to the relation

$$\widehat{\varphi}(\lambda) = \left[\sum_{k=-\infty}^{+\infty} |\widehat{g}(\lambda + k)|^2\right]^{-\frac{1}{2}} \widehat{g}(\lambda). \tag{42.19}$$

The following theorem is due to Meyer.

42.5.3 Theorem *Assume that $g \in V_0$ and that $\{\tau_k g\}$ is an unconditional basis for V_0. If φ is defined by (42.19), then $\{\tau_k \varphi\}$ is an orthonormal basis for V_0.*

Proof. The proof follows directly from the abstract argument, but we are going to give a more explicit argument.

We must first show that φ is well-defined and that $\{\tau_k \varphi\}$ is an orthonormal family in $L^2(\mathbb{R})$. We use the assumption about g to prove that

$$0 < C \leq \sum_{k=-\infty}^{+\infty} |\widehat{g}(\lambda + k)|^2 \leq D$$

for some positive constants C and D.

For any sequence (a_k) in $l^2(\mathbb{Z})$ the function

$$f(t) = \sum_{k=-\infty}^{+\infty} a_k g(t - k)$$

is in $L^2(\mathbb{R})$, and, as in the proof of Proposition 42.5.2,

$$\widehat{f}(\lambda) = \sum_{k=-\infty}^{+\infty} a_k e^{-2i\pi k\lambda} \widehat{g}(\lambda) = m(\lambda)\widehat{g}(\lambda),$$

where $m \in L^2_{\mathrm{p}}(0, 1)$. The hypothesis that $\{\tau_k \varphi\}$ is an unconditional basis implies that

$$0 < C \sum_{k=-\infty}^{+\infty} |a_k|^2 \leq \|f\|^2 \leq D \sum_{k=-\infty}^{+\infty} |a_k|^2$$

for some strictly positive constants C and D, and in Fourier space,

$$0 < C \int_0^1 |m(\lambda)|^2 \, d\lambda \leq \int_{-\infty}^{+\infty} |m(\lambda)|^2 |\widehat{g}(\lambda)|^2 \, d\lambda \leq D \int_0^1 |m(\lambda)|^2 \, d\lambda.$$

The middle integral is equal to $\displaystyle\int_0^1 |m(\lambda)|^2 \sum_{k=-\infty}^{+\infty} |\widehat{g}(\lambda + k)|^2 \, d\lambda$, so we have

$$0 < C \int_0^1 |m(\lambda)|^2 \, d\lambda \leq \int_0^1 |m(\lambda)|^2 \sum_{k=-\infty}^{+\infty} |\widehat{g}(\lambda + k)|^2 \, d\lambda \leq D \int_0^1 |m(\lambda)|^2 \, d\lambda$$

for all (a_k) in $l^2(\mathbb{Z})$, which is to say, for all $m \in L^2_{\mathrm{p}}(0,1)$. But this can be true if and only if

$$0 < C \leq \sum_{k=-\infty}^{+\infty} |\widehat{g}(\lambda + k)|^2 \leq D$$

for almost every $\lambda \in \mathbb{R}$, which is what we wished to prove. For convenience we write

$$M(\lambda) = \left[\sum_{k=-\infty}^{+\infty} |\widehat{g}(\lambda + k)|^2 \right]^{-\frac{1}{2}} \quad \text{and} \quad N(\lambda) = \left[\sum_{k=-\infty}^{+\infty} |\widehat{g}(\lambda + k)|^2 \right]^{\frac{1}{2}}.$$

Then

$$\sqrt{C} \leq N(\lambda) \leq \sqrt{D} \quad \text{and} \quad \frac{1}{\sqrt{D}} \leq M(\lambda) \leq \frac{1}{\sqrt{C}},$$

and we conclude that both M and N are in $L^\infty_{\mathrm{p}}(0,1)$ and hence in $L^2_{\mathrm{p}}(0,1)$. This implies that $\widehat{\varphi}(\lambda) = M(\lambda)\widehat{g}(\lambda)$ is well-defined as an element of $L^2(\mathbb{R})$. It is clear from this definition that $\widehat{\varphi}$ satisfies (42.16); hence $\{\tau_k \varphi\}$ is an orthonormal family.

It remains to show that $\{\tau_k \varphi\}$ spans V_0, which is by now close to obvious: Since $\widehat{g}(\lambda) = N(\lambda)\widehat{\varphi}(\lambda)$ and $N \in L^2_{\mathrm{p}}(0,1)$, there is a sequence (n_k) in $l^2(\mathbb{Z})$ such that

$$g(t) = \sum_{k=-\infty}^{+\infty} n_k \varphi(t-k).$$

Since the functions $g(t-k)$ span V_0, this shows that the functions $\varphi(t-k)$ span V_0 and completes the proof. □

42.5.4 Finding an orthonormal wavelet basis for $L^2(\mathbb{R})$

Starting with a multiresolution analysis of $L^2(\mathbb{R})$, we have managed to construct a basis $\varphi_k(t) = \varphi(t-k)$ for the space V_0 and thus for the spaces V_j. However, we have yet to find an orthonormal wavelet basis for $L^2(\mathbb{R})$. For this, let W_j be the orthogonal complement of V_j in V_{j+1},

$$V_{j+1} = V_j \bigoplus W_j.$$

The spaces W_j provide decompositions of the spaces V_n (since $V_j \downarrow \{0\}$) and of L^2 (since $V_n \uparrow L^2$) as direct sums of orthogonal subspaces:

$$V_n = \bigoplus_{j=-\infty}^{n-1} W_j \quad \text{and} \quad L^2 = \bigoplus_{j=-\infty}^{+\infty} W_j.$$

Thus, if we find an orthonormal basis for each W_j, then we will have an orthonormal basis for $L^2(\mathbb{R})$. As was the case with the V_j, it is sufficient to solve the problem for W_0, since

$$v(t) \in W_0 \iff v(2^j t) \in W_j.$$

The plan is to look for a function $\psi \in W_0$ such that the functions $\tau_k\psi$ form an orthonormal basis for W_0.

42.5.5 Proposition *Assume that φ is defined by (42.19). Then there exists a function $A \in L^2_{\mathrm{p}}(0,1)$ such that, for almost all λ,*

$$\widehat{\varphi}(2\lambda) = A(\lambda)\widehat{\varphi}(\lambda)$$

and

$$|A(\lambda)|^2 + |A(\lambda + 1/2)|^2 = 1. \tag{42.20}$$

Proof. The function $\frac{1}{2}\varphi\left(\frac{t}{2}\right)$ is in $V_{-1} \subset V_0$. Thus there exists a sequence (a_k) in $l^2(\mathbb{Z})$ such that

$$\frac{1}{2}\varphi\left(\frac{t}{2}\right) = \sum_{k=-\infty}^{+\infty} a_k\varphi(t-k).$$

Taking the Fourier transform, this becomes

$$\widehat{\varphi}(2\lambda) = \sum_{k=-\infty}^{+\infty} a_k e^{-2i\pi k\lambda}\widehat{\varphi}(\lambda) = A(\lambda)\widehat{\varphi}(\lambda),$$

and A is clearly in $L^2_{\mathrm{p}}(0,1)$. From (42.16) we see that

$$\sum_{k=-\infty}^{+\infty} |\widehat{\varphi}(2\lambda + 2k)|^2 = |A(\lambda)|^2 \sum_{k=-\infty}^{+\infty} |\widehat{\varphi}(\lambda + k)|^2 = |A(\lambda)|^2.$$

Replacing λ with $\lambda + 1/2$ we have

$$\begin{aligned}\sum_{k=-\infty}^{+\infty} |\widehat{\varphi}(2\lambda + 2k + 1)|^2 &= |A(\lambda + 1/2)|^2 \sum_{k=-\infty}^{+\infty} |\widehat{\varphi}(\lambda + k + 1/2)|^2 \\ &= |A(\lambda + 1/2)|^2.\end{aligned}$$

The result is obtained by adding the last two equations. □

We next investigate the conditions that ψ must satisfy.

42.5.6 Proposition *If ψ exists, then there exists a function B in $L^2_{\mathrm{p}}(0,1)$ that satisfies the following conditions:*

(i) $\widehat{\psi}(2\lambda) = B(\lambda)\widehat{\varphi}(\lambda)$. (42.21)

(ii) $|B(\lambda)|^2 + |B(\lambda + 1/2)|^2 = 1$. (42.22)

(iii) $A(\lambda)\overline{B}(\lambda) + A(\lambda + 1/2)\overline{B}(\lambda + 1/2) = 0$. (42.23)

Proof. First note that as for φ in Proposition 42.5.1, if the $\psi(t-k)$ form an orthonormal basis for W_0, then

$$\sum_{k=-\infty}^{+\infty} |\widehat{\psi}(\lambda+k)|^2 = 1. \tag{42.24}$$

On the other hand, all of the functions $\varphi(t-k)$ are orthogonal to ψ, and therefore

$$\int_{-\infty}^{+\infty} \varphi(t-k)\overline{\psi}(t)\,dt = \int_{-\infty}^{+\infty} \widehat{\varphi}(\lambda)\overline{\widehat{\psi}}(\lambda)e^{-2i\pi k\lambda}\,d\lambda = 0$$

for all $k \in \mathbb{Z}$. By Poisson's formula (or direct manipulation), this implies that

$$\sum_{k=-\infty}^{+\infty} \widehat{\varphi}(\lambda+k)\overline{\widehat{\psi}}(\lambda+k) = 0. \tag{42.25}$$

The existence of B satisfying (42.21) follows from the same argument we used in Proposition 42.5.5 to show the existence of A for φ. Thus

$$B(\lambda) = \sum_{k=-\infty}^{+\infty} b_k e^{-2i\pi k\lambda}$$

is derived from the relation

$$\frac{1}{2}\psi\left(\frac{t}{2}\right) = \sum_{k=-\infty}^{+\infty} b_k\varphi(t-k). \tag{42.26}$$

Similarly, identity (42.22) follows from (42.24). We can write (42.25) as

$$\sum_{k=-\infty}^{+\infty} \left[\widehat{\varphi}(2\lambda+2k)\overline{\widehat{\psi}}(2\lambda+2k) + \widehat{\varphi}(2\lambda+2k+1)\overline{\widehat{\psi}}(2\lambda+2k+1)\right] = 0.$$

Using the relations $\widehat{\varphi}(2\lambda) = A(\lambda)\widehat{\varphi}(\lambda)$ and $\widehat{\psi}(2\lambda) = B(\lambda)\widehat{\varphi}(\lambda)$ shows that

$$A(\lambda)\overline{B}(\lambda) \sum_{k=-\infty}^{+\infty} \widehat{\varphi}(\lambda+k)\overline{\widehat{\varphi}}(\lambda+k) +$$
$$A(\lambda+1/2)\overline{B}(\lambda+1/2) \sum_{k=-\infty}^{+\infty} \widehat{\varphi}(\lambda+k+1/2)\overline{\widehat{\varphi}}(\lambda+k+1/2) = 0,$$

which is

$$A(\lambda)\overline{B}(\lambda) \sum_{k=-\infty}^{+\infty} |\widehat{\varphi}(\lambda+k)|^2 +$$
$$A(\lambda+1/2)\overline{B}(\lambda+1/2) \sum_{k=-\infty}^{+\infty} |\widehat{\varphi}(\lambda+k+1/2)|^2 = 0.$$

This and (42.16) yield (42.23). □

42.5.7 Computing the functions B and ψ

To find ψ, we look for a B that satisfies (42.22) and (42.23). By solving equations (42.20) and (42.23) for $B(\lambda)$ we see that

$$B(\lambda) = -\overline{A}(\lambda + 1/2)[A(\lambda)B(\lambda + 1/2) - A(\lambda + 1/2)B(\lambda)].$$

We write this as

$$B(\lambda) = e^{-2i\pi\lambda}\overline{A}(\lambda + 1/2)\theta(\lambda)$$

with

$$\theta(\lambda) = -e^{2i\pi\lambda}[A(\lambda)B(\lambda + 1/2) - A(\lambda + 1/2)B(\lambda)]$$

and make two observations. Note that $\theta(\lambda + 1/2) = \theta(\lambda)$ and, since A and B must satisfy (42.20) and (42.22), that

$$|\theta(\lambda)| = 1. \tag{42.27}$$

Conversely, it is easy to show that any function θ with period 1/2 satisfying (42.27) will work. A simple family of functions θ is

$$\theta_\alpha(\lambda) = e^{-i\pi\alpha}, \quad \alpha \in \mathbb{R},$$

but in practice one usually takes $\alpha = 0$. Thus let B be defined by

$$B(\lambda) = e^{-2i\pi\lambda}\overline{A}(\lambda + 1/2); \tag{42.28}$$

then ψ is defined in terms of its Fourier transform by (42.21).

42.5.8 Theorem *If ψ is defined by (42.21) and (42.28), then the set of functions $\{\tau_k\psi\}_{k\in\mathbb{Z}}$ is an orthonormal basis for W_0 and the functions*

$$\psi_{jk} = 2^{j/2}\psi(2^j - k), \quad j, k \in \mathbb{Z},$$

are an orthonormal basis for $L^2(\mathbb{R})$.

Proof. The first task is to sort out what has been proved and what remains to be proved. We assume that we have a multiresolution analysis of $L^2(\mathbb{R})$ and that we have in hand a function φ such that the $\varphi_k(t) = \varphi(t-k)$ form an orthonormal basis for V_0. By definition, W_0 is the orthogonal complement of V_0 in V_1, so $V_1 = V_0 \bigoplus W_0$, and this implies by a change of scale that $V_{j+1} = V_j \bigoplus W_j$ for all $j \in \mathbb{Z}$. The fact that

$$L^2 = \bigoplus_{j=-\infty}^{+\infty} W_j$$

is then a direct consequence of Definition 42.4.2(iii). Thus, to prove that $\{\psi_{jk}\}_{j,k\in\mathbb{Z}}$ is an orthonormal basis for $L^2(\mathbb{R})$, it is sufficient to show that the $\psi_{0k} = \psi_k$ form an orthonormal basis for W_0.

This is how we proceed: Define B by (42.28) and ψ by (42.21); then work the arguments of Proposition 42.5.6 backwards to show that B satisfies (42.22) and (42.23) and that ψ satisfies relations (42.24) and (42.25). These are straightforward computations, and we leave this part as an exercise. This proves that the functions ψ_k are in W_0 and that they form an orthonormal family in W_0. What remains to be shown is that the ψ_k span W_0. To do this, we will show the existence of sequences $(c_k), (d_k), (e_k), (f_k) \in l^2(\mathbb{Z})$ such that the functions $\varphi(t)$, $\varphi(2t)$, and $\psi(t)$ are related by the following equations:

$$\begin{aligned} \varphi(2t) &= \sum_{k=-\infty}^{+\infty} c_k \varphi(t-k) + \sum_{k=-\infty}^{+\infty} d_k \psi(t-k), \\ \varphi(2t-1) &= \sum_{k=-\infty}^{+\infty} e_k \varphi(t-k) + \sum_{k=-\infty}^{+\infty} f_k \psi(t-k). \end{aligned} \tag{42.29}$$

Once we have (42.29) we have the result: These relations show that for each $n \in \mathbb{Z}$, the function $\varphi(2t-n)$ can be expressed as linear a combination of the φ_k and the ψ_k. If the ψ_k do not span W_0, there is a nonzero element $h \in W_0$ such that $(h, \psi_k) = 0$ for all k. Since $W_0 \perp V_0$, $(h, \varphi_k) = 0$ for all k. But $h \in V_1 = V_0 \bigoplus W_0$ and the $\varphi(2t-k))$ form a basis for V_1; this and (49.29) imply that $h = 0$. Hence the ψ_k must span W_0.

The last step is to show that we do indeed have (42.29). In the Fourier domain, (42.29) is equivalent to the existence of four functions C, D, E, F belonging to $L^2_{\mathrm{p}}(0,1)$ such that

$$\begin{aligned} \widehat{\varphi}\Big(\frac{\lambda}{2}\Big) &= C(\lambda)\widehat{\varphi}(\lambda) + D(\lambda)\widehat{\psi}(\lambda), \\ e^{-i\pi\lambda}\widehat{\varphi}\Big(\frac{\lambda}{2}\Big) &= E(\lambda)\widehat{\varphi}(\lambda) + F(\lambda)\widehat{\psi}(\lambda). \end{aligned} \tag{42.30}$$

Using the properties of A and the definitions of B and ψ it is easy to show that the following system satisfies the requirements. (It slightly more difficult to find these relations "from scratch.")

$$\begin{aligned} C(\lambda) &= \overline{A}(\lambda/2) + \overline{A}(\lambda/2 + 1/2), \\ D(\lambda) &= \overline{B}(\lambda/2) + \overline{B}(\lambda/2 + 1/2), \\ E(\lambda) &= e^{-i\pi\lambda}[\overline{A}(\lambda/2) - \overline{A}(\lambda/2 + 1/2)], \\ F(\lambda) &= e^{-i\pi\lambda}[\overline{B}(\lambda/2) - \overline{B}(\lambda/2 + 1/2)]. \end{aligned}$$

While this completes the proof, much can be said about this result and the questions it raises. A few comments are given below. □

42.5.9 Remarks

(a) Formula (42.28) provides a relation between the Fourier coefficients a_k and b_k of A and B:

$$b_k = (-1)^{1-k}\,\overline{a}_{1-k}.$$

This allows one to obtain ψ in terms of φ without a Fourier transform, since by (42.26),

$$\frac{1}{2}\psi\left(\frac{t}{2}\right) = \sum_{k=-\infty}^{+\infty} (-1)^k\,\overline{a}_k\varphi(t-1+k).$$

However, if one starts with a multiresolution analysis with

$$\sum_{k=-\infty}^{+\infty} |\widehat{g}(\lambda+k)|^2 \neq 1,$$

then a Fourier transform is needed to construct φ via (42.19).

Once we have an orthonormal basis for each W_j, a signal f is decomposed as the sum of its projections on the spaces W_j:

$$f = \sum_{j=-\infty}^{+\infty} f_j,$$

with

$$f_j = \sum_{k=-\infty}^{+\infty} (f, \psi_{jk})\psi_{jk},$$

and the approximation of f at the resolution 2^{-n} is given by its orthogonal projection on V_n, which is

$$F_n = \sum_{j=-\infty}^{1-n} f_j.$$

(b) The approach has been to start with a multiresolution analysis of $L^2(\mathbb{R})$ and to construct φ and ψ. It is possible to begin with a function φ in $L^2(\mathbb{R})$ and to consider the closed subspace V_0 spanned by the translates of φ. A natural question arises: What assumptions about φ will guarantee that $\{V_j\}_{j\in\mathbb{Z}}$ is a multiresolution analysis of $L^2(\mathbb{R})$? Clearly, we want $\{\varphi(t-k)\}$ to be an orthonormal family, so we must assume that φ satisfies (42.16). (Otherwise, we must assume $0 < C \leq \sum_{k\in\mathbb{Z}} |\widehat{\varphi}(\lambda+k)|^2 \leq D$ and transform $\{\varphi(t-k)\}$ into an orthonormal family.) The function φ must also satisfy

$$\frac{1}{2}\varphi\left(\frac{t}{2}\right) = \sum_{k=-\infty}^{+\infty} a_k\varphi(t-k). \tag{42.31}$$

for some sequence (a_k) in $l^2(\mathbb{Z})$. With these assumptions it is easy to show that the closed subspaces V_j generated by the orthonormal families $\{\varphi_{jk}\}_{k\in\mathbb{Z}} = \{2^{j/2}\varphi(2^j t - k)\}_{k\in\mathbb{Z}}$ fulfill conditions (i), (ii), and (iv) of Definition 42.4.1. It is also not difficult to prove that $\cap_{j\in\mathbb{Z}} V_j = \{0\}$. The difficult part is to show that $\cup_{j\in\mathbb{Z}} V_j$ is dense in L^2. This can be proved with the additional assumptions that $\widehat{\varphi}(\lambda)$ is bounded for all λ and that it is continuous near $\lambda = 0$ with $|\widehat{\varphi}(0)| = 1$ (see [Dau92, p. 142]). These conditions are fulfilled, for example, if $\varphi \in L^1$ and $|\int \varphi(t)\,dt| = 1$.

If $\widehat{\varphi}$ is continuous near $\lambda = 0$ with $|\widehat{\varphi}(0)| = 1$, then by Proposition 42.5.5, $|A(0)| = 1$ and $A(1/2) = 0$, and from the definition of B, we have $B(0) = 0$ and $|B(1/2)| = 1$. If A and B are interpreted as transfer functions of filters, then A passes frequencies near $\lambda = 1$ and attenuates frequencies near $\lambda = 1/2$. Thus A acts like a low-pass filter. Similarly, B acts like a high-pass filter. The impulse responses of the two filters are (a_k) and (b_k) respectively. Filters A and B that satisfy the relations of Propositions 42.5.5 and 42.5.6, which can be summarized by saying that the matrix

$$\begin{bmatrix} A(\lambda) & B(\lambda) \\ A(\lambda + 1/2) & B(\lambda + 1/2) \end{bmatrix}$$

is unitary for almost all λ, are called conjugate quadrature filters.

(c) As indicated several times, regularity and localization of the scaling function φ and the wavelets ψ_{jk} are necessary for efficient numerical computation. Thus the "minimal" assumptions made about φ in (b) do not lead to practical wavelets. If we assume, however, that

$$\int_{\mathbb{R}} (1 + |t|^m)|\varphi(t)|\,dt < +\infty$$

for all $m \in \mathbb{Z}$, in addition to assuming that φ satisfies (42.16) and (42.31) and that $|\widehat{\varphi}(0)| = 1$, the whole situation becomes much "smoother." In this case, the coefficients a_k decrease rapidly at infinity and $A \in C^\infty$. Furthermore, not only do assumptions about the regularity of φ and the localization of its derivatives lead to the regularity of ψ, but they also imply that ψ has vanishing moments. This analysis can be found in [CR95].

42.5.10 Spline wavelets

We began the discussion of multiresolution analysis by describing the multiresolution analysis of $L^2(\mathbb{R})$ based on the space V_0 that is spanned by the integer translates of the "hat" function (Figure 42.14). We also mentioned that this example could be generalized by takin V_0 to be the space spanned by the cardinal splines of degree n. More precisely, if r is the characteristic function of the interval $[-1/2, 1/2]$ and if g_n denotes the convolution $r * r * \cdots * r$ containing $n+1$ terms, the functions $\tau_k g_n$, $k \in \mathbb{Z}$, form a Riesz basis for V_0, and the nested spaces V_j constitute a multiresolution analysis of $L^2(\mathbb{R})$. We wish to continue this example in light of what we now know about the wavelets associated with a multiresolution analysis. For simplicity, we limit the discussion to the case $n = 1$ and write $g = g_0 = r * r$. It is clear from Figure 42.14 that the functions $\tau_k g$ are not orthogonal. However, it is easy to see that g satisfies the equation

$$g(t) = \frac{1}{2}g(2t+1) + g(2t) + \frac{1}{2}g(2t-1). \tag{42.32}$$

This relation implies that the $V_0 \subset V_1$ and thus that $V_j \subset V_{j+1}$by a change of scale. It was argued following Definition 42.4.2 that $\{V_j\}$ is a multiresolution analysis of $L^2(\mathbb{R})$.

For later use, we take the Fourier transform of both sides of (42.32) and write

$$\widehat{g}(2\lambda) = G(\lambda)\widehat{g}(\lambda), \tag{42.33}$$

where $G(\lambda) = (1 + \cos 2\pi\lambda)/2$. Note that G is real and even.

Since the translates $\tau_k g$, $k \neq 0$, are not orthogonal to g, it is necessary to transform the $\tau_k g$ into an orthonormal family. For this, we use Theorem 42.5.3 and define $\widehat{\varphi}$ by (42.19):

$$\widehat{\varphi}(\lambda) = \left[\sum_{k=-\infty}^{+\infty} |\widehat{g}(\lambda+k)|^2\right]^{-\frac{1}{2}} \widehat{g}(\lambda).$$

Since $\widehat{r}(\lambda) = \sin \pi\lambda/(\pi\lambda)$, we have

$$\widehat{g}(\lambda) = \frac{\sin^2 \pi\lambda}{\pi^2\lambda^2}.$$

An expression for the function $\sum_{k=-\infty}^{+\infty} |\widehat{g}(\lambda+k)|^2$ is computed by evaluating its Fourier coefficients as follows:

$$\begin{aligned}
\int_0^1 \sum_{k=-\infty}^{+\infty} |\widehat{g}(\lambda+k)|^2 e^{2i\pi n\lambda}\, d\lambda &= \int_{\mathbb{R}} |\widehat{g}(\lambda)|^2 e^{2i\pi n\lambda}\, d\lambda \\
&= \int_{\mathbb{R}} \widehat{g}(\lambda)\overline{\widehat{g}}(\lambda) e^{2i\pi n\lambda}\, d\lambda \\
&= \int_{\mathbb{R}} g(t)g(t-n)\, dt.
\end{aligned}$$

A simple computation shows that

$$\int_{\mathbb{R}} g(t)g(t-n)\, dt = \begin{cases} \frac{2}{3} & \text{if } n = 0, \\ \frac{1}{6} & \text{if } n = \pm 1, \\ 0 & \text{otherwise.} \end{cases}$$

Thus, this infinite sum has the simple expression

$$\sum_{k=-\infty}^{+\infty} |\widehat{g}(\lambda + k)|^2 = \frac{2}{3} + \frac{1}{3}\cos 2\pi\lambda = \frac{1}{3}[1 + 2\cos^2 \pi\lambda],$$

and we can write $\widehat{\varphi}$ (42.19) as

$$\widehat{\varphi}(\lambda) = M(\lambda)\widehat{g}(\lambda) = \frac{\sqrt{3}}{[1 + 2\cos^2 \pi\lambda]^{1/2}} \frac{\sin^2 \pi\lambda}{\pi^2\lambda^2}. \tag{42.34}$$

The function $M(\lambda) = \sqrt{3}[1 + 2\cos^2 \pi\lambda]^{-1/2}$ is in $C^\infty(\mathbb{R}) \cap L^1_{\text{p}}(0,1)$. It is a periodic tempered distribution, and by Theorem 36.2.2, $\overline{\mathscr{F}}M$, which we denote by m, can be expressed as

$$\overline{\mathscr{F}}M = m = \sum_{n=-\infty}^{+\infty} \alpha_n \delta_n,$$

where (α_n) is a slowly increasing sequence. In fact, since $M \in L^1_{\text{p}}(0,1)$, α_n tends to zero as $|n| \to +\infty$. An application of Proposition 33.2.1 shows that

$$g * \widehat{\overline{\mathscr{F}}M} = \widehat{g} \cdot M,$$

and since the Fourier transform in 1-to-1 on $\mathscr{S}'$, we must have

$$\varphi(t) = g * m(t). \tag{42.35}$$

We can draw several conclusions from this representation of φ. First, it is clear from (42.32) that φ is a cardinal spline of degree 1. To be precise, φ is the spline one obtains by connecting the points (n, α_n) with straight lines. The second observation is that φ does not have compact support, or more to the point, m does not have compact support. If it did, then M would be a trigonometric polynomial, but this assumption leads quickly to a contradiction. The conclusion is that the support of the scaling function φ is all of $\mathbb{R}$. Finally, note that (α_n) is real and even, since M is real and even.

As a step toward construction the wavelet ψ, we need to describe the filter A that appears in Proposition 42.5.5. From (42.33) and (42.44), it follows that

$$A(\lambda) = \frac{M(2\lambda)}{M(\lambda)} G(\lambda),$$

and from what we know about M and G, it is not difficult to see that A is real and even. A quick computation shows that $A(\lambda + 1/2)$ is also even.

The wavelet ψ is defined by $\psi(2\lambda) = e^{-2i\pi\lambda} A(\lambda + 1/2)\widehat{\varphi}(\lambda)$ ((42.21 and (42.28)), and from what we have seen so far, this can be written as

$$\psi(2\lambda) = e^{-2i\pi\lambda} A(\lambda + 1/2) G(\lambda)\widehat{g}(\lambda). \tag{42.36}$$

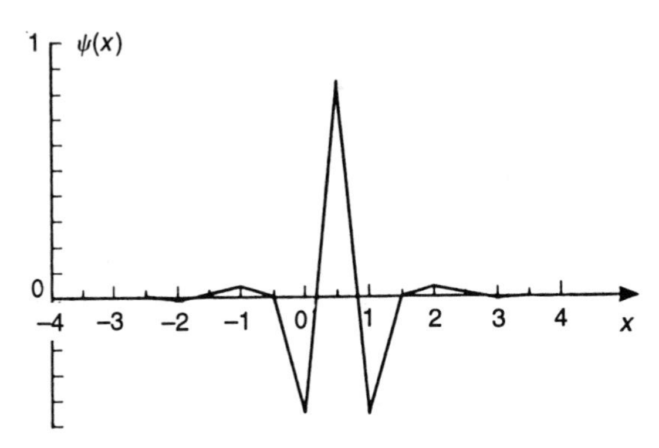

FIGURE 42.15. Spline wavelet of degree 1 (Lemarié–Battle).

As a last step, we wish to show that ψ is a real-valued cardinal spline of degree 1 and that it is symmetric about $t = 1/2$. For ease of notation, write $S(\lambda) = A(\lambda + 1/2)G(\lambda)$. The argument regarding M applies to S, and consequently $\overline{\mathscr{F}}\ S$ can be expressed as

$$\overline{\mathscr{F}}\ S = s = \sum_{n=-\infty}^{+\infty} \beta_n \delta_n, \tag{42.37}$$

where β_n tends to zero as $|n| \to +\infty$. Equation (42.36), written as

$$e^{2i\pi\lambda}\psi(2\lambda) = S(\lambda)\widehat{g}(\lambda),$$

implies that

$$\psi\Big(\frac{t+1}{2}\Big) = (s * g)(t).$$

Both s and g are even, and it follows that $\psi((t+1)/2)$ is even. The function $\psi(t + 1/2)$ obtained by replacing t with $2t$ is also even; thus its translate $\psi(t)$ is symmetric around $1/2$.

To summarize, starting with a multiresolution of $L^2(\mathbb{R})$ generated by the function $g = r * r$, we have used the constructions described in this lesson to generate the scaling function φ and the spline wavelet ψ. We have shown that φ and ψ are real spline functions of degree 1, that φ is even, and that ψ is symmetric about $t = 1/2$. We also argued that the support of φ is $\mathbb{R}$; similarly, since S cannot be a polynomial, the support of ψ is $\mathbb{R}$. However, both functions decay exponentially (for a proof see [Dau92]). These results generalize to the multiresolution generated by g_n. For a systematic discussion of spline wavelets, we suggest the article by Charles Chui in [RBC$^+$92]. The spline wavelet ψ and the modulus of its spectrum are shown, respectively, in Figures 42.15 and 42.16. The spline wavelet of degree 3 is shown in Figure 42.17 and the modulus of its spectrum is illustrated in Figure 42.18.

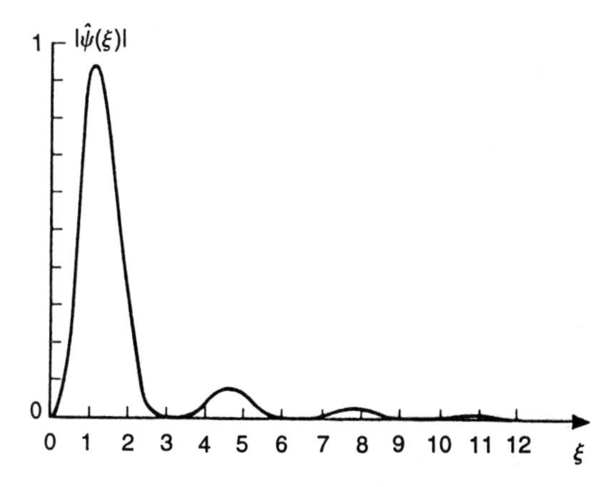

FIGURE 42.16. Amplitude of the spectrum of spline wavelet of degree 1.

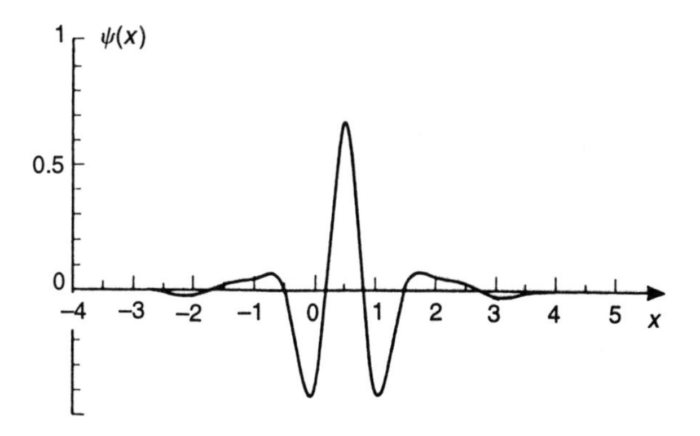

FIGURE 42.17. Spline wavelet of degree 3.

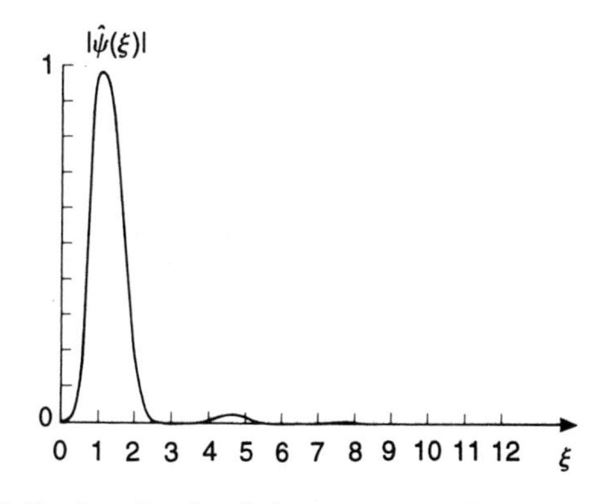

FIGURE 42.18. Amplitude of the spectrum of spline wavelet of degree 3.

42.6 Afternotes

This lesson has been but a brief introduction to the theory and applications of wavelets. We have presented only a few topics from what has become a dynamic and productive area of research with a rich theory and a wide range of applications. In this last section we indicate some other aspects of the field and provide a few pointers to the literature, which is now substantial.

A first point concerns history and the sociology of science. Since the beginning in the 1980s of what we call "modern wavelet theory," the field has been characterized by a healthy interplay between theory and applications. Simply put, mathematicians have worked in close collaboration with researchers from other areas of science and engineering, and wavelet theory has been strongly influenced by applied problems. These revolve naturally around signal and image processing, but the signals and images arrive from diverse fields: astronomy, biology, medicine, hydrodynamics, geophysics, and, of course, telecommunications—to mention but a few. The challenge is to find a field of science or engineering where wavelet techniques have not been applied, or at least tried. This was not always the case. As mentioned at the beginning of the lesson, we now see many older results in mathematics and in signal processing that are now interpreted in the language of wavelet theory. These results were for the most part unknown outside their respective communities. Since the initial collaboration between Morlet and Grossmann, the tradition of interaction and cross-fertilization among disciplines continues, and there is a resonance in this when we recall that Fourier was motivated by problems in heat conduction.

We have introduce two kinds of wavelet analysis: continuous wavelet analysis associated with a family of the form

$$\psi_{ab}(t) = \frac{1}{\sqrt{a}}\psi\left(\frac{t-b}{a}\right), \quad b \in \mathbb{R},\ a > 0,$$

and discrete wavelet analysis using wavelets of the form

$$\psi_{jk}(t) = 2^{j/2}\psi(2^j t - k), \quad j, k \in \mathbb{Z}. \tag{42.38}$$

Both of these analyses can be extended to higher dimensions; this is particularly important in two dimensions for image processing.

Continuous wavelet analysis, including variations involving the modulus of the wavelet transform, has been developed as a sensitive tool for analyzing local properties of a signals. These techniques have been used to analyze the singularities of "mathematical signals" such as the celebrated continuous, "nowhere-differentiable" function

$$R(x) = \sum_{n=1}^{\infty} \frac{\sin(\pi n^2 x)}{n^2}$$

attributed to Riemann as well as various "experimental signals," particularly fully developed turbulence. General information on the continuous point of view can be found in [Dau92] and [Tor95]. Applications to the analysis of fractal objects in physics can be found in [AAB+95]. A great deal of work on continuous wavelet analysis has been done by the group at Marseille under the general guidance of Alex Grossmann, who from the very beginning has been a leader in the field.

We proved the reconstruction formula using the same wavelet that was used for the analysis. It is possible, however, to use different wavelets for the analysis and synthesis. This technique was used profitably by Matthias Holschneider and Philippe Tchamitchian for their analysis of Riemann's function [HT91].

On the discrete side, the discovery by Ingrid Daubechies of wavelets having compact support stands as a landmark in the theory. Remarkably, given $r \in \mathbb{N}$, there exists an orthonormal basis for $L^2(\mathbb{R})$ of the form

$$2^{j/2}\psi_r(2^j t - k), \quad j, k \in \mathbb{Z},$$

such that the support of ψ_r is in $[0, 2r+1]$, the moments $\int t^n \psi_r \, dt = 0$ for $0 \le n \le r$, and ψ_r has about $r/5$ continuous derivatives. A complete account can be found in Daubechies's book [Dau92]. Another significant step was the discovery by Daubechies, Cohen, and Feauveau of a general way to generate biorthogonal wavelet bases. (A particular example had previously been constructed by Philippe Tchamitchian.) This means there are two families $\{\psi_{jk}\}$ and $\{\widetilde{\psi}_{jk}\}$, each of the form (42.32), that are unconditional bases for $L^2(\mathbb{R})$ and such that

$$(\psi_{jk}, \widetilde{\psi}_{j'k'}) = 0$$

except when $j = j'$ and $k = k'$, in which case it equals 1. A complete discussion of this construction and of why biorthogonal wavelets are interesting for applications is given in [CR95].

The original French version of this lesson appeared in 1990 at a time when the only book on wavelets was Yves Meyer's *Ondelettes et opérateurs I: Ondelettes* [Mey90]. Professor Meyer and his students have played a central role in the development of wavelet theory, and Meyer's books, both the technical work cited above and his more widely accessible account [Mey93], have had an influence on both sides of the Atlantic.

Ten Lectures on Wavelets [Dau92] by Ingrid Daubechies was the first book in English, and it has deservedly become a "best seller." Full accounts of most of the material in this lesson can be found there.

There are now many books on wavelets in English. Furthermore, all are accessible to anyone who has understood the material of these 42 lessons. We have included several books in the References, usually annotated, that have not been cited in the text.

Finally, there is a large amount of information and software available via the Internet. The *Wavelet Digest* is a free monthly news letter edited by Wim Sweldens that provides general information on publications, conferences, software, etc. A subscription is available by visiting the Web page `http://www.wavelet.org.` Information about software in the public domain can be found in the article *Wavelet analysis* by Andrew Bruce, David Donoho, and Hong-Ye Gao, IEEE Spectrum, October 1996.

42.7 Exercises

Exercise 42.1 With the notation and hypotheses of Theorem 42.2.1, show that

$$\frac{1}{K}\iint_{\mathbb{R}^2} C_f(a,b)\,\overline{C}_g(a,b)\,\frac{da\,db}{a^2} = \int_{\mathbb{R}} f(t)\,\overline{g}(t)\,dt$$

for f and g in $L^2(\mathbb{R})$.

Hint: Show that

$$C_k(a,b) = \sqrt{|a|}\,\overline{\mathscr{F}}_{\lambda}\big[\widehat{h}(\lambda)\,\overline{\widehat{\psi}}(a\lambda)\big](b)$$

for $h \in L^2(\mathbb{R})$ and use the proof of Exercise 41.1.

Exercise 42.2 Consider the Haar system ψ_{jk} defined by

$$\psi_{jk}(x) = 2^{j/2}\psi(2^j x - k), \quad x \in \mathbb{R}, \quad j,k \in \mathbb{Z},$$

where $\psi(x) = 1$ on $[0, 1/2)$; $\psi(x) = -1$ on $[1/2, 1)$; and $\psi(x) = 0$ otherwise.

(1) Show that $\{\psi_{jk}\}_{j,k\in\mathbb{Z}}$ is an orthonormal system in $L^2(\mathbb{R})$.

(2) We know that $\{\psi_{jk}\}$ is an orthonormal basis for $L^2(\mathbb{R})$. Consider the scaling function $\varphi = \chi_{[0,1)}$ associated with the wavelet basis $\{\psi_{jk}\}$. If $n \in \mathbb{N}^*$ and

$$A = \begin{bmatrix} a_0 \\ \vdots \\ a_{2^n-1} \end{bmatrix},$$

we define a scaling function e associated with A by

$$e(x) = \sum_{k=0}^{2^n-1} a_k\,\chi_{[k2^{-n},(k+1)2^{-n})}(x), \quad x \in \mathbb{R}.$$

Verify that $e \in L^2(\mathbb{R})$ and that the wavelet decomposition of e is of the form

$$e(x) = d_{00}\varphi(x) + \sum_{j=0}^{n-1}\sum_{k=0}^{2^j-1} c_{jk}\psi_{jk}(x). \tag{1}$$

(3) For $n = 2$ take

$$A = \begin{bmatrix} 1 \\ 0 \\ -1 \\ 2 \end{bmatrix} \quad \text{and write} \quad B = \begin{bmatrix} d_{00} \\ c_{00} \\ c_{10} \\ c_{11} \end{bmatrix}.$$

(a) Draw the graph of e.

(b) Find the matrix $M \in M_{\mathbb{R}}(4,4)$ such that $A = MB$.

(c) Find B.

(d) Show explicitly that one indeed has the solution by computing the values of the two terms of (1) for each x.

(4) Treat explicitly the case $n = 3$ for an A of your choice.

References

[AAB+95] A. Arneodo, F. Argoul, E. Bacry, J. Elezgaray, and J.-F. Muzy. *Ondelettes, multifractales et turbulences, de l'ADN aux croissances cristallines.* Diderot Editeur, Arts et Sciences, Paris, 1995. English translation, Diderot Publishers, New York, 1997.

[Bas78] J. Bass. *Cours de Mathématiques*, volume I. Masson, Paris, 1978.

[Bel81] M. Bellanger. *Traitement numérique du signal.* Masson, Paris, 1981.

[Ber70] J.P. Bertrandias. *Analyse fonctionnelle.* Armand Colin, Paris, 1970.

[BL80] R. Boite and H. Leich. *Les filtres numériques.* Masson, Paris, 1980.

[Bre83] H. Brezis. *Analyse fonctionnelle. Théorie et applications.* Masson, Paris, 1983.

[Car63] H. Cartan. *Théorie élémentaire des fonctions analytiques d'une ou plusieures variables complexes.* Hermann, Paris, 1963.

[CLW67] J.W. Cooley, P.A.W. Lewis, and P.D. Welch. The Fast Fourier Transform algorithm and its applications. Technical report, I.B.M. Research, 1967.

[CLW70] J.W. Cooley, P.A.W. Lewis, and P.D. Welch. The Fast Fourier Transform algorithm. Programming considerations in the calculation of sine, cosine and Laplace transforms. *J. Sound Vibrations*, 12(3):315–337, 1970.

[Cou84] F. De Coulon. *Théorie et traitement des signaux.* Dunod, Paris, 1984.

[CR95] A. Cohen and R.D. Ryan. *Wavelets and Multiscale Signal Processing.* Chapman & Hall, London, 1995.

[Dau92] I. Daubechies. *Ten Lectures on Wavelets.* Society for Industrial and Applied Mathematics, Philadelphia, PA, 1992.

[DH82] P.J. Davis and R. Hersh. *The Mathematical Experience.* Houghton Mifflin, Boston, 1982.

[Ebe70] A. Eberhard. *Algorithmes de l'analyse harmonique numérique.* PhD thesis, Univesity of Grenoble, June 1970.

[Gab46] D. Gabor. Theory of communication. *J. Inst. Elec. Eng. (London)*, 93:429–457, 1946.

[GM84] A. Grossmann and J. Morlet. Decomposition of Hardy functions into square integrable wavelets of constant shape. *SIAM J. Math.*, 15:723–736, 1984.

[Haa10] A. Haar. Zur theorie der orthogonalen funktionen-systeme. *Math. Ann.*, 69:331–337, 1910.

[Hal64] P.R. Halmos. *Measure Theory.* D. Van Norstrand Company, Inc., New York, 1964.

[Her86] M. Hervé. *Distributions et transformée de Fourier.* P.U.F., Paris, 1986.

[HT91] M. Holschneider and Ph. Tchamitchian. Pointwise regularity of Riemann's "nowhere differentiable" function. *Inventiones Mathematicae*, 105:157–175, 1991.

[Hub96] B.B. Hubbard. *The World According to Wavelets.* A.K Peters, Wellesley, MA, 1996. A popular account of the basic ideas of wavelets, their history, and the people involved.

[Jac63] D. Jackson. *Fourier Series and Orthogonal Polynomials.* Number 6 in Carus Mathematical Monographs. Mathematical Association of America, Washington, D.C., 1963.

[KF74] A. Kolmogorov and S. Fomine. *Eléments de la théorie des fonctions et de l'analyse fonctionnelle.* Editions du Moscou, 1974.

[Kho72] Vo Khac Khoan. *Distributions, Analyse de Fourier. Opérateurs aux dérivées partielles.* Vuibert, 1972.

[Kun84] M. Kunt. *Traitement numérique des signaux.* Dunod, Paris, 1984.

[Lau72] P.J. Laurent. *Approximation et optimisation.* Hermann, Paris, 1972.

[Lip81] J.D. Lipson. *Elements of algebra and algebraic computing.* Addison-Wesley, 1981.

[LM86] P.G. Lemarié and Y. Meyer. Ondelettes et bases hilbertiennes. *Revista Ibero-Americana*, 2:1–18, 1986.

[Mal89] S. Mallat. A theory for multiresolution signal decomposition: The wavelet representation. *IEEE Trans. Pattern Anal. Machine Intell.*, 11:674–693, 1989.

[Mey90] Y. Meyer. *Ondelettes et Opérateurs I: Ondelettes.* Masson, Paris, 1990. English translation, *Wavelets and operators*, Cambridge University Press, 1992.

[Mey93] Y. Meyer. *Wavelets: Algorithms & Applications.* SIAM, Philadelphia, 1993.

[MJR87] Y. Meyer, S. Jaffard, and O. Rioul. L'analyse par ondelettes. *Pour la Science*, Sept. 1987.

[Nus81] H.J. Nussbaumer. *Fast Fourier Transform and Convolution Algorithms.* Springer-Verlag, 1981.

[RBC+92] M. B. Ruskai, G. Beylkin, R. Coifman, I. Daubechies, S. Mallat, Y. Meyer, and L. Raphael, editors. *Wavelets and their Applications.* Jones and Bartlett, Boston, 1992.

[Roy63] H.L. Royden. *Real Analysis.* The Macmillan Company, New York, 1963.

[Sch65a] L. Schwartz. *Méthodes mathématiques pour les sciences physiques.* Hermann, Paris, 1965.

[Sch65b] L. Schwartz. *Théorie des distributions.* Dunod, Paris, 1965.

[SN96] G. Strang and T. Nguyen. *Wavelets and Filter Banks.* Wellesley-Cambridge Press, Wellesley, MA, 1996.

[Sze59] G. Szegő. *Orthogonal polynomials*, volume 23. A.M.S. Colloquium Publications, 1959.

[Tor95] B. Torrésani. *Analyse continue par ondelettes.* InterEditions/CNRS Editions, Paris, 1995.

[VK95] M. Vetterli and J. Kovačević. *Wavelets and Subband Coding.* Prentice Hall, Englewood Cliffs, NJ, 1995. Written in the language of signal processing, this book presents an integrated view of wavelets and subband coding.

[Wic94] M.V. Wickerhauser. *Adapted Wavelet Analysis from Theory to Software.* A.K Peters, Wellesley, MA, 1994. A detailed treatment for engineers and applied mathematicians with an emphasis on the analysis of real signals. A good place to learn about wavelet packets.

Index